Phase-Locked and Frequency-Feedback Systems

ELECTRICAL SCIENCE
A Series of Monographs and Texts

Editors: Henry G. Booker and Nicholas DeClaris
A complete list of titles in this series appears at the end of this volume

Phase-Locked and Frequency-Feedback Systems

PRINCIPLES AND TECHNIQUES

JACOB KLAPPER
*Department of Electrical Engineering
Newark College of Engineering
Newark, New Jersey*

JOHN T. FRANKLE
*Engineering Department
RCA Global Communications Inc.
New York, New York*

ACADEMIC PRESS   New York and London   1972

COPYRIGHT © 1972, BY ACADEMIC PRESS, INC.
ALL RIGHTS RESERVED
NO PART OF THIS BOOK MAY BE REPRODUCED IN ANY FORM,
BY PHOTOSTAT, MICROFILM, RETRIEVAL SYSTEM, OR ANY
OTHER MEANS, WITHOUT WRITTEN PERMISSION FROM
THE PUBLISHERS.

ACADEMIC PRESS, INC.
111 Fifth Avenue, New York, New York 10003

United Kingdom Edition published by
ACADEMIC PRESS, INC. (LONDON) LTD.
24/28 Oval Road, London NW1 7DD

LIBRARY OF CONGRESS CATALOG CARD NUMBER: 72-76546

PRINTED IN THE UNITED STATES OF AMERICA

To

ROXANA   **RENA**
Rachelle   *Andy*
Robert   *Susan*
             Jill

AND OUR PARENTS

Contents

Preface xi
Acknowledgments xiii
Glossary of Terms xv

1 Introduction

1.1. Historical Development 1
1.2. Organization of the Book 2
References 3

2 Review of Concepts

2.1. Network Theory 4
2.2. Feedback Theory 11
2.3. Representation and Properties of Noise 14
2.4. Trigonometric Identities 21
References 21

3 Loop Components and Systems Considerations

3.1. Introduction 23
3.2. Limiter-Discriminator 24
3.3. Low-Threshold Demodulation 37

3.4. Filtering of FM Carriers	40
3.5. Voltage-Controlled Oscillator	42
3.6. Phase Detectors	44
References	46

4 FM Feedback Loop Principles

4.1. Operational Principles	49
4.2. The Linear Equivalent Circuit	53
4.3. Linear Operation of the Loop	57
4.4. Nonlinear Operation	65
4.5. Effect of Excess Delay	73
References	75

5 Phase-Locked Loop Principles

5.1. Operational Principles	76
5.2. The Linear Equivalent Circuit	80
5.3. Linear Operation	84
5.4. Nonlinear Operation	95
5.5. Excess Delay and Minimum Noise Bandwidth	110
References	112

6 Design of Phase-Locked Loops for FM Demodulation

6.1. FM Improvement Region	113
6.2. Distortion-Limited Region	122
6.3. Threshold-Limited Region	132
6.4. Step-by-Step Design Procedure and Examples	154
References	168

7 Design of Frequency-Feedback Loops for FM Demodulation

7.1. FM Improvement Region	170
7.2. Distortion-Limited Region	177
7.3. Threshold-Limited Region	195
7.4. Step-by-Step Design Procedure and Examples	215
References	224

8 Design of Compound and Multiple Loops for Low-Threshold Demodulation

8.1. General Design Considerations	227
8.2. The FMFB-PLL Compound Loop	231

8.3. The FMFB-FMFB Compound Loop	239
8.4. Extended-Range Phase-Locked Demodulator (ERPLD) and the FMFB-ERPLD	247
References	264

9 Digital FM and Other PLL Applications

9.1. Introduction to Digital FM Systems	266
9.2. Binary Error Rates with Limiter-Discriminator Demodulation	274
9.3. Binary FM Demodulation with Angular-Feedback Demodulators	283
9.4. Other PLL Applications	295
References	300

10 Testing and Evaluation Procedures

10.1. Component Tests	304
10.2. Loop Tests	313
10.3. Systems Tests of FM Demodulators	323
References	338

Appendix A	**Derivation of Angle-Modulation Improvement Equations**	339
Appendix B	**Derivation of the Discriminator Baseband Response**	344
Appendix C	**The Ideal Demodulator**	346
Appendix D	**Varactor VCO Distortion**	348
Appendix E	**Baseband-Equivalent Response of a Single-Tuned Circuit to a Small-Index Off-Tuned FM Carrier**	351
Appendix F	**Calculation of Minimum Noise Bandwidth and Loop Filter Zero Constant for the FMFB**	353

Bibliography 355

Author Index 381
Subject Index 385

Preface

Phase-locked and frequency-feedback systems are now in widespread use. They are prominent in the fields of telemetry, communications, control, and instrumentation, and wherever it is necessary to extract frequency or phase information from signals embedded in noise or other interference. This text presents the operating principles and methods of design of these systems giving special emphasis to their application as FM demodulators with lowered thresholds. Much of the material presented herein has never appeared in book form before, and some of it is published here for the first time.

The authors feel that in view of the increasing use of angular-feedback techniques, there is a growing interest in the material presented in this text. It is the subject of a graduate level course given by one of the authors at Newark College of Engineering. The required background of the reader is essentially equivalent to that obtained in undergraduate courses on linear systems, feedback theory, communication systems, and stochastic processes. A brief review of these principles is given in Chapter 2. More advanced analytical tools are introduced where they are needed.

It is hoped that the material will be found especially useful by the practicing engineer. For this purpose the text contains detailed step-by-step design procedures and graphical aids, with illustrations bearing on real problems taken from the authors' varied experience in this field. Engineering approaches and approximations are prominent, often bypassing difficult theoretical formulizations. A chapter describing detailed test procedures and an extensive bibliography on theory and applications are included.

Acknowledgments

This book is an outgrowth of an early manuscript developed by the authors under the sponsorship of the U.S. Department of Defense. In connection with this early manuscript, the authors are particularly indebted to Drs. J. Birch and W. Poe, DOD, for their guidance and review, and to Mr. S. J. Mehlman, Manager, Advanced Communications Techniques, RCA, for his interest and encouragement. The authors are also indebted to A. Acampora and F. Lefrak for their contributions to Chapter 10, to A. Newton for his constructive review, and M. P. Rosenthal for editing and styling the early manuscript.

The administrations of Newark College of Engineering, RCA Corporation, and GTE Laboratories provided support and encouragement in the preparation of the book. Dean L. B. Anderson, Dr. F. A. Russell, and Professor R. E. Anderson, of NCE, and Dr. John E. Fulenwider of GTE Laboratories deserve particular mention for their interest in this endeavor. The students of NCE were not only stimulating critics but also helped actively. W. Novick read the manuscript with care. Much typing was done by Mrs. Bernice Thompson, Miss J. Bulbach, and Miss B. Walker. Much of the art was prepared by Mrs. N. Bronstein and Mr. M. Temes. Lastly, the authors found the association with the publisher's personnel most pleasant.

To our wives Roxana and Rena who participated with pleasure and devotion, our sincere thanks.

Glossary of Terms

α	Input CNR referred to PLL noise bandwidth, B_n
α_b	Entire baseband threshold CNR
α_{CH}	Channel threshold CNR in FDM
δ_2, δ_3	Distortion coefficient
Δ	Deviation percentage error
Δf_p	Peak signal frequency deviation (hertz)
$\Delta\omega$	VCO frequency deviation (radians per second) about quiescent center frequency
$\Delta\omega_p$	Peak signal frequency deviation (radians)
θ_n	Noise angle
$(\Delta\omega_{rms})^2$	Mean-square signal frequency deviation (radians per second)^2
$(\dot{\Delta}\omega)_{max}$	Maximum sweep frequency (radians per second squared)
η	Spectral density level (watts per hertz)
η_m	Modulation density in terms of phase modulation (radians squared per hertz)
λ	Limiter-discriminator sensitivity (also K_1)
Λ	Normalized phase margin

Glossary of Terms

μ	VCO sensitivity (radians per volt-second (also K_3)
μs	Microsecond
ξ	Damping factor (also z)
ρ	Carrier-to-noise ratio (CNR)
σ	Root-mean-square (rms) modulation index; also standard deviation
τ	Delay (seconds)
Φ	Phase angle, frequency domain
ϕ	Phase angle, time domain (radians)
ϕ_b	Excess phase shift per base bandwidth (radians)
ϕ_d	Distortion generator in terms of phase modulation (radians)
ϕ_e	Loop phase error (radians)
ϕ_{es}	Modulation-induced phase error component (radians)
ϕ_{en}	Noise-induced phase error component (radians)
ϕ_i	Input signal phase modulation (radians)
ϕ_{is}	Received signal phase modulation (radians)
ϕ_{np}	Equivalent peak phase error, noise component (radians)
ϕ_r	VCO phase modulation (radians)
ϕ_{rd}	Loop response distortion component in terms of phase modulation (radians)
ψ	Phase angle (radians)
ω	Radian frequency (radians per second)
ω_a	Bottom frequency of speech spectrum model or of FDM baseband (radians per second)
ω_b	Top baseband frequency of transmission (radians per second)
ω_c	Signal center frequency (radians per second)
ω_{CH}	Channel center frequency (radians per second)
ω_e	IF center frequency (FMFB) (radians per second)
ω_i	Input signal center frequency (radians per second)
ω_n	Loop natural frequency (radians per second)
ω_r	VCO center frequency (radians per second)
ω_T	Test-tone frequency (radians per second)
a, b	Loop zero constants
b'	Ratio of predetection semibandwidth to base bandwidth
dB	Decibels

Glossary of Terms

e_n	Equivalent noise input generator that accounts for voltage-controlled oscillator internal noise
e_r	Demodulated signal or loop output signal (volts)
f	Frequency (hertz)
f_b	Base-bandwidth or top baseband transmission frequency (hertz)
h	Impulse response of closed loop
h_1	Impulse response of IF filter
h_{b1}	Equivalent baseband inpulse response of internal IF filter
kHz	Kilohertz (kilocycles per second)
mH	Millihenry
m_p	Peak modulation index (also $\Delta\omega_p/\omega_b$)
$n(t)$	Noise
nsec	Nanosecond
r	Radius of gyration
rad	Radian
rms	Root-mean-square
sec	Second
⊛	Convolution
A	Signal amplitude or constant carrier amplitude (volts)
AFC	Automatic frequency control
AGC	Automatic gain control
AM	Amplitude modulation
B	Channel noise bandwidth in frequency-division multiplex (hertz); also, a bandwidth-related parameter in ERPLD
B_p	Predetection bandwidth (hertz)
B_{CR}	Carson's rule bandwidth (hertz)
BER	Bit error rate
B_{IF}	IF filter 3-dB bandwidth (hertz)
BINR	Baseband intrinsic noise ratio
B_n	Equivalent noise bandwidth (hertz)
B_R	Bit rate
BWR	Base-bandwidth to channel-bandwidth ratio
CNR	Carrier-to-noise ratio (ρ)
$(CNR)_{AM}$	Input CNR referred to twice base bandwidth ($2f_b$)
$(CNR_{AM})_{TH}$	Threshold CNR referred to twice base bandwidth ($2f_b$)

Glossary of Terms

D	Deviation index
DC	Direct current
D_e	rms frequency deviation (hertz)
DPSK	Differentially coherent phase-shift keying
E/η	Energy ratio
ERPLD	Extended-range phase-locked demodulator
F	Frequency modulation feedback factor
FDM	Frequency-division multiplex
FM	Frequency modulation
FMFB	FM feedback loop
FMFB-ERPLD	Compound-loop FMFB demodulator, ERPLD as internal demodulator
FMFB-FMFB	Compound-loop FMFB demodulator, FMFB as internal demodulator
FMFB-PLL	Compound-loop FMFB demodulator, phase-locked loop as internal demodulator
FSK	Frequency shift keying
GHz	Gigahertz (gigacycles per second)
$G(s)$	Open-loop transfer function
$H_2(s)$	Transfer function of loop baseband filter
$H_D(s)$	Transfer function of internal demodulator
$H(j\omega)$	Closed-loop response
$H(s)$	Closed-loop transfer function
$H_1(s)$	Transfer function of IF filter (FMFB)
$H_{L1}(s)$	Lowpass equivalent of $H_1(s)$
$H_{b1}(s)$	Baseband equivalent of $H_1(s)$
.Hz	hertz (cycles per second)
IF	Intermediate frequency
K	Loop gain constant (product of phase detector sensitivity, amplifier gain, and voltage-controlled oscillator sensitivity in the phase-locked loop)
K_0	Closed-loop gain in FMFB
K_1	Phase-locked loop phase detector sensitivity (volts per radian); also LD sensitivity in FMFB
K_2	Amplifier gain (PLL and FMFB)
K_3	VCO sensitivity (radians per volt-second) (PLL and FMFB)

Glossary of Terms

L	Multiplex noise loading ratio
LD	Limiter-discriminator
LLI	Loss-of-lock impulses
MHz	Megahertz (megacycles per second)
$|M(j\omega)|$	Magnitude of predetection filter response (PLL)
$|M_p(j\omega)|$	Magnitude of predetection filter response (FMFB)
MLR	Maximum likelihood receiver
MTBF	Mean time between failure
N	Spike rate, cycle skipping rate
$N(t)$	Additive noise
NCR	Noise-to-carrier ratio
NIF	Noise improvement factor
N_{po}	Output noise power from postdetection filter following the PLL (watts)
NPR	Noise power ratio
NT	Average number of encirclements per bit
PCM	Pulse code modulation
P	Noise penalty factor
P_e	Probability of error
P_s	Signal power
P_n	Noise power
PLD	Phase-locked demodulator
PLL	Phase-locked loop
PM	Phase modulation
rf	Radio frequency
$|R(j\omega)|$	Magnitude of postdetection filter response
RLC	Resistance-inductance-capacitance (network)
SNR	Signal-to-noise ratio
SNR_{TT}	Test-tone signal-to-noise ratio
S_p	Speech power
S_{po}	Postdetection filter output signal power (watts)
$S(\omega)$	Power spectral density of transmitted baseband (watts per hertz)
ThI	Threshold impulses
TT	Test-tone
V	Voltage

VCO	Voltage-controlled oscillator
VCXO	Voltage-controlled crystal oscillator
VSWR	Voltage standing wave ratio
$W(f)$	Power spectral density function
$\cong$	Approximately equal

CHAPTER

1

Introduction

1.1. Historical Development

The phase-locked loop (PLL) has in the past received much more attention than the frequency-feedback loop because of its wider range of applicability and greater simplicity.† Invented as a technique to realize coherent AM reception,[1] the PLL received serious study and application as a synchronization circuit of television receivers.[2,3] Much further application impetus came with the space age and the requirement to realize low-level signal-reception, -tracking, -phase extraction, -filtering, and frequency synchronization. While the basic circuitry is very simple, which adds to its desirability, the inherent nonlinear behavior of the loop, further complicated by the inclusion of external random processes (such as input additive noise), have been substantial obstacles in the analytical treatment and optimization of the system.

The first analysis and optimization of the PLL, as a demodulator of FM signals with additive input noise, was reported by Jaffe and Rechtin.[4] They used a linear equivalent circuit for the PLL and optimized the loop for transient signals and additive Gaussian noise interference. More recently, Develet[5] presented an approximate nonlinear analysis applicable to the case where both signal and noise are of a Gaussian distribution. He derived in this manner the best FM threshold performance for an optimum PLL and for an optimum second-order PLL. Other approximate nonlinear analyses have also appeared in the literature.[6,7]

† Compare, e.g., the respective bibliographies at the end of the book.

An exact analysis exists for the first-order loop,[8] and more recently, for a generalized tracker.[9] However, practical design is generally based on a linear model, with the effect of nonlinearities included by way of experimental parameters.

The FM feedback (FMFB) demodulator was first described by Chaffee in 1939.[10] Its low-threshold capabilities were used by Morita and Ito in 1953.[11] A widely used design procedure was first formulated by Enloe[12]; this was followed by the appearance of competitive design procedures in the literature.[13]

A considerable effort was made recently, and is still under way, to extend the FM demodulator threshold beyond the capabilities of the simple FMFB or PLL. Interesting demodulators were recently analyzed and described by Frankle[14] and Acampora and Newton.[15] These advanced demodulators use either combinations of FMFB and PLL or "range extension" of the PLL. The term *angular feedback loop* will refer generically to any of the loops treated in this text.

1.2. Organization of the Book

Chapter 2 reviews briefly the elements of linear systems, feedback theory, and noise, providing the minimum background for the material presented in the remainder of the text. Chapter 3 reviews the characteristics of the major components that comprise the loops. Also discussed is the performance of the conventional FM demodulator (limited-discriminator), and a performance preview of the multiple-loop demodulators (discussed more fully in Chapter 8). Loop principles, with the basic describing equations, are presented in Chapters 4 and 5 for the FMFB and PLL, respectively. Chapters 6 and 7 cover PLL and FMFB design. Here, step-by-step design procedures with performance characteristics for low-threshold angle demodulation are illustrated by typical design examples. The design principles are then extended in Chapter 8 to the design of advanced demodulators which feature demodulation thresholds lower than those of the simple PLL or FMFB. Digital FM demodulation is featured in Chapter 9, along with a list and discussion of PLL applications other than FM demodulation. Finally, Chapter 10 presents methods of testing and evaluating loop performance.

References

1. H. de Bellescize, La Reception Synchrone. *Onde Elec.* **11**, 209–272 (1932).
2. K. R. Wendt and G. L. Fredendall, Automatic frequency and phase control of synchronization in television receivers. *Proc. IRE* **31**, 7–15 (1943).
3. D. Richman, Color-carrier reference phase synchronization accuracy in NTSC color television. *Proc. IRE* **42**, 106–133 (1954).
4. R. Jaffe and E. Rechtin, Design and performance of phase-lock loops capable of near-optimum performance over a wide range of input signal and noise levels. *Trans. IRE* **IT-1**, 66–76 (1955).
5. J. A. Develet, A threshold criterion for phase-lock demodulation. *Proc. IEEE* **51**, 349 (1963); Correction *Proc. IEEE* **51**, 580 (1963).
6. H. L. Van Trees, Functional techniques for the analysis of the nonlinear behavior of phase-locked loops. *WESCON* (1963).
7. D. L. Schilling, The response of an automatic phase control system to FM signals and noise. *Proc. IEEE* **51**, 1306–1315 (1963).
8. A. J. Viterbi, Phase-locked loop dynamics in the presence of noise by Fokker-Planck techniques. *Proc. IEEE* **51**, 1737 (1963).
9. W. C. Lindsey, Nonlinear analysis of generalized tracking systems. *Proc. IEEE* **57**, No. 10, 1705–1722 (1969).
10. J. G. Chaffee, The application of negative feedback to frequency modulation systems. *Bell Syst. Tech. J.* **18**, 403–437 (1939).
11. M. Morita and S. Ito, High sensitivity receiving system for frequency modulated waves. *IRE Int. Conv. Rec.* **8**, Pt. 5, 228 (1960).
12. L. H. Enloe, Decreasing the threshold in FM by frequency feedback. *Proc. IRE* **50**, No. 1, 18–30 (1962); The synthesis of frequency feedback demodulators. *Proc. Nat. Electron. Conf.* **18**, 477–497 (1962).
13. P. Frutiger, Noise in FM receivers with negative frequency feedback. *Proc. IEEE* **54**, No. 11, 1506–1520 (1966).
14. J. Frankle, Threshold performance of analog FM demodulators. *RCA Rev.* **27**, No. 4, 521–562 (1966).
15. A. Acampora and A. Newton, Use of phase subtraction to extend the range of a phase-locked demodulator. *RCA Rev.* **27**, No. 4, 577–599 (1966).

CHAPTER

2

Review of Concepts

The theory and application of phase-lock and frequency-feedback principles require an understanding of several engineering disciplines. This chapter briefly reviews the appropriate elements of network analysis, feedback theory, and stochastic processes without a rigorous development. The purpose of the chapter is to acquaint the reader with formulas, concepts, terms, and "language" required by the subsequent text. The reader is referred to the references for a more complete treatment of these subjects.

2.1. Network Theory

A. Laplace and Fourier Transforms[1-3]

In the design and analysis of networks it is convenient to represent network and signal characteristics in either the time or the frequency domain. An analytical tool for transforming functions from one domain to the other is the *Laplace transform*, which is defined in terms of the complex frequency variable s (where $s = \sigma + j\omega$) by

$$\mathscr{L}[x(t)] = X(s) = \int_{-\infty}^{\infty} x(t) \exp(-st)\, dt \quad \text{(transform)} \quad (2\text{-}1)$$

$$\mathscr{L}^{-1}[X(s)] = x(t) = \frac{1}{2\pi j} \int_{c-j\infty}^{c+j\infty} X(s) \exp(st)\, ds \quad \text{(inverse transform)}$$

$$(2\text{-}2)$$

2.1. Network Theory

where $x(t)$ is a physical function of time, such as a voltage appearing in an electrical network, $X(s)$ is the transform function in terms of s, c is a constant forming a convergent path of integration, and $\mathscr{L}[\ ]$, $\mathscr{L}^{-1}[\ ]$ are the transform operation and the inversion operation, respectively.

The above formulation is in terms of the *bilateral* Laplace transform, i.e., the integration is for both positive and negative time. The *unilateral* Laplace transform, often used in transient analysis with signals existing for positive time only, is obtained simply by letting the lower limit of integration in Eq. (2-1) be $t = 0$. Clearly, for signals existing for positive time only, the bilateral and unilateral transforms are identical. The only exception lies in the handling of *initial conditions*, i.e., the energy stored in reactances prior to the application of the signal. The introduction of initial conditions will not be of much interest in this text. On the other hand, the bilateral transform will be found more natural in a number of instances.

The bilateral Laplace transform formula reduces to the very useful Fourier transform formula if the complex frequency s is replaced by $j\omega$, i.e.,

$$\mathscr{F}[x(t)] \equiv X_F(j\omega) = \int_{-\infty}^{\infty} x(t) \exp(-j\omega t)\, dt \quad \text{(transform)} \quad (2\text{-}1a)$$

$$x(t) = \mathscr{F}^{-1}[X_F(j\omega)] = \frac{1}{2\pi j} \int_{-\infty}^{\infty} X_F(j\omega) \exp(j\omega t)\, d(j\omega)$$

$$\text{(inverse transform)} \quad (2\text{-}2a)$$

The restriction of the integration path in Eq. (2-2a) to the $j\omega$ axis leads to some differences in the convergence properties of the integral so that some functions may be handled by one transform and not by the other. However, in our cases it will be generally possible to obtain the Fourier transform $X_F(j\omega)$ by simply replacing every s in $X(s)$ by $j\omega$, i.e.,

$$X_F(j\omega) = X(s)\big|_{s=j\omega} = X(j\omega)$$

and we shall do so freely. The variable $2\pi f = \omega$, where f is in hertz, is freely substituted in the text for ω in radians per second. In terms of f, Eq. (2-2a) becomes

$$x(t) = \int_{-\infty}^{\infty} X(j\omega) \exp(j\omega t)\, df \quad (2\text{-}2b)$$

The inverse Laplace transform formula, Eq. (2-2), is seldom used by the practicing engineer. Instead, he refers to readily available tables of transform pairs. Important Laplace transform pairs used repeatedly in this text are given in Table I.

TABLE I

Laplace Transforms

Description	Function of time $x(t)$	Transform $X(s)$
1. Definition	$x(t) = \dfrac{1}{2\pi j}\displaystyle\int_{c-j\infty}^{c+j\infty} X(s)\exp(st)\,ds$	$X(s) = \displaystyle\int_{-\infty}^{\infty} x(t)\exp(-st)\,dt$
2. Amplitude scaling	$ax(t)$	$aX(s)$
3. Addition	$x_1(t) + x_2(t)$	$X_1(s) + X_2(s)$
4. Multiplication	$x_1(t)\cdot x_2(t)$	$\displaystyle\int_{-\infty}^{\infty} X_1(v)\cdot X_2(s-v)\,dv$
5. Damping	$[\exp(-s_0 t)]\cdot x(t)$	$X(s + s_0)$
6. Derivative[a]	$\dfrac{d}{dt}[x(t)]$	$s\cdot X(s)$
7. Integral[a]	$\displaystyle\int^{t} x(v)\,dv$	$X(s)/s$
8. Multiple derivative[a]	$\dfrac{d^n}{dt^n}[x(t)]$	$s^n X(s)$
9. Multiply by time	$tx(t)$	$-\dfrac{dX(s)}{ds}$
10. Convolution	$\displaystyle\int_{-\infty}^{\infty} x_1(\tau)x_2(t-\tau)\,d\tau$	$X_1(s)X_2(s)$
11. Delay	$x(t-\tau)$	$[\exp(-\tau s)]\cdot X(s)$
12. Scaling	$x(at)$	$(1/a)X(s/a)$
13. Final value[b]	$\lim_{t\to\infty} x(t)$	$\lim_{s\to 0} sX(s)$
14. Initial value[b]	$\lim_{t\to 0+} x(t)$	$\lim_{s\to\infty} sX(s)$
15. Unit impulse	$\delta(t)$	1
16. Unit step	$u(t) = \begin{cases} 0 & t<0 \\ 1 & t>0 \end{cases}$	$1/s$
17. Unit ramp	$tu(t)$	$1/s^2$
18. High-order function	$\dfrac{t^{n-1}}{(n-1)!}u(t)$	$1/s^n$
19. Cosinusoid	$\cos\omega_0 t\, u(t)$	$s/(s^2 + \omega_0^2)$
20. Sinusoid	$\dfrac{1}{\omega_0}\sin\omega_0 t\, u(t)$	$1/(s^2 + \omega_0^2)$
21. Exponential	$\exp(-s_0 t)$	$1/(s + s_0)$
22. Damped sinusoid	$e^{-\sigma t}\sin\omega_0 t\, u(t)$	$\omega_0/[(s+\sigma)^2 + \omega_0^2]$

[a] These formulas apply for the unilateral Laplace transform only if the initial conditions are zero or are considered part of the source.
[b] For unilateral time functions.

B. Input–Output Relationships[4,5]

Consider an electrical network comprised of linear, non-time-varying elements. For an input $x_i(t)$ the network responds with an output $x_o(t)$, as shown diagrammatically in Fig. 2-1(a). The action of the network on the input

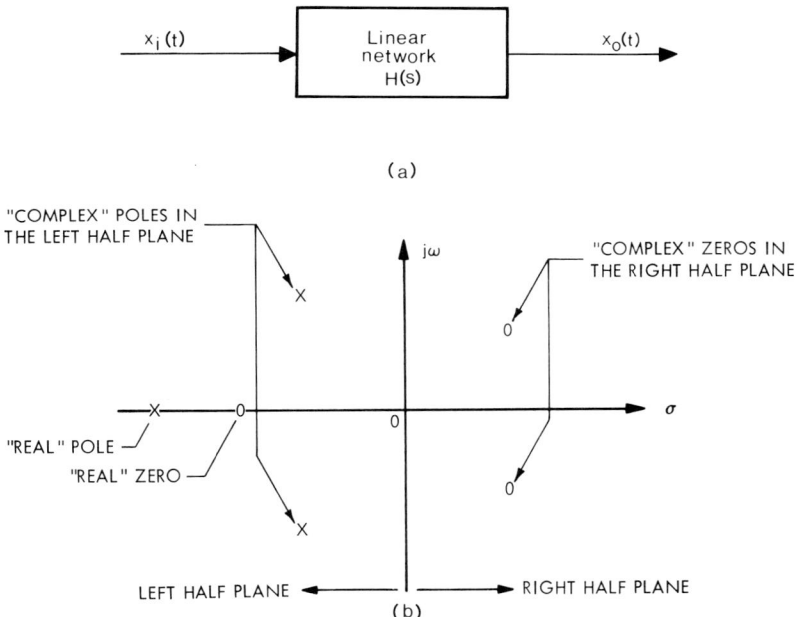

Fig. 2-1. (a) Two-port network; (b) s-plane representation of its transfer function (0 : zeros; × : poles).

signal to produce the output signal is expressed by a linear differential equation of the form

$$a_m \frac{d^m}{dt^m} x_i(t) + a_{m-1} \frac{d^{m-1}}{dt^{m-1}} x_i(t) + \cdots + a_0 x_i(t)$$
$$= b_n \frac{d^n}{dt^n} x_o(t) + b_{n-1} \frac{d^{n-1}}{dt^{n-1}} x_o(t) + \cdots + b_0 x_o(t) \qquad (2\text{-}3)$$

where the a_m and b_n are constant coefficients determined by the network elements. The largest value of m or n specifies the *order of the network*. Symbolically we may write $p = d/dt$. Thus,

$$d^j/dt^j = p^j \qquad (2\text{-}4)$$

and (2-3) may be written as

$$[a_m p^m + a_{m-1} p^{m-1} + \cdots + a_0]x_i(t) = [b_n p^n + b_{n-1} p^{n-1} + \cdots + b_0]x_o(t) \tag{2-5}$$

Continuing in this manner, the ratio of the output to the input $H(p)$ is

$$\frac{x_o(t)}{x_i(t)} = \frac{a_m p^m + a_{m-1} p^{m-1} + \cdots + a_0}{b_n p^n + b_{n-1} p^{n-1} + \cdots + b_0} \equiv H(p) \tag{2-6}$$

It is emphasized that this is only a symbolic representation of the differential equation. It is of interest, however, to note that the Laplace transform of Eq. (2-5) or (2-6) is obtained simply by replacing in that equation the operator p by the complex frequency variable s and the input and output quantities $[x_i(t), x_o(t)]$ by their respective Laplace transforms $[X_i(s), X_o(s)]$. Conversely, one may go from the "s" to the "time" domain by reversing the above steps. We may also obtain the Laplace transform directly from Eq. (2-3) and the "derivative relationships" of Table I. Thus

$$[a_m s^m + a_{m-1} s^{m-1} + \cdots + a_0]X_i(s) = [b_n s^n + b_{n-1} s^{n-1} + \cdots + b_0]X_o(s) \tag{2-7}$$

The *transfer function* of the network $H(s)$ is defined by the ratio of the output to the input in the s-domain

$$H(s) \triangleq \frac{X_o(s)}{X_i(s)} = \frac{a_m s^m + a_{m-1} s^{m-1} + \cdots + a_0}{b_n s^n + b_{n-1} s^{n-1} + \cdots + b_n} \tag{2-8}$$

For a *passive network* (no amplifiers) m is restricted so that $0 \le m \le n + 1$. The polynomial numerator and denominator of $H(s)$ may be factored so that $H(s)$ may be expressed as a ratio of the product of roots

$$H(s) = \frac{a_m(s - z_m)(s - z_{m-1}) \cdots (s - z_1)}{b_n(s - p_n)(s - p_{n-1}) \cdots (s - p_1)} \tag{2-9}$$

Roots of the numerator are called *zeros*, while roots of the denominator are called *poles* and, in general, they are complex. The roots may be represented diagrammatically by an s-plane plot, such as shown in Fig. 2-1(b). Complex roots always appear in conjugate pairs. For the network to be stable and realizable the poles are restricted to the left half plane, $\sigma \le 0$. The zeros, however, may appear anywhere in the s-plane. Networks with zeros confined to the left half plane are called *minimum-phase networks*.

The main advantage of the s-domain is that the output of a linear system is obtained as a product of the input and the transfer function, viz.

$$X_o(s) = H(s)X_i(s) \tag{2-10}$$

2.1. Network Theory

In the time domain, $H(s)$ becomes $h(t)$, which is the response of the system to a unit impulse applied at $t = 0$

$$\mathscr{L}^{-1}[H(s)] = h(t) \qquad (2\text{-}11)$$

and the input–output relationship in terms of $h(t)$ is given by the convolution integral

$$x_o(t) = \int_{-\infty}^{t} x_i(\tau) h(t - \tau) \, d\tau \qquad (2\text{-}12)$$

or symbolically,

$$x_o(t) = x_i(t) \circledast h(t) \qquad (2\text{-}13)$$

In conclusion, the input–output relationship of a linear system may be expressed in the differential equation form of Eq. (2-3) [equivalently Eq. (2-5)], or in the s-domain product of Eq. (2-10), or, finally, as a convolution integral given by Eq. (2-13).

C. Network Frequency Response[5]

The response of a linear network to a sinusoidal input is an important tool in network calculations. If it is known for all possible frequencies of the sinusoid, then the network is fully described and the response to any other waveshape can be calculated. Suppose the input to the network $x_i(t)$ is

$$x_i(t) = \sin \omega t \qquad (2\text{-}14)$$

Then the output of the network $x_o(t)$ is of the form

$$x_o(t) = A(\omega) \sin[\omega t + \phi(\omega)] \qquad (2\text{-}15)$$

i.e., the output is a sinusoid of the identical frequency, but its amplitude and phase may be altered depending upon the input frequency ω. Thus, $A(\omega)$ described for all ω is termed the *amplitude response*, and $\phi(\omega)$ the *phase response* of the network. Together, $A(\omega)$ arg $\phi(\omega)$ define the *frequency response* of the network.

It is of interest to note that the frequency response is easily obtainable from the network's transfer function, simply by replacing every s in $H(s)$ by $j\omega$ (i.e., $\sigma = 0$). It is generally written in polar form

$$H(j\omega) = A(\omega) e^{j\phi(\omega)} \qquad (2\text{-}16)$$

where $A(\omega)$ is the magnitude and $\phi(\omega)$ is the argument of $H(s)|_{s=j\omega}$. The amplitude response is most frequently given in decibels, i.e., as $20 \log_{10} A(\omega)$, and $\phi(\omega)$ in degrees. As an example, Fig. 2-2 shows the typical frequency response of a second-order PLL, plotted (as is generally done) on a normalized logarithmic frequency scale.

A *rectangular filter* is a filter whose amplitude response is a rectangle, having a constant transmission in the passband and infinite attenuation outside. It is generally considered to have an associated linear phase response. This filter cannot be fully realized in practice. Chebyshev and Butterworth filters are essentially approximations to the rectangular type.

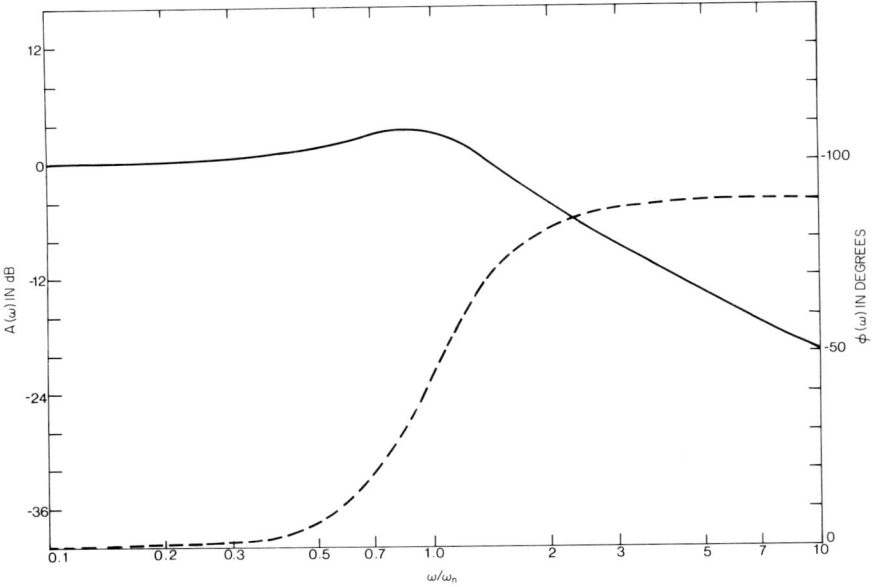

Fig. 2-2. Typical frequency response characteristic of a second-order PLL (from Fig. 5-5 with $a/\omega_n = 1$) (—: magnitude; – – –: phase).

D. Lowpass Equivalent Frequency Response[2]

The *lowpass equivalent* frequency response $H_L(j\omega)$ of a *symmetrical bandpass* frequency response $H(j\omega)$ is defined as the response formed by shifting the upper half of $H(j\omega)$ to zero frequency. This transformation, illustrated in Fig. 2-3, is useful in calculations with modulated carriers.

Caution: $H_L(j\omega)$ is not the same as the lowpass prototype of a bandpass network used in network synthesis. In particular, the lowpass prototype is formed from a geometrically symmetric bandpass response and has twice the bandwidth of the lowpass equivalent response.

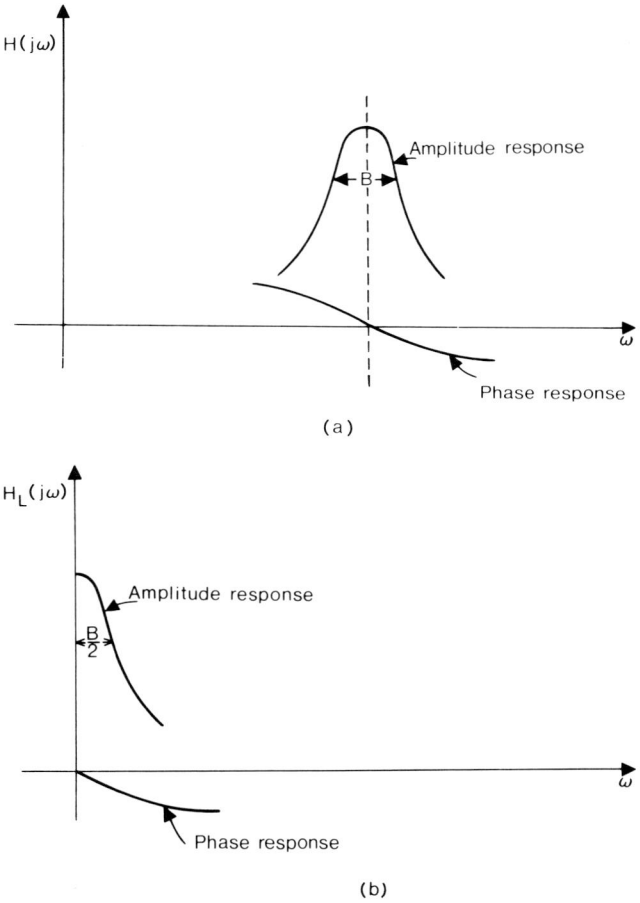

Fig. 2-3. Bandpass to lowpass equivalent transformation. (a) Bandpass response; (b) lowpass equivalent response.

2.2. Feedback Theory[4]

Phase-locked and frequency-feedback loops are feedback systems in which the fed-back parameter is the angle (or its derivative) of a sine wave. While these systems are inherently nonlinear, we shall deal largely with linear equivalent loops to which conventional linear system feedback theory directly applies. Therefore, a review of classical linear feedback theory is given here. Some techniques for nonlinear analysis are introduced in the text where they are used.

A. Fundamental Relations

A basic feedback loop is shown in Fig. 2-4. It consists of a forward gain network with transfer function $A(s)$, a feedback network with transfer function $B(s)$, and a subtractor network that forms the error signal $y_e(t)$ from the difference between the input signal $y_i(t)$ and the return signal $y_r(t)$. The output of the loop is $y_o(t)$.

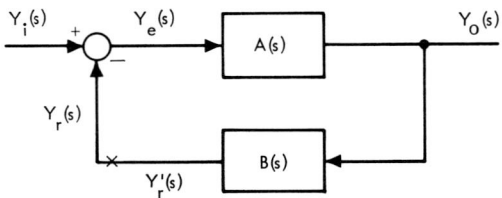

Fig. 2-4. Basic feedback loop block diagram.

The applicable relations in the s-domain are

$$Y_e(s) = Y_i(s) - Y_r(s) \tag{2-17}$$

$$Y_o(s) = A(s) \cdot Y_e(s) \tag{2-18}$$

$$Y_r(s) = B(s) \cdot Y_o(s) \tag{2-19}$$

The signals y_i, y_e, and y_o are known as the input, error, and output signals, respectively. From these we obtain the closed-loop transfer function

$$H(s) = Y_o(s)/Y_i(s) = A(s)/[1 + A(s)B(s)] \tag{2-20}$$

and the closed-loop error response

$$Y_e(s)/Y_i(s) = 1/[1 + A(s)B(s)] \tag{2-21}$$

We shall find occasions in which the *open-loop response* will be useful in describing the system. It may be obtained as follows: A break is made at the point within the loop designated by × in Fig. 2-4, assuring that block $B(s)$ is terminated in the same impedance as before. With an applied signal $y_i(t)$, signal $y_r'(t)$ is measured. The open-loop response is defined by

$$G(s) = \frac{Y_r'(s)}{Y_i(s)} = A(s)B(s) \tag{2-22}$$

The angular feedback loops require special techniques for this measurement, as described in Chapter 10.

Often the transfer function of the feedback network B is unity, reducing the closed-loop transfer function to

$$H(s) = A(s)/[1 + A(s)] \qquad (2\text{-}23)$$

the error response to

$$Y_e(s)/Y_i(s) = 1 - H(s) = 1/[1 + A(s)] \qquad (2\text{-}24)$$

and the open-loop response to $A(s)$.

B. Stability Tests[4, 6]

In any feedback system, care must be taken to assure that the system is not prone to oscillations at *any* frequency, i.e., that the system is *stable*. While sustained oscillations necessarily involve nonlinearities, analytical tests to check the system's stability are generally made on a small-signal linear-system basis.

There are a number of classical analytical tests that are used to check the stability of a linear system. Observe from Eq. (2-20) that the frequency response of a feedback system $H(j\omega)$ becomes unbounded (indicating oscillations) if the open-loop response $A(j\omega)B(j\omega) = -1$. Thus one criterion of stability is that the *open-loop amplitude response fall below unity before the phase response reaches* 180°. This criterion is relatively stringent, but is handy whenever the open-loop frequency response is known graphically. Furthermore, for minimum phase networks, which include most of our common circuits (with some exceptions such as the lattice type), the amplitude and phase responses are intimately related, and the stability criterion can be given in terms of the amplitude response only. Thus, a slope of 6 dB/octave of frequency causes a maximum phase shift of 90°, a slope of 12 dB/octave—a maximum phase shift of 180°, etc. A minimum phase system in which the open-loop amplitude response never has a greater slope than 12 dB/octave of frequency, and over some region of frequency it has a smaller slope, never reaches a phase shift of 180° and is unconditionally stable. In practice some allowance must be made for extraneous phase shifts not accounted for.

Our systems will generally be designed to meet the stability requirement just described, as illustrated by the typical open-loop PLL amplitude response shown in Fig. 2-5. Actually, the curve consists of the asymptotes to the response, referred to as a *Bode plot*. We shall find, however, that inherent modulation and noise characteristics, and inherent nonlinearities in the systems, will result in various interesting types of instabilities.

Another very popular and useful stability test is in terms of the s-plane representation of the closed-loop transfer function $H(s)$. Oscillations occur whenever there are poles in the right half plane. We shall find occasions in which this test will be more convenient.

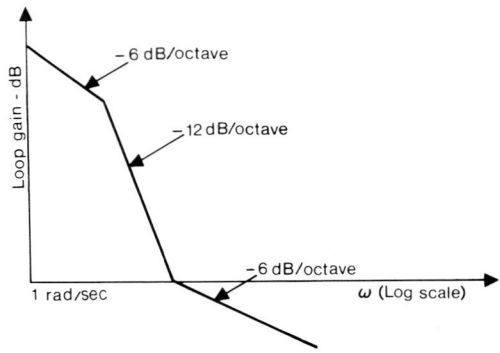

Fig. 2-5. Typical open-loop amplitude response of second-order PLL.

2.3. Representation and Properties of Noise[7, 8]

Much of this text is concerned with the improved extraction of signals embedded in noise, by use of phase-locked and frequency-feedback systems. It will, therefore, be necessary to be familiar with the analytical properties and representation of noise and random signals.

Consider a random noise voltage $n(t)$. Its actual waveshape is unpredictable and can be described only in a *statistical* manner. We shall generally assume that $n(t)$ is *stationary*; that is, all statistical properties are constant with time. Often, the full statistical data are not required. Instead, some characteristic numbers representing certain *averages* are sufficient. In addition to the familiar averaging of a parameter of a known waveshape in time, there is also the *statistical average*, defined analytically below. We shall generally assume that the random process is *ergodic*, i.e., that statistical or time averaging gives identical results, and we will therefore conveniently designate any averaging by an overbar.

A. Time Averages

A. AVERAGE OR DC VALUE

$$\bar{n}(t) = \lim_{T \to \infty} \frac{1}{2T} \int_{-T}^{T} n(t)\, dt \qquad (2\text{-}25)$$

B. MEAN-SQUARE VALUE OR POWER

$$\overline{n^2(t)} = \lim_{T \to \infty} \frac{1}{2T} \int_{-T}^{T} n^2(t)\, dt \qquad (2\text{-}26)$$

2.3. Representation and Properties of Noise

C. rms VALUE

$$\text{rms} = \left[\overline{n^2(t)}\right]^{1/2} \tag{2-27}$$

D. AUTOCORRELATION FUNCTION

$$R(\tau) = \lim_{T \to \infty} \frac{1}{2T} \int_{-T}^{T} n(t)n(t+\tau)\,dt \tag{2-28}$$

B. Statistical Description and Averages

A. PROBABILITY DENSITY FUNCTION (PDF): $p(n)$

We define $p(n)$ in terms of its integral, viz., $\int_{n_1}^{n_2} p(n)\,dn$ is the probability of n having a value between n_1 and n_2. This is illustrated in Fig. 2-6. The three important properties of $p(n)$ on which all of probability theory may be based are

$$p(n) \geq 0 \quad \text{for all } n \tag{2-29}$$

$$\int_{-\infty}^{\infty} p(n)\,dn = 1 \tag{2-30}$$

and that the probability of two mutually exclusive events equals the sum of the individual probabilities.

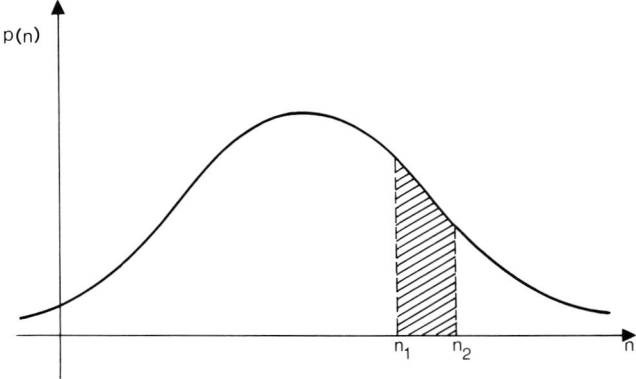

Fig. 2-6. Illustration of probability density function. (Shaded area equals the probability that n is between n_1 and n_2. Note that $p(n)$ is always positive and the total area under the curve equals unity.)

B. Expected Value, Mean, or Statistical Average

$$\bar{n} = \int_{-\infty}^{\infty} np(n)\, dn \qquad (2\text{-}31)$$

Physically, statistical averaging may be visualized as follows: A large number of waveforms having the same statistics are stacked vertically. Statistical averaging is along a vertical line (i.e., at a particular time t_1), picking a value from each waveform, while time averaging is performed horizontally along a single waveform.

C. Mean-Square Value (MSV)

$$\overline{n^2} = \int_{-\infty}^{\infty} n^2 p(n)\, dn \qquad (2\text{-}32)$$

D. Variance σ^2

$$\sigma^2 = \overline{(n-\bar{n})^2} = \overline{n^2} - \bar{n}^2 \qquad (2\text{-}33)$$

where σ is termed the *standard deviation*.

E. Autocorrelation Function

$$R(t_1, t_2) = \overline{n(t_1)n(t_2)} \qquad (2\text{-}34)$$

F. Ergodicity

Ergodicity implies stationarity and that all time and statistical averages are the same, viz.,

$$\overline{n(t)} = \bar{n} \qquad (2\text{-}35)$$

$$\overline{n^2(t)} = \overline{n^2} \qquad (2\text{-}36)$$

Now σ^2 is power in time-varying component, σ is rms of $n(t)$, and

$$R(t_1, t_2) = R(\tau) \qquad (2\text{-}37)$$

We shall generally assume that the noise is of zero mean, resulting in

$$\overline{n^2} = \sigma^2 \qquad (2\text{-}38)$$

G. Gaussian Distribution

It will be frequently assumed that the random function is of a Gaussian distribution, defined in terms of the PDF as

$$p(n) = \frac{1}{(2\pi\sigma^2)^{1/2}} \exp\left[\frac{-(n-\bar{n})^2}{2\sigma^2}\right] \qquad (2\text{-}39)$$

2.3. Representation and Properties of Noise

A useful property of a Gaussian random process is that its PDF remains unchanged after linear processing such as differentiation, integration, or linear filtering. This is generally not true with other distributions.

H. Poisson Distribution

The Poisson distribution is representative of a discrete random process and is defined by

$$P(n) = e^{-\bar{n}T}(\bar{n}T)^n/n! \tag{2-40}$$

where $P(n)$ is the probability of exactly n events in time interval T, and $\bar{n}$ is the average number of events per second. It will be used to describe the statistics of certain spikes at the output of FM discriminators.

C. Power Spectra and Filtering

A.

The *Power spectral density* [PSD or $W(f)$] is defined as the Fourier transform of the autocorrelation function

$$W_2(f) = \int_{-\infty}^{\infty} R(\tau)e^{-j\omega\tau}\, d\tau, \qquad \omega = 2\pi f \tag{2-41}$$

with the inverse relationship

$$R(\tau) = \int_{-\infty}^{\infty} W_2(f)e^{j\omega\tau}\, df \tag{2-42}$$

The subscript two is to identify the PSD as two-sided, having values for both positive and negative frequencies. Equations (2-41) and (2-42) are known as the *Wiener–Khintchine theorem*.

B.

For real time waveforms, $W_2(f)$ has even symmetry about $f = 0$, i.e.,

$$W_2(-f) = W_2(f) \tag{2-43}$$

It is therefore possible to define a one-sided PSD $W_1(f)$ having twice the value but ranging only over positive frequencies, i.e.,

$$W_1(f) = \begin{cases} 2W_2(f), & f \geq 0 \\ 0, & f < 0 \end{cases} \tag{2-44}$$

We shall generally use the one-sided representation and drop the subscript one. In the isolated instances where a two-sided PSD is used, it is made evident in the text.

C.

The PSD represents average power per cycle of bandwidth (watts per hertz) and when integrated over all frequencies yields the msv (or average power) of the wave, assuming as is generally done, that the waveform is across a 1-Ω resistance

$$\int_0^\infty W(f)\,df = \overline{n^2(t)} \tag{2-45}$$

D.

Input–output relationships: The PSD at the output of a linear filter $W_o(f)$ is obtained from the PSD at the input $W_i(f)$ and the filter amplitude response $|H(j\omega)|$ as

$$W_o(f) = |H(j\omega)|^2 W_i(f) \tag{2-46}$$

Note that the phase response of the filter is of no significance. The term $|H(j\omega)|^2$ is also referred to as the *power response*.

The autocorrelation function at the output of a linear filter $R_o(\tau)$ is

$$R_o(\tau) = \int_{-\infty}^\infty h(\lambda) \left[\int_{-\infty}^\infty h(t) R_i(\tau + \lambda - t)\,dt \right] d\lambda \tag{2-47}$$

where $R_i(\tau)$ is the autocorrelation function at the input, and λ is a dummy variable.

E.

A PSD is termed *white* or *flat* if it has a constant value over all frequencies. In practice, it is sufficient for the noise PSD to be flat over a bandwidth substantially wider than the passband of the system to be termed white. Noise with a nonflat PSD is referred to as *colored*.

D. Narrowband Noise

A bandpass noise is narrowband† if its bandwidth is small compared to its center frequency ω_i. Such a noise is conveniently written as a modulated wave

$$N(t) = x(t)\cos\omega_i t - y(t)\sin\omega_i t \tag{2-48}$$

---

† This representation is also valid for wideband noise. However, the representation by an amplitude- and phase-modulated sine wave is more meaningful for a narrowband process, as can be observed on an oscilloscope. The noise bandwidth can be no wider than twice the center frequency in order to have symmetry about the center frequency.

2.3. Representation and Properties of Noise

or

$$N(t) = [x^2(t) + y^2(t)]^{1/2} \cos\left[\omega_i t + \tan^{-1}\frac{y(t)}{x(t)}\right] \quad (2\text{-}49)$$

with $x(t)$ and $y(t)$ having the following properties:

A.

The spectra of $x(t)$ and $y(t)$ are lowpass in nature, and can be derived from the PSD of $N(t)$: If $W(f)$ has arithmetic symmetry about ω_i, then $x(t)$ and $y(t)$ have spectra of the lowpass equivalent shape with twice the amplitude. For example, if the spectrum of $N(t)$ is a bandpass rectangle of width B, height η, and centered about ω_i, i.e.,

$$W(f) = \begin{cases} \eta, & f_i - \tfrac{1}{2}B \le f \le f_i + \tfrac{1}{2}B \\ 0, & \text{elsewhere} \end{cases} \quad (2\text{-}50)$$

then $x(t)$ and $y(t)$ have the spectra $W_x(f)$ and $W_y(f)$, respectively,

$$W_x(f) = W_y(f) = \begin{cases} 2\eta, & f \le \tfrac{1}{2}B \\ 0, & f > \tfrac{1}{2}B \end{cases} \quad (2\text{-}51)$$

B.

If $N(t)$ is Gaussian, then $x(t)$ and $y(t)$ are also Gaussian.

C.

If $N(t)$ has zero mean, then $x(t)$ and $y(t)$ also have zero mean values.

D.

The variance of $x(t)$ and $y(t)$ is the same as that of $N(t)$

$$\sigma^2 = \sigma_x^2 = \sigma_y^2 \quad (2\text{-}52)$$

E.

The functions $x(t)$ and $y(t)$ are statistically independent.

E. Equivalent Noise Bandwidth

The *equivalent noise bandwidth* of a system B_n is defined in terms of the power response as

$$B_n = \frac{1}{H_0^2} \int_0^\infty |H(j\omega)|^2 \, df \quad (\text{hertz}) \quad (2\text{-}53)$$

where H_0^2 is generally taken as the power response at DC for a lowpass wave or at the center frequency for a bandpass wave. As illustrated in Fig. 2-7, B_n equals the bandwidth of a (fictitious) rectangular filter that would pass as much white-noise power as the system in question. We shall often drop the adjective "equivalent" and refer to it simply as the noise bandwidth.

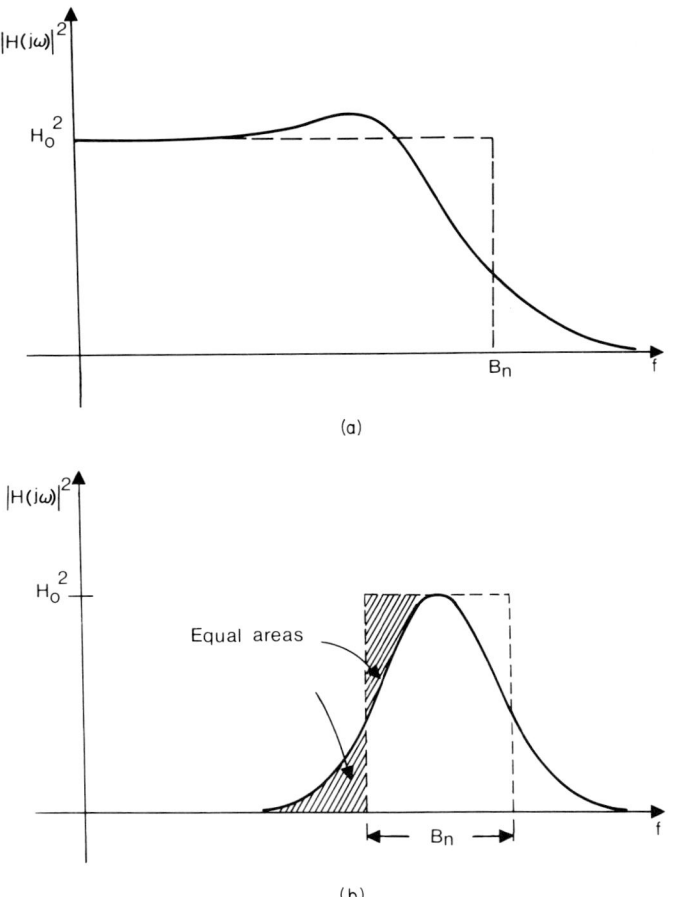

Fig. 2-7. Illustration of the equivalent noise bandwidth (solid and dashed lines enclose equal areas). (a) Lowpass response; (b) bandpass response.

Why use it? It permits a quick determination of the noise power P_n at the output of a system, if the input power spectral density is white, say of value η

$$P_n = \eta B_n H_0^2 \qquad (2\text{-}54)$$

Thus system noise rejection is judged by the value of B_n.

To calculate the equivalent noise bandwidth using Eq. (2-53), it is generally required to solve a definite integral of the form

$$I_n = \int_0^\infty \left| \frac{c_{n-1}(j\omega)^{n-1} + \cdots + c_0}{d_n(j\omega)^n + \cdots + d_0} \right|^2 df \tag{2-55}$$

As an aid, the solutions to these integrals[1] are tabulated below for n, up to $n = 4$.

$$I_1 = \frac{c_0^2}{4d_0 d_1}$$

$$I_2 = \frac{c_1^2 d_0 + c_0^2 d_2}{4d_0 d_1 d_2}$$

$$I_3 = \frac{c_2^2 d_0 d_1 + (c_1^2 - 2c_0 c_2)d_0 d_3 + c_0^2 d_2 d_3}{4d_0 d_3 (d_1 d_2 - d_0 d_3)} \tag{2-56}$$

$$I_4 = \frac{c_3^2(d_0 d_1 d_2 - d_0^2 d_3) + (c_2^2 - 2c_1 c_3)d_0 d_1 d_4 + (c_1^2 - 2c_0 c_2)d_0 d_3 d_4 + (d_2 d_3 d_4 - d_1 d_4^2)c_0^2}{4d_0 d_4 (d_1 d_2 d_3 - d_0 d_3^2 - d_1^2 d_4)}$$

2.4. Trigonometric Identities

The following trigonometric identities are repeatedly used in the text. They are, therefore, listed here for convenience:

$$\cos(\alpha \pm \beta) = \cos \alpha \cos \beta \mp \sin \alpha \sin \beta \tag{2-57}$$

$$\sin(\alpha \pm \beta) = \sin \alpha \cos \beta \pm \cos \alpha \sin \beta \tag{2-58}$$

$$\cos \alpha \cos \beta = \tfrac{1}{2}[\cos(\alpha + \beta) + \cos(\alpha - \beta)] \tag{2-59}$$

$$\sin \alpha \sin \beta = \tfrac{1}{2}[\cos(\alpha - \beta) - \cos(\alpha + \beta)] \tag{2-60}$$

$$\sin \alpha \cos \beta = \tfrac{1}{2}[\sin(\alpha + \beta) + \sin(\alpha - \beta)] \tag{2-61}$$

$$\cos \alpha \sin \beta = \tfrac{1}{2}[\sin(\alpha + \beta) - \sin(\alpha - \beta)] \tag{2-62}$$

References

1. W. W. Seifert and C. W. Steeg, "Control Systems Engineering." McGraw-Hill, New York, 1962.
2. A. Papoulis, "The Fourier Integral and Its Applications." McGraw-Hill, New York, 1962.
3. B. Van der Pol and H. Bremmer, "Operational Calculus Based on the Two-Sided Laplace Integral." Cambridge Univ. Press, London and New York, 1955.

4. H. Chestnut and R. W. Mayer, "Servomechanisms and Regulating System Design." Wiley, New York, 1951.
5. M. E. Van Valkenburg, "Network Analysis." Prentice-Hall, Englewood Cliffs, New Jersey, 1964.
6. H. W. Bode, "Network Analysis and Feedback Amplifier Design." Van Nostrand-Reinhold, Princeton, New Jersey, 1945.
7. W. B. Davenport and W. L. Root, "Random Signals and Noise." McGraw-Hill, New York, 1958.
8. A. Papoulis, "Probability, Random Variables, and Stochastic Processes." McGraw-Hill, New York, 1965.

CHAPTER

3

Loop Components and Systems Considerations

This chapter describes and defines the salient components of the PLL and FMFB systems and introduces some FM system considerations. The limiter-discriminator is discussed in greater detail because it is both a critical element in the FMFB loop and it is the "conventional FM demodulator," which serves as a yardstick against which improved performance of the angular feedback loops is measured. The major application of the PLL and FMFB systems stressed in this text is in FM demodulation with *lowered thresholds*. (The term "threshold" is defined presently.) The reasons for the desirability of low-threshold demodulation are discussed, and an indication is given as to the extent of improved performance available from the advanced FM demodulators described in the succeeding chapters. The detailed design of the loop components is not given, and analytical results are sometimes stated without derivations. For details, the reader is referred to the references. Some elementary knowledge of modulation theory is assumed.

3.1. Introduction

FM reception using any demodulator is characterized by a performance curve such as that shown in Fig. 3-1. The curve gives the output fidelity as a function of the purity of the received FM carrier. Three distinct regions are indicated: (1) the FM improvement region, (2) the below-threshold region, and (3) the distortion-limited region.

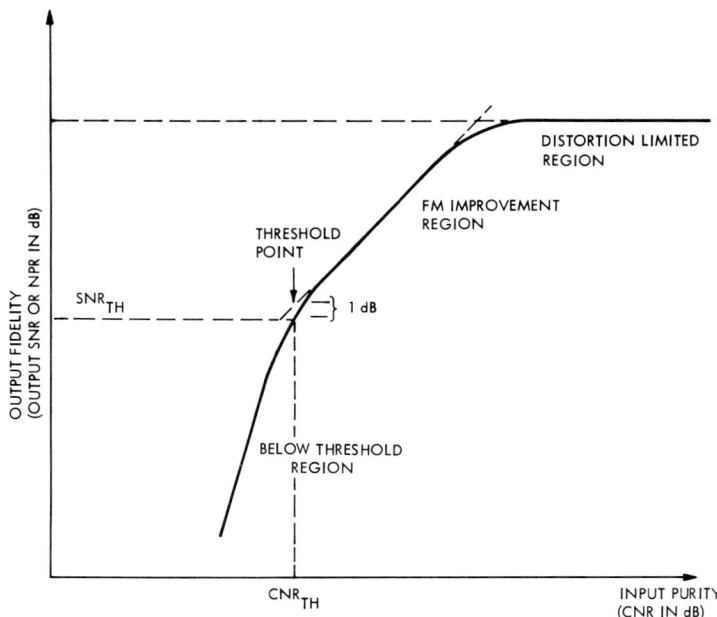

Fig. 3-1. Typical performance characteristics of FM receiving systems. (SNR is signal-to-noise ratio, NPR is noise power ratio, CNR is carrier-to-noise ratio, SNR_{TH} and CNR_{TH} are SNR and CNR values at threshold, respectively.)

The FM *improvement region* encompasses the usual operating conditions. Its name is adapted from the fact that the quality of reception for a given transmitter power can be improved at the cost of bandwidth. This is illustrated in Fig. 3-5, where a family of such curves is presented with the modulation index as a parameter.

The *below-threshold region* is marked by a rapid deterioration of output fidelity. The physical basis of the poor performance is discussed presently, as are the means of reducing the "threshold point" through the application of angular feedback.

Finally, for a high received carrier-to-noise ratio (CNR) the output fidelity is limited by the system distortions, resulting in a *distortion-limited region*. Introductory remarks on the sources and control of distortion are given in this chapter. The details have been left for the design chapters.

3.2. Limiter-Discriminator†

The limiter-discriminator (LD) is a basic element in FM reception. In the FM improvement region, its performance is essentially the same as that of

† The limiter-discriminator (LD) is used here as a generic term for all known non-compressive types of FM demodulators that are insensitive to AM. Their performance in terms of noise is about the same, and they include the Foster–Seeley, pulse counting, delay line, zero crossing, or the ratio detector types.

the low-threshold, angular-feedback demodulators. It is a critical element in the FMFB loop, and its performance serves as a yardstick against which improvements by angular feedback are measured.

The noise performance of an LD in the FM improvement region presents little problem in analysis. However, when noise peaks become sufficiently frequent to cause a "threshold" condition in analog FM reception or errors in digital FM demodulation, its exact theory is generally too cumbersome. Therefore, the material on threshold is based on an approximate analysis, which holds sufficiently well for engineering purposes and provides an intuitive description of LD operation. Furthermore, the theory presented serves as a useful introduction to the operational principles of the phase-locked and FMFB loops.

The actual design of discriminators is not treated herein. A design for low distortion is described in Section 7.2. For further material, the reader is referred to the literature.[1-3]

A. Operational Principles

Operationally, the LD can be represented as consisting of (a) a limiter, (b) a differentiator, and (c) an envelope detector. In its usual mode of operation, the LD is preceded by band-limiting circuitry (or predetection filter) and is followed by a baseband filter (or postdetection filter). This demodulator, referred to as the "conventional" FM demodulator, is shown in block diagram form in Fig. 3-2.

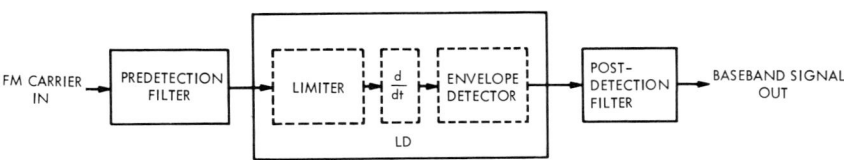

Fig. 3-2. Conventional FM demodulator.

An FM carrier can be written as

$$e_i(t) = A \cos[\omega_i t + \phi_i(t)] \qquad (3\text{-}1)$$

where A is the constant amplitude of the carrier, ω_i is the unmodulated frequency, and $\phi_i(t)$ is proportional to the time-integral of the modulating signal $e_m(t)$, or

$$e_m(t) = \frac{d}{dt}\phi_i(t) \equiv \dot{\phi}_i(t) \qquad (3\text{-}2)$$

Angle $\phi_i(t)$ is the *phase deviation*, and $\dot{\phi}_i(t)$ is the *instantaneous frequency deviation* of the carrier due to the modulation. Clearly, the limiter is not needed to demodulate an uncorrupted FM wave described by Eq. (3-1). The output of the differentiator is

$$[\omega_i + \dot{\phi}_i(t)]A' \sin[\omega_i t + \phi_i(t) + \pi] \qquad (3\text{-}3)$$

i.e., the differentiator transforms the FM into AM, with the percentage of AM modulation given by the ratio of the maximum frequency deviation to the carrier frequency

$$\%\text{AM} = \frac{|\dot{\phi}_i|_{\text{peak}}}{\omega_i} \times 100 \qquad (3\text{-}4)$$

The constant A was replaced by A' to allow for amplitude changes by the limiter and for a proportionality constant in differentiation. Finally, the desired baseband signal is obtained at the output of the envelope detector as $\lambda\dot{\phi}_i(t)$, where λ is the *LD sensitivity* in volt-seconds per radian. The LD is usually of the balanced type, so that the DC term due to ω_i is canceled out.

In a practical LD, it may not be possible to break the circuitry into these functional parts. However, it is useful to keep the three functions in mind in order to appreciate both the operational principles of an LD and the sources of possible malfunctions. For example, the choice of low-frequency time constants in the envelope detector is made in the same manner as for an AM wave with the percentage of modulation given by Eq. (3-4). To minimize corruption of the output by noise, the predetection and postdetection filters are inserted as shown in Fig. 3-2.

If the received signal $e_r(t)$ is corrupted by additive noise $N(t)$, then it may be written as

$$e_r(t) = e_i(t) + N(t) = R(t) \cos[\omega_i t + \phi_i(t) + \theta_n(t)] \qquad (3\text{-}5)$$

The resultant of the sum of an FM carrier and additive noise is a carrier which is both amplitude and frequency modulated, with $R(t)$ and $\theta_n(t)$ as stochastic quantities. The latter form of Eq. (3-5) assumes that the noise is a bandpass type, written in the form

$$N(t) = x(t) \cos \omega_i t - y(t) \sin \omega_i t \qquad (3\text{-}6)$$

where the first component is in phase and the second term is in quadrature with the signal carrier [see Eq. (2-48)]. It may also be written as

$$N(t) = r(t) \cos[\omega_i t + \psi_n(t)] \qquad (3\text{-}7)$$

The properties of $x(t)$ and $y(t)$ are described in Section 2.3. The distributions of R and θ_n are functions of the CNR.

It is helpful to represent by phasors the carrier and noise,[4] as shown in

3.2. Limiter-Discriminator   27

Fig. 3-3. First, let the carrier be unmodulated and be considered the reference in phase. The phasor addition of carrier and noise produces a resultant having a time-varying amplitude $R(t)$ and angle $\theta_n(t)$, per Eq. (3-5). It is assumed that the limiter removes completely the AM from $R(t)$; therefore, only the angle $\theta_n(t)$ and its time derivative $\dot{\theta}_n(t)$ are of interest. It is observed from Fig. 3-3a that the angle $\theta_n(t)$ is given by

$$\theta_n(t) = \tan^{-1}\frac{y(t)}{[A + x(t)]} \qquad [\phi_i(t) = 0] \qquad (3\text{-}8)$$

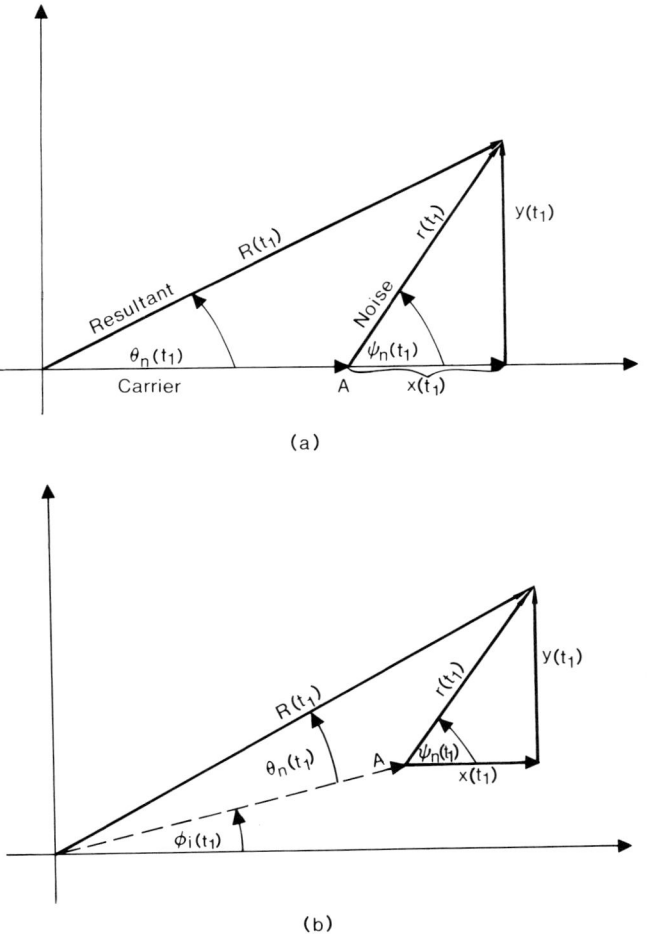

Fig. 3-3. Phasor representation of carrier and noise. (a) No modulation. (b) With modulation.

3. Loop Components and Systems Considerations

At present we shall discuss only the case where x and y are small compared to A most of the time, so that the angle due to the noise can be approximated by

$$\theta_n(t) \cong y(t)/A \tag{3-9}$$

This approximation holds very well above "threshold." The output noise is proportional to the derivative of the angle, viz.,

$$\dot{\theta}_n(t) \cong \dot{y}(t)/A \tag{3-10}$$

It is of interest to note that the output noise given by Eq. (3-10) is of a Gaussian distribution, since y is Gaussian, and the result of a linear operation (derivative) on a Gaussian process is a Gaussian process. For the case when the carrier is modulated, the noise angle becomes a function of the modulation and is given from Fig. 3-3(b) by[5]

$$\theta_n(t) = \tan^{-1} \frac{r \sin(\psi_n - \phi_i)}{A + r \cos(\psi_n - \phi_i)} = \tan^{-1} \frac{y \cos \phi_i - x \sin \phi_i}{A + x \cos \phi_i + y \sin \phi_i} \tag{3-11}$$

The dependence on time is assumed but not indicated, to simplify the notation. Under modulation, the equivalent of Eq. (3-9) is

$$\theta_n(t) \cong (y \cos \phi_i - x \sin \phi_i)/A \tag{3-12}$$

Generally, the inclusion of modulation heavily complicates the analysis. Fortunately, Eq. (3-9) usually represents sufficiently well the output noise above threshold even with a modulated carrier. Thus, for high CNR, the input wave can, from Eqs. (3-5) and (3-9), be approximated by

$$e_r \cong R \cos[\omega_i t + \phi_i + (y/A)] \tag{3-13}$$

and the output of the LD is

$$e_o \cong \dot{\phi}_i + (\dot{y}/A) \tag{3-14}$$

where the demodulation sensitivity λ was taken as unity.

In order to obtain a relation between the output signal-to-noise ratio (SNR) and the input CNR, it is necessary to invoke the ratio of the noise bandwidth of the predetection to the postdetection filters (Fig. 3-2), as well as the consideration of the power spectra. Assuming that both the predetection and postdetection filters are rectangular types, the modulation is a sinusoidal tone of radian frequency ω_T, and the input noise is white, we obtain the well-known relationship[6]

$$(SNR)_{TT} = 3m_p^2 (CNR)_{AM} \tag{3-15}$$

where $(CNR)_{AM}$, also written as $(CNR)_{2f_b}$, is the carrier-to-noise power ratio with the noise measured in a filter bandwidth of twice the baseband (twice the

3.2. Limiter-Discriminator

message bandwidth), $(SNR)_{TT}$ is the output signal-to-noise power ratio with the noise measured in base bandwidth equal to the test-tone frequency f_T, and m_P is the modulation index defined as the ratio of the peak deviation to the top baseband frequency.

Why is the CNR treated in terms of the noise power in twice the baseband? To begin with, it permits a direct comparison against AM whose rf bandwidth is actually twice the baseband. More importantly, however, it permits a comparison of the performance as a function of m_P for the same interference condition, i.e., for the same noise power density at the receiver input.

It is instructive to derive Eq. (3-15) for it presents insight into the process by which FM gives noise improvement. Let the peak frequency deviation be $\Delta\omega_P$; then the average output signal power P_S is

$$P_S = (\lambda \Delta\omega_P)^2/2 \tag{3-16}$$

(without loss of generality the load impedance was assumed unity.) Now, the output noise PSD is from Eq. (3-10), (2-46), and Rule 6 of Table I

$$W_o(f) = |\lambda j\omega|^2 \, W_y(f)/A^2 = \lambda^2\omega^2 2\eta/A^2 \tag{3-17}$$

where η is the input noise power spectral density. The total noise power at the output is

$$\overline{n_o^2(t)} = \int_0^{f_T} (2\lambda^2\omega^2\eta/A^2) \, df = 2\lambda^2\eta f_T^3 (2\pi)^2/(3A^2) \tag{3-18}$$

The output SNR is simply the ratio of (3-16) and (3-18)

$$(SNR)_{TT} = 3(\Delta\omega_P/\omega_T)^2 (A^2/2)/(\eta 2 f_T) = 3m_P^2 \, (CNR)_{AM} \tag{3-15'}$$

Following the steps, and as illustrated in Fig. 3-4, one observes that the FM demodulator transforms a flat power spectral density into a parabolic type, and that the output filter passes only the noise from the region near the beginning of the parabola where the values are small. This is the mechanism by which FM has greater noise immunity than PM [see Eq. (3-21)]. Both FM and PM offer an output SNR proportional to the square of the modulation index, traced to the fact that the noise in these systems enters through the angle. These systems also are capable of greater transmitter efficiency over AM because class-C amplification is permitted.

Other performance equations are similarly derived from Eq. (3-14), in terms of appropriate system parameters, as given below (for derivations see Appendix A).

A. SINGLE-CHANNEL SPEECH [frequency modulation (FM) and phase modulation (PM)]

$$SNR_{SP} = 6\sigma^2 (CNR)_{AM} \quad\quad (FM) \tag{3-19}$$

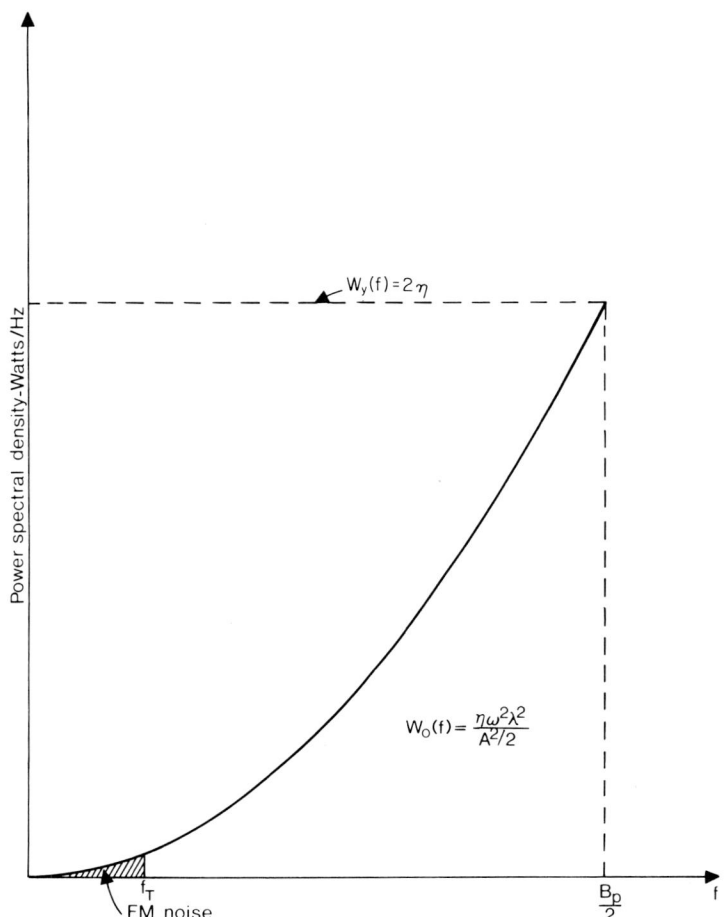

Fig. 3-4. Illustration of noise power spectra in an FM receiver (B_p is predetection bandwidth; curve assumes $\lambda \pi B_p = A$).

$$\text{SNR}_{\text{SP}} = 2(\omega_b/\omega_a)\sigma^2(\text{CNR})_{\text{AM}} \qquad \text{(PM)} \qquad (3\text{-}20)$$

$$\text{SNR}_{\text{TT}} = m_P^2(\omega_b/\omega_T)^2(\text{CNR})_{\text{AM}} \qquad \text{(PM)} \qquad (3\text{-}21)$$

B. Frequency Division Multiplex† (FM and PM)

$$\text{NPR} = \frac{2(\omega_b/\omega_{\text{CH}})^2 \sigma^2}{[1 - \omega_a/\omega_b]} (\text{CNR})_{\text{AM}} \qquad \text{(FM)} \qquad (3\text{-}22)$$

† High-Q channels are assumed ($2\pi B \ll \omega_{\text{CH}}$), i.e., the ratio of power spectral densities adequately determines channel NPR.

3.2. Limiter-Discriminator

$$\text{NPR} = \frac{6\sigma^2}{[1-(\omega_a/\omega_b)^3]}(\text{CNR})_{\text{AM}} \qquad \text{(PM)} \qquad (3\text{-}23)$$

$$\text{SNR}_{\text{TT}} = (f_b/B)(\omega_b/\omega_{\text{CH}})^2 m_P{}^2 (\text{CNR})_{\text{AM}} \qquad \text{(FM or PM)} \quad (3\text{-}24)$$

with the symbols defined as follows:

SNR = output SNR for speech waveform (SP) or test-tone (TT)
NPR† = noise power ratio = the power ratio of signal-to-noise tested with a noise-like signal in a narrow FDM channel
$(\text{CNR})_{\text{AM}}$ = input CNR referred to twice base bandwidth ($2f_b$)
σ = rms modulation index = $\dfrac{\text{rms carrier frequency deviation}}{\text{top baseband frequency}} = \dfrac{\Delta\omega_{\text{rms}}}{\omega_b}$
m_P = peak modulation index = $\dfrac{\text{peak carrier frequency deviation}}{\text{top baseband frequency}} = \dfrac{\Delta\omega_P}{\omega_b}$
ω_a = bottom frequency of speech spectrum model (nominally $2\pi \times 10^3$ radians per second) or bottom frequency of frequency division multiplex (FDM) baseband (radians per second)
ω_b = top baseband frequency of transmission (radians per second)
ω_{ch} = channel center frequency (radians per second)
ω_T = test-tone (TT) frequency (radians per second, nominally $2\pi \times 10^3$ radians per second for speech)
B = channel noise bandwidth (FDM) in hertz

The assumptions are the same as before; namely, the system is in the FM improvement region, the modulation has a negligible effect on the noise power output, the input noise is white, and the pre- and postdetection filters are rectangular. The speech and FDM signals were modeled as described in Appendix A.

A relationship between the required predetection bandwidth and the modulation parameters found useful in practice is that given by *Carson's rule*,

$$B_{\text{CR}} = 2(1 + m_P)f_b \qquad (3\text{-}25)$$

where B_{CR} is the Carson's rule bandwidth. Returning to tone modulation, the measured improvement in noise from the LD input to the postdetection filter output is, from Eqs. (3-15) and (3-25),

$$(\text{SNR})_{f_b}/(\text{CNR})_{\text{CR}} = 3m_P{}^2(1 + m_P) \qquad (3\text{-}26)$$

where $(\text{SNR})_{f_b}$ is the output SNR measured in base bandwidth, and $(\text{CNR})_{\text{CR}}$ is the input CNR measured in Carson's rule bandwidth.

Another measure is the improvement in SNR due to the use of FM rather than AM. For 100% modulation, the output SNR of an AM receiver is the same as its unmodulated input CNR. Since AM requires a predetection filter

† For a more detailed definition and method of measurement see Section 10.3.

of twice the baseband, the improvement due to FM is, from Eq. (3-15), simply

$$(\text{SNR})_{\text{FM}}/(\text{SNR})_{\text{AM}} = 3m_\text{P}^2 \qquad (3\text{-}27)$$

which is also the improvement of FM over a suppressed-carrier AM system. The above equations show quantitatively the most important property of FM, namely, that transmitter power can be traded for spectrum occupancy. However, as we shall see, the threshold mechanism sets a limit to this tradeoff, and a basic advantage of the low-threshold demodulators described presently is that they provide an increase in the range over which this tradeoff is possible.

Figure 3-5 gives a plot of SNR (in decibels) vs $(\text{CNR})_{2f_b}$ (in decibels) with m_P as a parameter[7] including the FM improvement and below threshold regions. For high CNR, the FM noise-improvement factor is readily apparent, and the SNR–CNR relation is linear. As the received CNR decreases, a region appears where the SNR drops off more rapidly than predicted by Eq. (3-15). The point at which the SNR has dropped by 1 dB more than

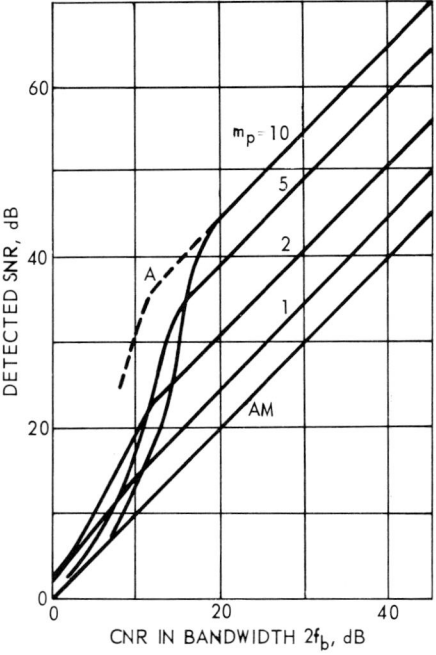

Fig. 3-5. Variation of detected SNR with input CNR level. For AM : Input CNR with no modulation, output SNR with 100% modulation; A is hypothetical low-threshold FM demodulator (adapted from Walsh[7]).

predicted by the high-CNR linear relation is defined as the *threshold point*. This definition of FM threshold is convenient and reasonably widespread, although other definitions have appeared in the literature. Ideally, the definition of threshold should incorporate the effect of the type of service and the results of subjective tests on the effect of the noise.

An important consideration is that the threshold CNR level increases as m_P is increased. This is the phenomenon which limits the power-bandwidth tradeoff. Referring to Fig. 3-5, consider the following example:

Problem. The system requires an output SNR of 30 dB, which is to be achieved with a minimum of transmitted power. It is permitted to trade transmitter power for rf bandwidth occupancy. The received signal is corrupted by additive noise, which may be considered to be of a Gaussian distribution with a power spectrum density, η, sufficiently flat over the rf bandwidth of interest. Select the best modulation index m_P.

Solution. In Fig. 3-5, follow the horizontal line of SNR = 30 dB and observe the intersections with the solid curves. Of the four modulation indexes indicated, $m_P = 5$ requires the least received carrier power. It is observed that due to the threshold effect, $m_P = 10$ requires a larger received carrier power.

The low-threshold FM demodulators, which are discussed in detail presently, feature thresholds at lower CNR's. It is apparent from Fig. 3-5 that for a demodulator with a lower threshold, a further tradeoff between received carrier power and bandwidth is possible.

Another, perhaps more obvious, advantage of a demodulator with a reduced threshold is that it permits reception of FM signals in fringe areas. Referring again to Fig. 3-5, for a $(CNR)_{2f_b}$ of 10 dB and $m_P = 10$, the conventional FM demodulator has an output SNR of about 13 dB, while the hypothetical low-threshold demodulator, "A" (dashed curve), has an SNR of 30 dB because its rapid dropoff began at a lower CNR.

B. Threshold

A phasor representation of the received carrier and noise was given in Fig. 3-3. On an instantaneous basis, the noise angle θ_n will generally behave as shown in (a) of Fig. 3-6. As the noise amplitude is small most of the time, angle θ_n will also be small, in the main. However, occasionally the noise is large enough to cause an encirclement of the origin by the resultant. This results in a 2π shift of the resultant's phase angle, and a new reference point is created about which the resultant again makes small undulations.

The output of the discriminator is illustrated in (b) of Fig. 3-6. It is apparent that the origin encirclements result in "spike" noise at the output.

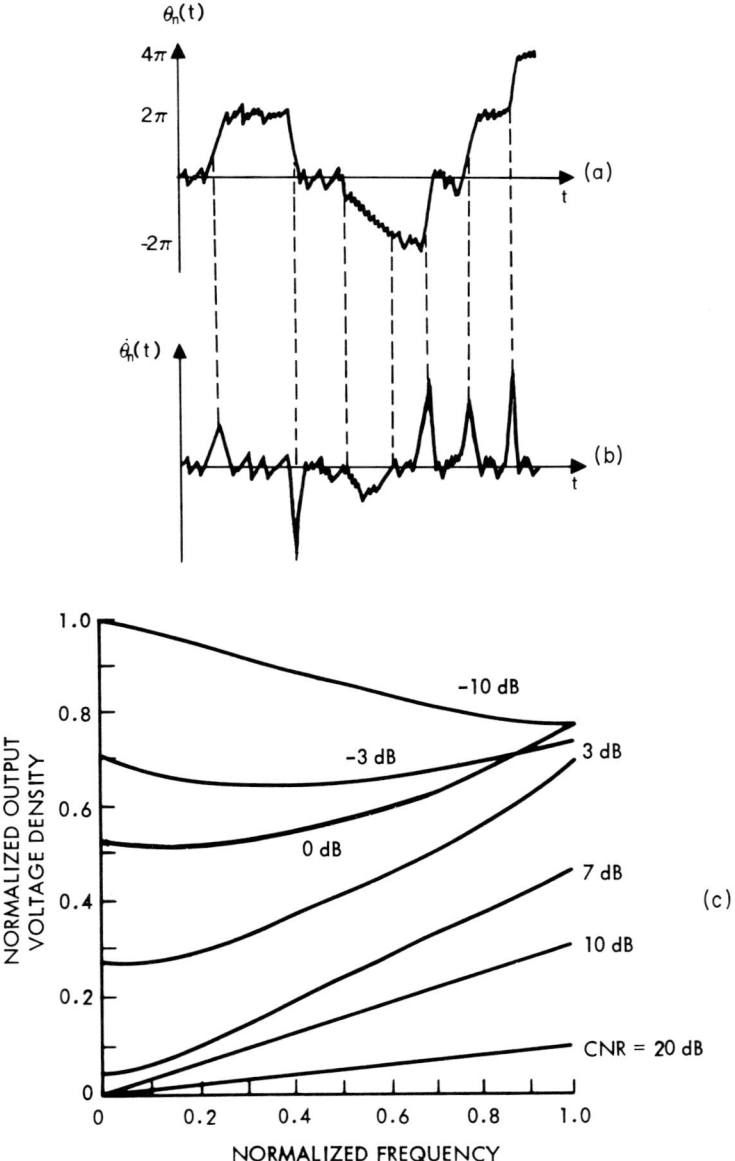

Fig. 3-6. Threshold effects. (a) Phase steps. (b) Noise spikes. (c) Detected noise voltage spectra. (The rms voltage of the noise in a small band of B Hz is $(B/B_p)^{1/2}$ times the value given by the curve; from Stumpers.[8])

3.2. Limiter-Discriminator

These spikes are generally sharp compared to the information signal and, therefore, have a spectral distribution that is essentially flat over the lowpass bandwidth of interest. It is these origin encirclements that essentially are the cause of threshold because of their large low-frequency content.† With the onset of threshold, the output noise spectrum alters to one having a higher low-frequency content, as illustrated in (c) of Fig. 3-6. The curves are from the classical work of Stumpers[8] and are given in terms of the voltage (not power) spectral density which is linear above threshold, for a rectangular predetection filter of bandwidth B_p Hz and white noise at its input.

That the onset of threshold is accompanied by spike noise and a change in spectral distribution was found early in practical FM history.[9] More recently, the spike noise model has been suggested[10] and utilized[4] in the calculation of the noise level in the threshold region, and it has been shown[4] that the results obtained in this manner are a good approximation to the exact results obtained from much more complex mathematical formulation.[11]

The model, then, for the output noise of a discriminator, is as follows: The output noise is comprised of (a) small noise for which the FM improvement holds and which has an essentially Gaussian probability density function, and (b) spike noise due to the origin encirclements by the resultant vector of carrier and noise. This general picture essentially holds for all known FM demodulators, but the source of spike noise may be different, as discussed in later chapters. The assumption that the *nonencirclement noise* is fully subject to the FM improvement appears to result in only a small error.[4]

The spike rate N for an unmodulated carrier corrupted by Gaussian noise having an arithmetically symmetric spectrum about the carrier is given by

$$N = r[1 - \text{erf}(\rho)^{1/2}] \quad \text{pulses per second} \tag{3-28}$$

where ρ is the CNR at the limiter input and r is the radius of gyration of the noise spectrum defined by

$$r^2 = \frac{\int_0^\infty (f - f_c)^2 W(f)\,df}{\int_0^\infty W(f)\,df} \tag{3-29}$$

where $W(f)$ is the power spectral density of the noise and f_c is its center frequency at the limiter input. For white noise at the receiver input, $W(f)$ is identical to the power response of the predetection filter. The symbol r is also referred to as the "typical" noise frequency.[12] With modulation, the

---

† Compare "bipolar" or "doublet" type of noise which has a parabolic spectral density and, therefore, does not affect the output SNR as much.

spikes occur mainly in the direction opposing the signal[9,13] and their rate is increased. For an offset carrier, or a similar modulation where the carrier spends much of the time at a certain deviation Δf, the spike rate is given approximately by[13]

$$N \cong \Delta f\, e^{-\rho} \tag{3-30}$$

Equation (3-30) holds sufficiently well when

$$\Delta f/r > 0.2 \tag{3-31}$$

This is further discussed in Chapter 9.

It has been shown[14] that the spike rate increases by an equivalent of 0.5 dB of received carrier power under noise modulation, and by an equivalent of 1 dB for a carrier offset to the edge of the rectangular predetection filter. Therefore, the effect of modulation can be ignored in many situations, since the error will be only a fraction of 1 dB.

Having calculated the rate of spikes N one obtains a relation for the noise response which holds also at threshold[14] and over some region below it, until further capture effects are noted[5]

$$(\text{SNR})_{f_b} = \frac{(\Delta f_p)^2/2}{\tfrac{1}{3}[0.5 f_b^{\,2}/(\text{CNR})_{2f_b}] + 8\pi^2 N f_b} \tag{3-32}$$

The difference between Eq. (3-32) and (3-15) lies only in the term $8\pi^2 N f_b$ which is the noise power contributed by the spikes. This noise term may be derived using the Wiener–Khintchine theorem [Eqs. (2-41) and (2-42)], or less rigorously as follows: An impulse of strength 2π has a Fourier transform of 2π (a constant) (see Rule 15, Table I), contributing a two-sided *energy spectral density* (joules per Hertz) of $(2\pi)^2$. Then N independent impulses per second contribute a two-sided power spectral density of $N(2\pi)^2$, or a one-sided power spectral density of $2N(2\pi)^2$. Thus, the noise power contributed by N spikes/sec in a bandwidth f_b is $2N(2\pi)^2 f_b$.

A more exact analytical procedure involves the expansion of Eq. (3-8) in a series. The first two terms are then given by[4]

$$\theta_n(t) \cong \frac{y}{A} + \frac{xy}{A^2} + \cdots \tag{3-33}$$

It is recognized that the first term in Eq. (3-33) is the approximation made in Eq. (3-9) for the above-threshold noise angle. When the second term becomes of significance, then the approach of threshold is indicated. It has been shown[4] that if the contribution of both terms in Eq. (3-33) is utilized in addition to the spike noise, then the resulting noise power is nearly identical with that predicted by the exact solution.[11]

C. Equivalent Baseband Response

In the design of wideband FMFB loops it is useful to know the equivalent baseband response of the discriminator in order that it may be appropriately considered and compensated for in the design. We shall now give such an approximate equivalent response.[15]

A simplified circuit diagram for the discriminator is shown in Fig. 3-7. An assumption of arithmetic symmetry in the frequency response and small modulation index operation is made to facilitate analysis. For this circuit, Zinn[16] derived the baseband response to a step in carrier phase, or impulse in frequency. If this resulting impulse response is subjected to Laplace transformation, the desired steady-state frequency response function $H_d(s)$ is obtained.

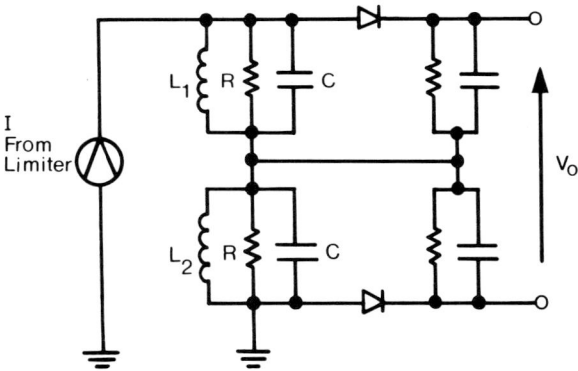

Fig. 3-7. Simplified discriminator circuit diagram.

The derivation carried out in Appendix B gives the function

$$H_d(s) = \frac{K\Delta}{s^2 + 2\alpha s + \alpha^2 + \Delta^2} \qquad (3\text{-}34)$$

having two complex poles. The quantity α is half the angular 3-dB bandwidth of each tuned circuit, Δ is the angular center-to-peak bandwidth, and K is a gain constant.

3.3. Low-Threshold Demodulation

Section 3.2 discussed the threshold performance of the limiter-discriminator and some advantages gained by low-threshold demodulation. The design and performance of low-threshold FM demodulators is the topic of the succeeding chapters. Only a brief summary of their performance is given here as a preview.

An illustration of the threshold performance of various FM demodulators is given in Fig. 3-8.[14] The system is shown in part (a) of the figure. The modulation signal is assumed to be a voice channel of rms modulation index σ, modeled by Gaussian noise with the power spectral density as shown in the figure, and a waveform peak-to-rms ratio of 10 dB. A Carson's rule rectangular predetection and a 4-kHz rectangular postdetection bandwidth are assumed. The bandwidth relation is thus

$$B_{CR} = 2(1 + 10^{1/2}\sigma)f_b \qquad (3\text{-}35)$$

In Fig. 3-8b, the ordinate is output SNR (in decibels) measured in base bandwidth, and the abscissa is input CNR (in decibels) measured in twice the base bandwidth. In the FM improvement region, the noise performance of the various demodulators, if properly designed, is the same and is essentially that given by the dashed lines. The solid lines are the loci of the threshold points of the various demodulators as a function of the modulation index. As before, threshold is defined as the point at which the noise at the output of the FM demodulator (after the lowpass filter) has increased by 1 dB more than predicted by the above-threshold linear relation. The curve at the far right gives the performance of the LD, or the "conventional demodulator," obtained using Eqs. (3-32), (3-28), and (3-35). The threshold performance curves of the angular feedback demodulators are for convenience grouped and approximated by three lines, giving an indication of the approximate advantage available from the various techniques. Thus, next in improved performance are the PLL demodulator and the FMFB demodulator whose performance is very similar. These "first generation" threshold extenders are already subject to substantial use in practice. The remaining demodulators shown in this figure are still partly in the development stage.

The "second generation" low-threshold demodulators offer further improved performance at the cost of some additional complexity. The extended-range phase-locked demodulator (ERPLD) is a phase-locked loop whose phase detector has an effective monotonic range greater than $\pm 90°$. The FMFB-PLL is an FMFB whose internal detector, normally an LD, is a PLL. Finally, the FMFB–ERPLD is an FMFB with the internal demodulator replaced by an ERPLD. The design principles for all of these demodulators are discussed in succeeding chapters.

Can the threshold of FM demodulators be reduced ad infinitum by increased complexity of the demodulator? The answer is negative, as can be shown from Shannon's theory.[14, 17, 18] Shannon's limit for "errorless" transmission leads to the linear demodulation boundary shown in Fig. 3-8, as derived in Appendix C. While the ideal demodulator boundary derived from Shannon's theory is not based on the 1-dB noise degradation point or, for that matter on FM, it does indicate from the information-theoretic point of view a limit of "linear" demodulation.

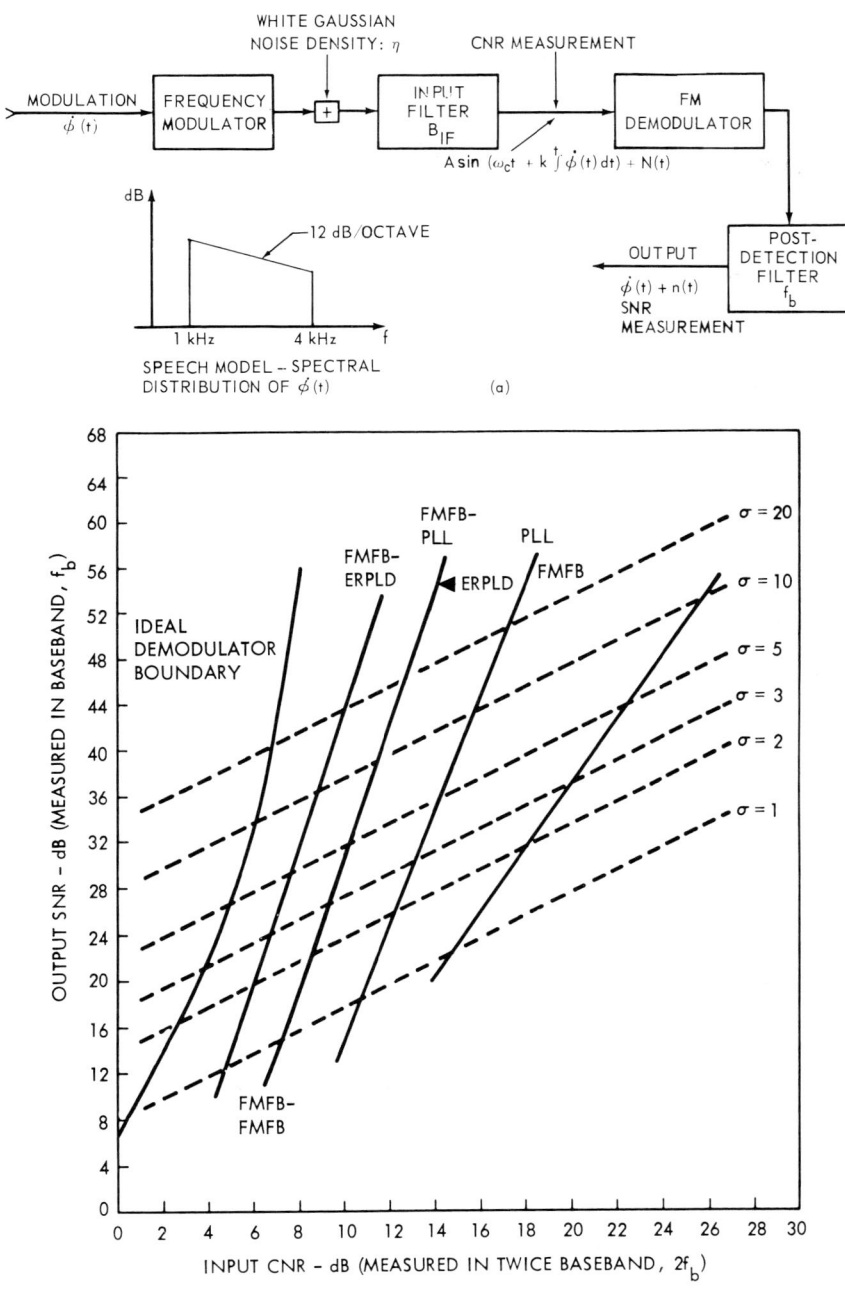

Fig. 3-8. Threshold performance of various FM demodulators (from Frankle[14]). (a) FM system. (b) Threshold curves (– –: linear demodulator performance above threshold; —: threshold boundary).

3.4. Filtering of FM Carriers

In the design of angular feedback loops we encounter two applications of FM carrier filtering. The first is the filtering of the wave before it is fed to the loop input, and the second is a narrower filter found within the FMFB loop proper. It is of interest to know the distortion of the signal caused by these filters, since it may determine the onset of the distortion-limited region or unduly alter the signal shape. In addition, for the filter within the loop, a baseband equivalent circuit is required for design purposes, as will be seen in Chapters 7 and 8. These considerations will now be briefly explored. Further discussion is given in Chapter 7.

The response of a linear filter to an FM carrier has been widely discussed in the literature.[19-27] No simple, general solution is as yet available. Practical solutions generally involve an approximation to the response plus correction terms. The approximation that is most applicable in a particular case depends largely on the modulation index of the FM wave. Oddly enough, the intermediate modulation indexes present the greatest difficulty in analysis. Small modulation indexes tend to make the FM carrier similar to an AM wave, as the spectrum tends to a "two sideband" kind. For large modulation indexes, the analysis is aided by the "slowness" of the frequency variations and leads to "quasi-stationary" solutions. The literature[19-27] contains the salient material for further study of the subject. We shall present here only some data that are particularly applicable to our loop designs.

A related problem is the required rf bandwidth for an FM carrier which, as expected, depends on the modulating signal, the deviation, and the amount of distortion that can be tolerated. Here one similarly resorts to practical approximations. The reader who wishes to pursue the subject further should see the literature.[28-32] In practice, a very popular and convenient rule of thumb for the required filter bandwidth which appears to be satisfactory for most situations of analog signal transmission is Carson's rule given earlier. Digital FM can tolerate much more distortion, and the filtering bandwidth that is usually taken is not much larger than the bit rate. It should be noted that phase linearity is important for FM carrier filtering, since phase nonlinearities, cause distortion products. Transient FM signals, as for example in digital FM, present further analytical difficulties.[22, 24] Generally, restrictions of the rf bandwidth leads to both AM and FM transients.

The filter within the FMFB loop is often restricted by design considerations to a single-tuned circuit.[33] For this case, useful experimental distortion data (for sinusoidal modulation) are given by Izatt.[27] Figure 3-9 gives the amplitude of the fundamental component of the modulation at the output normalized with respect to the same component at the input to the filter, plotted as a

3.4. Filtering of FM Carriers

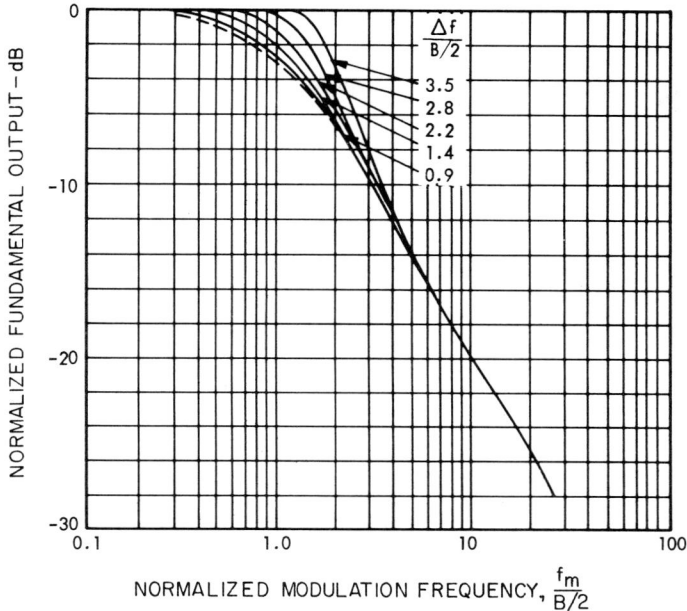

Fig. 3-9. Amplitude of the fundamental component of the demodulated signal (parallel resonant circuit; – – : low-pass equivalent; from Izatt[27]).

function of the modulation frequency normalized with respect to one-half the filter 3-dB bandwidth. The normalized peak deviation is a parameter. The dashed line represents the response of the lowpass equivalent filter defined in Section 2.1. It is observed that, for deviations less than one-half bandwidth (the usual case in practice), the response is essentially that of the lowpass equivalent. Deviations from this response characteristic occur for larger deviations, and especially for modulating frequencies of the order of the half bandwidth. For an FM carrier centered in the filter, the dominating distortion component at the output is the third harmonic of the signal. Figure 3-10 presents experimental data for this distortion component.[27] The plot gives the third harmonic output relative to the fundamental component as a function of the normalized modulation frequency. Further distortion information is given in Chapter 7 where actual calculations are carried out.

In the design of FMFB loops, a baseband equivalent circuit for the loop intermediate frequency (IF) filter is of interest. The IF filter affects both the amplitude and the phase of the FM carrier. Only the effect on the phase related to the modulation is required in FMFB design. Consequently, we desire to find a lowpass filter transfer function which adequately represents at baseband the effect on the phase of the FM carrier by the IF filter. Such

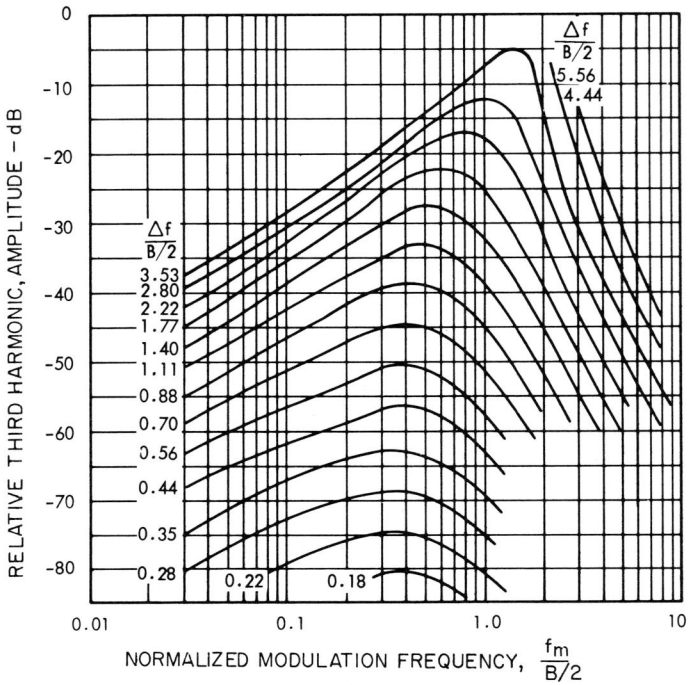

Fig. 3-10. Third harmonic in the demodulated signal (parallel resonant circuit, from Izatt[27]).

an equivalent response was derived by Enloe[34] for a signal representable by a small-index modulated FM wave with a slowly-varying carrier center frequency. The formulas and their use are discussed in Chapter 4. The application to the single-tuned circuit is given in Appendix E. For a small-index carrier which is not offset from the filter's center frequency, the conventional lowpass equivalent response holds sufficiently well.

3.5. Voltage-Controlled Oscillator

The voltage-controlled oscillator (VCO) is an oscillator whose instantaneous frequency is varied in accordance with an applied signal. Generally, it is desired that the oscillator's deviation from its center frequency be linearly related to the control voltage. The defining equation is

$$e_r(t) = A \cos\left(\omega_i t + \mu \int^t e_c(t)\, dt\right) \qquad (3\text{-}36)$$

where $e_r(t)$ is the output signal of the VCO, A is its constant amplitude, ω_i is

References

17. C. E. Shannon, Communication in the presence of noise. *Proc. IRE* **37**, 10–21 (1949).
18. J. A. Develet, A threshold criterion for phase-lock demodulation. *Proc. IEEE* **51**, 349 (1963); Correction. *Proc. IEEE* **51**, 580 (1963).
19. J. R. Carson and T. C. Fry, Variable frequency electric circuit theory with application to the theory of frequency modulation. *Bell Syst. Tech. J.* **16**, No. 4, 513–540 (1937).
20. B. van der Pol, The fundamental principles of frequency modulation. *J. Inst. Elec. Eng. Part 3* **13**, 153–158 (1946).
21. F. L. H. M. Stumpers, Distortion of frequency-modulated signals in electrical networks. *Commun. News (Phillips)* **9**, No. 3, 82–92 (1948).
22. D. D. Weiner and B. J. Leon, The quasi-stationary response of linear systems to modulated waveforms. *Proc. IEEE* **53**, No. 6, 564–574 (1965).
23. E. Bedrosian and S. O. Rice, Distortion and crosstalk of linearly filtered, angle-modulated signals. *Proc. IEEE* **56**, No. 1, 2–13 (1968).
24. T. T. N. Bucher, Network response to transient frequency modulation inputs. *Trans. (Communications and Electronics) Amer. Inst. Elec. Eng.* **78**, 1017–1022 (1960).
25. References 4, 8, 9, 11, 13, 18–23, 28, 34 as well as a bibliography for further reading are contained in: J. Klapper, "Selected Papers on Frequency Modulation." Dover, New York, 1970.
26. P. F. Panter, "Modulation, Noise, and Spectral Analysis." McGraw-Hill, New York, 1965.
27. J. B. Izatt, The distortion produced when frequency-modulated signals pass through certain networks. *Proc. IEE (Brit.)* **110**, No. 1, 149–156 (1963).
28. M. S. Corrington, Variation of bandwidth with modulation index in frequency modulation. *Proc. IRE* **35**, No. 10, 1013–1020 (1947).
29. J. L. Stewart, The power spectrum of a carrier frequency-modulated by Gaussian noise. *Proc. IRE* **42**, 1539–1542 (1954).
30. R. G. Medhurst, RF bandwidth of frequency-division multiplex systems using frequency modulation. *Proc. IRE* **44**, 189–199 (1956).
31. J. A. Mullen and D. Middleton, Limiting forms of FM noise spectra. *Proc. IRE* **45**, 874–877 (1957).
32. H. E. Rowe, "Signals and Noise in Communication Systems." Van Nostrand-Reinhold, Princeton, New Jersey, 1965.
33. L. H. Enloe, The synthesis of frequency feedback demodulators. *Proc. Nat. Electron. Conf.* **18**, 477–497 (1962).
34. L. H. Enloe, Decreasing the threshold in FM by frequency feedback. *Proc. IRE* **50**, 18–30 (1962).
35. F. M. Gardner, "Phaselock Techniques." Wiley, New York, 1966; F. M. Gardner, S. S. Kent, and R. D. Dasenbrock, Theory of phaselock techniques, NASA Rep. N66-10515. Resdel Eng. Corp., Pasadena, California, 1966.
36. Oscillator Catalog. Greenray Ind., Mechanicsburg, Pennsylvania, 1969.
37. W. A. Edson, "Vacuum Tube Oscillators." Wiley, New York, 1953.
38. A. W. Warner, Design and performance of an ultra precise 2.5 Mc quartz-crystal unit. *Bell Syst. Tech. J.* **34**, 1193–1217 (1960).
39. H. J. Reich, "Functional Circuits and Oscillators." Van Nostrand-Reinhold, Princeton, New Jersey, 1961.
40. K. H. Sann, Phase stability of oscillators. *Proc. IRE* **49**, 527–528 (1961).
41. L. R. Malling, Phase-stable oscillators for space communications, including the relationship between the phase noise, the spectrum, the short-term stability and Q of the oscillator. *Proc. IRE* **50**, 1656–1664 (1962).
42. R. R. Real, Direct frequency modulation of crystal controlled transistor oscillator. *IEEE Trans.* **CS-10**, 459 (1962).

43. J. Graszkowski, "Frequency of Self Oscillations." Macmillan, New York, 1964.
44. A. Luna and R. Cafissi, Transistorized frequency modulator for wideband radio links. *Alta Freq.* **34**, 534–542 (1965).
45. M. H. Norwood, Voltage variable capacitor tuning: A review. *Proc. IEEE* **56**, No. 5, 788–798 (1968).
46. K. Noda, Design for frequency modulator circuit using variable capacitance diode. *Electron. Commun. Jap.* **51-B**, No. 6, 107–112 (1968).
47. G. M. Pelchat, S. B. Boor, and D. B. Allen, Distortion in varicap FM oscillators. *IEEE Trans.* **COM-17**, No. 1, 49–53 (1969).
48. H. R. Camenzind and A. B. Grebene, An integrated FM multiplex generator. *Nat. Telemetering Conf. Rec.* pp. 228–233 (1969).
49. W. A. Gardner, Modulation rate distortion in frequency modulators. *IEEE Trans.* **CT-16**, No. 3, 295–302 (1969).
50. S. Krishnan, Diode phase detectors. *Electron. and Radio Eng.* **36**, No. 2, 45–50 (1959).
51. *Instruction Manual,* Hewlett Packard Balanced Mixer Type 10534, Hewlett Packard Co., Palo Alto, California, 1970.

CHAPTER

4

FM Feedback Loop Principles

In this chapter we present a qualitative description and the basic equations governing the FMFB loop. The pertinent equations are first developed for the ideal loop in the absence of input noise; they are then extended to include the effects due to input noise and nonideal functional components. Acquisition as well as tracking characteristics are considered. The results presented in this chapter serve as the basis for the design procedures found in Chapter 7.

4.1. Operational Principles

The essential elements of the FMFB circuit are the *input mixer*, the *internal IF amplifier and filter*, the *limiter-discriminator* (LD), the *baseband filter and amplifier*, and the *VCO*, as shown in Fig. 4-1. A basic operational requirement for the loop is to track the frequency of the input signal. This is accomplished by forming in the input mixer an IF signal which contains the frequency difference (or error) between the VCO signal and the input signal. This IF signal is first filtered to remove excess noise and other undesirable components and is then applied to the LD. The LD detects the frequency error, which is then amplified and further filtered prior to application to the VCO. The feedback is such as to effect frequency tracking (or following) of the input signal by the VCO signal.

One of two distinct tracking objectives may be desired, and each naturally results in a different loop design. One objective is that of automatic frequency

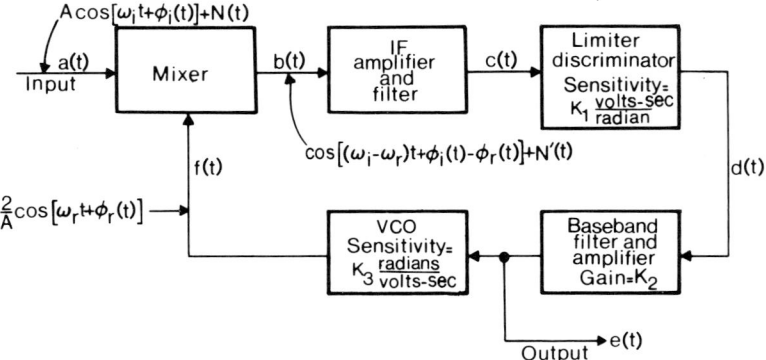

Fig. 4-1. Block diagram of basic FMFB loop.

control (AFC), where it is desired to track the average (or center) frequency of the input signal with minimum interference from the modulation components. This is accomplished by utilizing a relatively wideband internal IF filter and a narrowband baseband filter. Another objective is to closely track the FM components to provide a reduced deviation signal within the IF of the loop. The IF filter may then be designed with a narrower bandwidth to provide noise immunity for the signal, and still adequately transmit the FM signal. Tracking fidelity is obtained by utilizing a relatively wideband baseband filter within the loop. This design results in a low-threshold FMFB demodulator for which design and performance considerations are presented in detail in Chapter 7. We shall generally consider the input to the VCO as the loop output; however, the output may be taken at any point in the loop, keeping in mind that the response, SNR, and distortion are a function of the output point.

We continue by developing the describing equations for an FMFB loop with idealized elements. Consider an input signal of the form

$$a(t) = A \cos[\omega_i t + \phi_i(t)] \qquad (4\text{-}1)$$

where A is peak signal amplitude (a constant), ω_i is input signal center frequency, and $\phi_i(t)$ is input signal phase modulation. The input noise $N(t)$ is neglected for the present. Similarly, the VCO output signal is of the form

$$f(t) = (2/A) \cos[\omega_r t + \phi_r(t) + \phi_0] \qquad (4\text{-}2)$$

where ω_r is VCO center frequency, $\phi_r(t)$ is VCO phase modulation, and ϕ_0 is constant phase shift (does not affect loop operation and is, therefore, neglected in Fig. 4-1 and for the remainder of the chapter). The amplitude $2/A$ in Eq. (4-2) is chosen for convenience, and does not restrict the discussion. The subscript r indicates the kinship to a "return" signal. Normally,

4.1. Operational Principles

$\phi_i(t)$ and $\phi_r(t)$ contain the information-bearing components of the signal as well as components due to drift and doppler shift. Frequencies ω_i and ω_r represent the quiescent condition $[\phi_i(t) = \phi_r(t) = 0, \dot{\phi}_i(t) = \dot{\phi}_r(t) = 0]$ about which the loop is centered.

The mixer is characterized as an ideal multiplier, resulting in an output of

$$b(t) = a(t) \cdot f(t)$$
$$= \cos[(\omega_i + \omega_r)t + \phi_i(t) + \phi_r(t)] + \cos[(\omega_i - \omega_r)t + \phi_i(t) - \phi_r(t)] \quad (4\text{-}3)$$

The IF filter then selects one of the sidebands† while rejecting the other, resulting in an output of

$$c(t) = h_1(t) \circledast \cos[(\omega_i - \omega_r)t + \phi_e(t)] \quad (4\text{-}4)$$
$$= E(t) \cos[\omega_e t + \hat{\phi}_e(t)] \quad (4\text{-}5)$$

where $h_1(t)$ is the impulse response of the IF filter, $E(t)$ is the envelope function of $c(t)$, $\omega_e = \omega_i - \omega_r$, chosen to coincide with the center frequency of the IF, $\phi_e(t) = \phi_i(t) - \phi_r(t)$ is the phase error, $\hat{\phi}_e(t)$ is the phase function of $c(t)$, and $\circledast$ denotes convolution.

The LD is assumed to measure the instantaneous frequency deviation of its input. Its output is

$$d(t) = K_1 \frac{d}{dt} \hat{\phi}_e(t) \quad (4\text{-}6)$$

where K_1 is the detection sensitivity in volt-seconds per radian also referred to as λ. After baseband filtering and amplification we have

$$e(t) = K_1 K_2 h_2(t) \circledast \frac{d}{dt} \hat{\phi}_e(t) \quad (4\text{-}7)$$

where K_2 is the baseband amplifier gain, and $h_2(t)$ is the impulse response of the baseband filter.

The VCO produces an FM signal with a frequency deviation proportional to the control voltage, and a phase deviation proportional to the integral of it.

$$\phi_r(t) = \int_{-\infty}^{t} K_1 K_2 K_3 \left\{ h_2(t) \circledast \frac{d}{dt} [\hat{\phi}_e(t)] \right\} dt \quad (4\text{-}8)$$

where K_3 is the VCO modulation sensitivity in radians per volt-second, also referred to as μ. Thus‡

---

† In this development the lower sideband is selected; however, it is possible to select the upper sideband and/or the relative magnitudes of ω_i and ω_r, with the proviso that the discriminator baseband gain polarity be selected to effect VCO tracking.

‡ The phase constant that may be lost when going from (4-8) to (4-9) can be ignored, as it carries no information.

$$\phi_r(t) = K_1 K_2 K_3 h_2(t) \circledast [\hat{\phi}_e(t)] \tag{4-9}$$

with the derivative due to the LD canceled by the integration effect of the VCO. Taking the Laplace transform of the foregoing equation the following describing equations are obtained in complex frequency notation for the FMFB loop

$$\Phi_r(s) = K_1 K_2 K_3 [H_2(s)] \hat{\Phi}_e(s) \tag{4-10}$$

and

$$\Phi_e(s) = \Phi_i(s) - \Phi_r(s) \tag{4-11}$$

Solution of these equations requires the relationship between ϕ_e and $\hat{\phi}_e$ [or $\Phi_e(s)$ and $\hat{\Phi}_e(s)$]. Obtaining this relationship, which is the FM response for the IF filter, is a complex and difficult problem (see Section 3.4). Some available results are applied below.

We begin by considering the case where the phase error is small, i.e., $\phi_e \ll 1$ rad, with the result that the IF filter response can be characterized by its equivalent lowpass filter $H_{L1}(s) = H_1^{+}(s + j\omega_e)$, where $H_1^{+}(s)$ is the Laplace transform of $h_1(t)$ for $\text{Im}(s) > 0$ and $H_1^{+}(s) = 0$ for $\text{Im}(s) < 0$.† Therefore,

$$\hat{\Phi}_e(s) = H_1^{+}(s + j\omega_e)\Phi_e(s) = H_{L1}(s)\Phi_e(s) \tag{4-12}$$

which results in a linear set of describing equations that permits a linear model representation for the loop.

We next consider a phase error ϕ_e composed of the sum of a constant-frequency offset term and a small time-varying phase term of the form

$$\phi_e = \Delta\omega_e t + \phi_e'(t) \tag{4-13}$$

where $\Delta\omega_e$ is the frequency offset in the IF and $\phi_e'(t)$ is the time-varying phase modulation. This representation is a convenient method of approximating the IF response to a modulation consisting of both low- and high-frequency components. The low-frequency modulation components, in general, are not compressed sufficiently by the loop to satisfy the small ϕ_e approximation made above, and are more suitably approximated by a constant frequency offset term $\Delta\omega_e$. This term, therefore, represents a quasi-stationary approximation to the low-frequency modulation components, and (as will be required later) may also include low-frequency noise components. It is again possible to represent the IF filter response in terms of an equivalent baseband filter, with, however, a dependence upon $\Delta\omega_e$

$$\hat{\phi}_e(t) = \Delta\omega_e t + \hat{\phi}_e'(t) \tag{4-14}$$

with

$$\hat{\Phi}_e'(s) = H_{b1}(s, \Delta\omega_e)\Phi_e'(s) \tag{4-15}$$

† See pages 10–11.

4.2. The Linear Equivalent Circuit

Naturally, for $\Delta\omega_e = 0$,

$$H_{b1}(s, 0) = H_{L1}(s) \tag{4-16}$$

The general results, describing the equivalent baseband filter,[1] are summarized as follows (for application see Appendix E):

$$H_{b1}(j\omega, \Delta\omega_e) = \frac{1}{2|H_{L1}(\Delta\omega_e)|} [H_x(j\omega) + H_x^*(-j\omega)] \tag{4-17}$$

where

$$H_x(j\omega) = H_{L1}[j(\omega + \Delta\omega_e)]e^{-j\theta_{L1}(\Delta\omega_e)}, \tag{4-18a}$$

$$H_x^*(-j\omega) = H_{L1}[j(\omega - \Delta\omega_e)]e^{j\theta_{L1}(\Delta\omega_e)} \tag{4-18b}$$

and

$$H_{L1}[j(\omega + \Delta\omega_e)] = |H_{L1}(\omega + \Delta\omega_e)|e^{j\theta_{L1}(\omega + \Delta\omega_e)} \tag{4-19}$$

With these results, we continue by considering the linear model.

4.2. The Linear Equivalent Circuit

A. The Linear Model

The linear model shown in Fig. 4-2 is derived from the operational equations presented in Section 4.1, with the signal taken in terms of phase modulation. The desirability of forming such a model resides in the ease with which further analysis can be carried forth, and provides a base from which synthesis procedures can be formulated. The validity of the model is determined

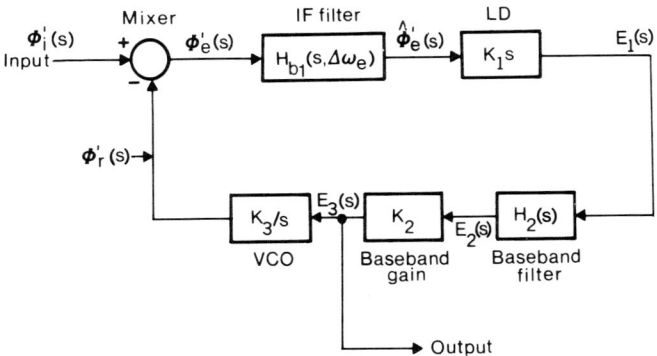

Fig. 4-2. Linear model representing FMFB. (Actual signals—Input: $A \cos[\omega_i t + \phi_i(t)]$ with $\phi_i(t) = \phi_i'(t) + \Delta\omega_i t$, $\phi_e(t) = \phi_e'(t) + \Delta\omega_e t$; VCO output: $(2/A)\cos[\omega_r t + \phi_r(t)]$ with $\phi_r(t) = \phi_r'(t) + \Delta\omega_r t$.)

basically by the representation for the IF filter, and depends on maintaining a small modulation phase error with a relatively unrestricted static (or quasi-static) frequency error. As previously discussed, this frequency error modifies the "equivalent" IF filter response.

The model elements are developed in the following manner: The mixer output signal component contains the frequency and phase differences between the input and the VCO signals and, therefore, can be represented by a subtraction element. The IF filter is represented by its "equivalent" baseband characteristic $H_{b1}(s, \Delta\omega_e)$. The LD is equivalent to a differentiation in terms of the signal phase, whereas the VCO, in performing the complementary function, is represented by an integration. The baseband gain and filter are transcribed directly from the basic model. This representation is, of course, in terms of ideal elements; methods for extending the model to include effects due to nonideal elements will be indicated later. The method for including input noise in the model will be considered next.

B. Effects Due to Input Noise

In the usual application, the FMFB loop is subjected to a signal with additive input noise interference. We consider, as a representative case, a condition that occurs often in practice; namely, corruption by bandpass Gaussian noise.

The loop exhibits a response to the input noise leading to a noise component of phase modulation in the VCO output. The mixer output will then consist of a signal-carrier component containing both signal and noise phase modulation, and an additive-noise component. The remainder of the analysis depends upon evaluating the internal IF filter response to both the signal and noise terms. The complexity of the problem at hand is underscored if we recall the difficulty in obtaining a general solution for the IF filter response in Section 4.1. The signal now contains an additional modulation component (the loop noise response), and is in the presence of an additive interfering noise component. We approach the problem in a manner similar to that followed in Section 4.1; assumptions and restrictions are placed on both the signal and the noise, with the result that a linear model representation is developed. This model then serves as a convenient means to perform noise calculations.

Let the input signal and noise be represented as

$$a(t) = A \cos[\omega_i t + \phi_i(t)] + N(t), \qquad (4\text{-}20)$$

with the noise written as (see Section 2.3)

$$N(t) = x(t) \cos \omega_i t - y(t) \sin \omega_i t \qquad (4\text{-}21)$$

4.2. The Linear Equivalent Circuit

Analytical combining of the signal and the noise results in

$$a(t) = R(t)\cos[\omega_i t + \phi_i(t) + \beta(t)] \tag{4-22}$$

where

$$R(t) = [A^2 + 2A(x\cos\phi_i + y\sin\phi_i) + x^2 + y^2]^{1/2} \tag{4-23}$$

and

$$\beta(t) = \tan^{-1}\frac{y\cos\phi_i - x\sin\phi_i}{A + x\cos\phi_i + y\sin\phi_i} \tag{4-24}$$

We shall now consider the high input CNR case, i.e., $A \gg x$ and $A \gg y$. The input for this case can be approximated by

$$R(t) \cong A[1 + (x_1/A)], \quad \text{where} \quad x_1 = x\cos\phi_i + y\sin\phi_i \tag{4-25}$$

and

$$\beta(t) \cong y_1/A, \quad \text{where} \quad y_1 = y\cos\phi_i - x\sin\phi_i \tag{4-26}$$

The response of the loop (VCO output) contains both a signal term and a noise term, $\phi_{rs}(t)$ and $\phi_{rn}(t)$, respectively, so that

$$f(t) = (2/A)\cos[\omega_r t + \phi_{rs}(t) + \phi_{rn}(t)] \tag{4-27}$$

At the output of the mixer, the difference-frequency component $b(t)$ is given by

$$b(t) \cong [1 + (x_1/A)]\cos(\omega_e t + \phi_{es} + \phi_{en}) \tag{4-28}$$

where

$$\phi_{es} = \phi_i(t) - \phi_{rs}(t) \quad \text{and} \quad \phi_{en} = \beta(t) - \phi_{rn}(t) \tag{4-29}$$

This IF component can be split into two parts, with one part being closer to a signal component and the other more representative of noise, viz.,

$$b(t) = \cos(\omega_e t + \phi_{es} + \phi_{en}) + (x_1/A)\cos(\omega_e t + \phi_{es} + \phi_{en}) \tag{4-30}$$

The last term will be referred to as $N'(t)$.

If we assume that the error components are small (ϕ_{es} and $\phi_{en} \ll 1$), then the response of the IF filter to the signal term is calculable by the method described earlier, and we have at the input to the LD a carrier signal containing phase noise and an interfering noise term

$$c(t) = E'(t)\cos[\omega_e t + \hat{\phi}_{es}(t) + \hat{\phi}_{en}(t)] + \hat{N}'(t) \tag{4-31}$$

with

$$\hat{\phi}_{es}(t) = \phi_{es}(t) \circledast h_{b1}(t) \tag{4-32}$$

$$\hat{\phi}_{en}(t) = \phi_{en}(t) \circledast h_{b1}(t) \tag{4-33}$$

$E'(t)$ = amplitude variations (removed by the limiter), and

$$\hat{N}'(t) = (x_1/A)\cos(\omega_e t + \phi_{es} + \phi_{en}) \circledast h_1(t) \qquad (4\text{-}34)$$

where $h_{b1}(t)$ is the equivalent-baseband impulse response of the internal IF filter, and $h_1(t)$ is the impulse response of the internal IF filter.

Let us examine $\hat{N}'(t)$. It can be observed from Eq. (4-30) that the term may be neglected for high CNR conditions ($x_1/A \ll 1$). However, for the benefit of later discussion, we shall analyze it further. Relying on the assumptions that $x/A, y/A, \phi_{es}$, and ϕ_{en} are significantly less than one, we can approximate $\hat{N}'(t)$ as follows:

$$(x_1/A)\cos(\omega_e t + \phi_{es} + \phi_{en}) \cong (x_1/A)\cos\omega_e t - (x_1/A)(\phi_{es} + \phi_{en})\sin\omega_e t \qquad (4\text{-}35)$$

with the significant term being $(x_1/A)\cos\omega_e t$. The resulting IF response is

$$\hat{N}'(t) \cong (\hat{x}_1/A)\cos\omega_e t \qquad (4\text{-}36)$$

with $\hat{x}_1(t)$ being interpreted as the envelope response of the IF filter to the driving envelope $x_1(t)$.

It is now possible to determine the influence of $\hat{N}'(t)$ upon the detected voltage. From Eqs. (4-31) and (4-36), $c(t)$ can be written as

$$c(t) \cong E'(t)\left[1 + \frac{2\hat{x}_1}{AE'(t)}\cos\theta_e + \left(\frac{\hat{x}_1}{AE'(t)}\right)^2\right]^{1/2}$$
$$\times \cos\left[\omega_e t + \theta_e - \tan^{-1}\frac{(\hat{x}_1/AE'(t))\sin\theta_e}{1 + (\hat{x}_1/AE'(t))\cos\theta_e}\right] \qquad (4\text{-}37)$$

$$\cong E'(t)\left(1 + \frac{2\hat{x}_1}{AE'(t)}\right)^{1/2}\cos\left(\omega_e t + \theta_e - \frac{\hat{x}_1}{AE'(t)}\cdot\theta_e\right) \qquad (4\text{-}38)$$

with $\theta_e = \hat{\phi}_{es}(t) + \hat{\phi}_{en}(t)$ = signal and noise induced phase error. Therefore, the detected voltage from the LD is

$$d(t) = \frac{d}{dt}\left[\left(1 - \frac{\hat{x}_1}{AE'(t)}\right)\theta_e(t)\right] = \dot{\theta}_e - \frac{1}{A}\frac{d}{dt}\left[\frac{\hat{\phi}_{es}(t)\hat{x}_1(t)}{E'(t)}\right] - \frac{1}{A}\frac{d}{dt}\left[\frac{\hat{\phi}_{en}(t)\hat{x}_1(t)}{E'(t)}\right]$$

$$(4\text{-}39)$$

The important term is $\dot{\theta}_e$ as predicted earlier. The remaining terms stemming from $\hat{N}'(t)$ are small, as they are inversely proportional to the signal amplitude. The term $\dot{\theta}_e$ consists of the linear sum of the signal and noise phase-error components. In order to achieve the same causal effect in $d(t)$, the noise may be considered as an additional modulation term and, thus, simply included in the linear model by $n(t)$ as indicated in Fig. 4-3. We recall that ϕ_{en}

is due to $\beta(t)$ [Eq. (4-29) and insert in Fig. 4-3] and that for high CNR, $\beta(t) \cong y_1/A$ [Eq. (4-26)]. Therefore, the equivalent noise modulation angle in our linear equivalent model is

$$n(t) = \beta(t) \cong y_1/A \qquad (4-40)$$

For no modulation,

$$n(t) \cong y/A \qquad (4-41)$$

having a power spectral density $2\eta/A^2$ over a lowpass frequency region equal to one-half the predetection bandwidth, where η is the spectral density of $N(t)$ (See Section 2.3).

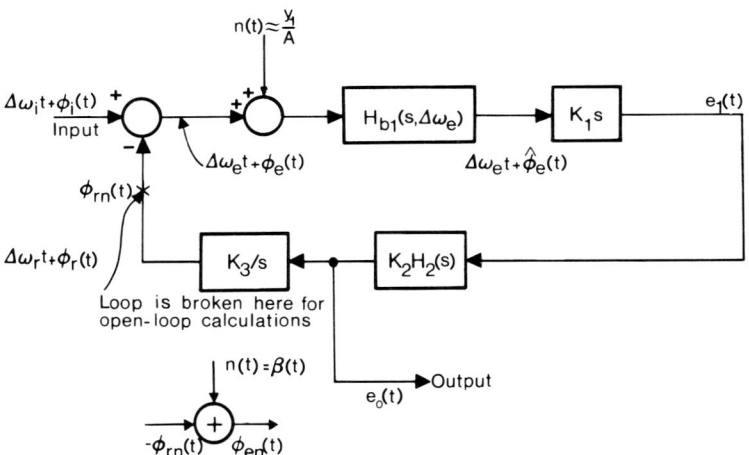

Fig. 4-3. FMFB linear model with equivalent-noise input. (Actual signals—Input: $A \cos[(\omega_i + \Delta\omega_i)t + \phi_i(t)] + N(t)$ with $N(t) = x(t) \cos \omega_i t - y(t) \sin \omega_i t$; VCO output: $(2/A) \sin[(\omega_r + \Delta\omega_r)t + \phi_r(t)]$.)

The conclusion to be drawn is that for high input CNR's the response of the system is essentially linear, and the effect of noise can be included by considering it to be an additional input term with density and bandwidth determined by the predetection filter.

4.3. Linear Operation of the Loop

Proceeding with the linear model, we calculate loop characteristics necessary to formulate and carry forth the loop design.

A. DC Tracking

The DC tracking analysis is particularly important in AFC applications, and for stability and noise response calculations in low-threshold FM detection. The DC analysis pertains to the loop response to the $\Delta\omega_i t$ term in the signal input, and represents the steady-state response to a "continuous wave" input offset from the center frequency by $\Delta\omega_i$. Under these conditions the loop will respond with a "tracking" offset $\Delta\omega_r$ producing a frequency error of $\Delta\omega_e$ within the loop. The relationships between these parameters are (see Section 2.2)

$$\Delta\omega_e/\Delta\omega_i = 1/[1 + K_1 K_2 K_3 H_2(0)] \tag{4-42}$$

and

$$\Delta\omega_r/\Delta\omega_i = K_1 K_2 K_3 H_2(0)/[1 + K_1 K_2 K_3 H_2(0)] \tag{4-43}$$

where $H_2(0)$ designates the DC response of the baseband filter. The effect of $H_{b1}(s, \Delta\omega_e)$ is, for a static signal, eliminated by the limiter. The discriminator output voltage, as well as the VCO input voltage are, therefore,

$$e_1/\Delta\omega_i = K_1/[1 + K_1 K_2 K_3 H_2(0)] \tag{4-44}$$

and

$$e_o/\Delta\omega_i = K_1 K_2 H_2(0)/[1 + K_1 K_2 K_3 H_2(0)] \tag{4-45}$$

Satisfying these equations depends on several factors in an actual loop. First, it is assumed that the loop is stable and in the tracking mode—the conditions for this are discussed further in Section 4.3. Second, the limiter employed within the loop restores the signal amplitude after IF filtering, even though the frequency error may place the signal on the IF skirt. Third, the frequency error is sufficiently small, so that the signal is within the central region of the discriminator "S" curve. Conversely, if any of these conditions are not met, pull-in (the acquisition of a signal) cannot be guaranteed, and a set of DC describing equations as well as the linear model may not necessarily exist. We continue by discussing the factors necessary to establish a stable loop.

B. Open- and Closed-Loop Response

The synthesis of a stable closed-loop system, and the conditions necessary to maintain stability can be described in terms of the open-loop response. We first consider the case of an unmodulated input signal at the loop center frequency in the absence of input noise. The linear model developed above assumes "ideal" elements for the LD and the VCO. With this assumption

4.3. Linear Operation of the Loop

and the application of a low-level test signal $\phi_i(t)$ (to maintain a small IF phase error), the open-loop response taken by figuratively breaking the loop at the VCO output is

$$\left.\frac{\Phi_r}{\Phi_i}(s)\right|_{\text{open loop}} = G(s) = K_1 K_2 K_3 H_{b1}(s, 0) H_2(s) \qquad (4\text{-}46)$$

The closed-loop stability of a linear feedback system is discussed in Section 2.2, where a convenient test for loop stability is given as follows: When

$$|H_{b1}(s, 0) \cdot H_2(s)|_{s=j\omega} \geq 1/(K_1 K_2 K_3) \qquad (4\text{-}47)$$

then

$$\arg[H_{b1}(s, 0) \cdot H_2(s)]_{s=j\omega} < 180° \qquad (4\text{-}48)$$

If we assume that the stability requirement is met, then the closed-loop transfer function is simply

$$\frac{\Phi_r}{\Phi_i}(s) = H(s) = \frac{G(s)}{1 + G(s)} = \frac{K_1 K_2 K_3 H_{b1}(s, 0) H_2(s)}{1 + K_1 K_2 K_3 H_{b1}(s, 0) H_2(s)} \qquad (4\text{-}49)$$

In subsequent applications, and in low-threshold FM demodulation in particular, the following normalized open-loop transfer function is a design goal; therefore, it was selected as an example for the remainder of the chapter. It is

$$G(s) = \frac{K[(s/a\omega_b) + 1]}{(s/\omega_b)^2 + 2\xi(s/\omega_b) + 1} \qquad (4\text{-}50)$$

It is seen to have a complex pole pair of natural frequency ω_b, a damping factor ξ, and a real zero at $s = -a\omega_b$. Note that, in general, to synthesize a particular $G(s)$, $H_{b1}(s, 0)$ and $H_2(s)$ must contain complementary poles and zeros. Then $H(s)$, from Eq. (4-50) is

$$H(s) = \frac{K[(s/a\omega_b) + 1]}{(s/\omega_b)^2 + 2[\xi + (K/2a)](s/\omega_b) + (1 + K)} \qquad (4\text{-}51)$$

By defining a new set of parameters it is possible to reexpress $H(s)$ in the normalized form

$$H(s) = \frac{K_0[(s/a_0 \omega_n) + 1]}{(s/\omega_n)^2 + 2\xi_0(s/\omega_n) + 1} \qquad (4\text{-}52)$$

where

$$\omega_n = \omega_b(1 + K)^{1/2} \qquad (4\text{-}53)$$

with

$$K_0 = K/(1 + K) \tag{4-54}$$

$$a_0 = a/(1 + K)^{1/2} \tag{4-55}$$

and

$$\xi_0 = [\xi + (K/2a)]/(1 + K)^{1/2} \tag{4-56}$$

It is also instructive to have available graphically the magnitude and phase of $H(j\omega)$ directly in terms of a, K, and ξ, to appreciate the variational effect in $H(j\omega)$ in terms of these parameters. Figures 4-4 through 4-6 present this magnitude and phase information.

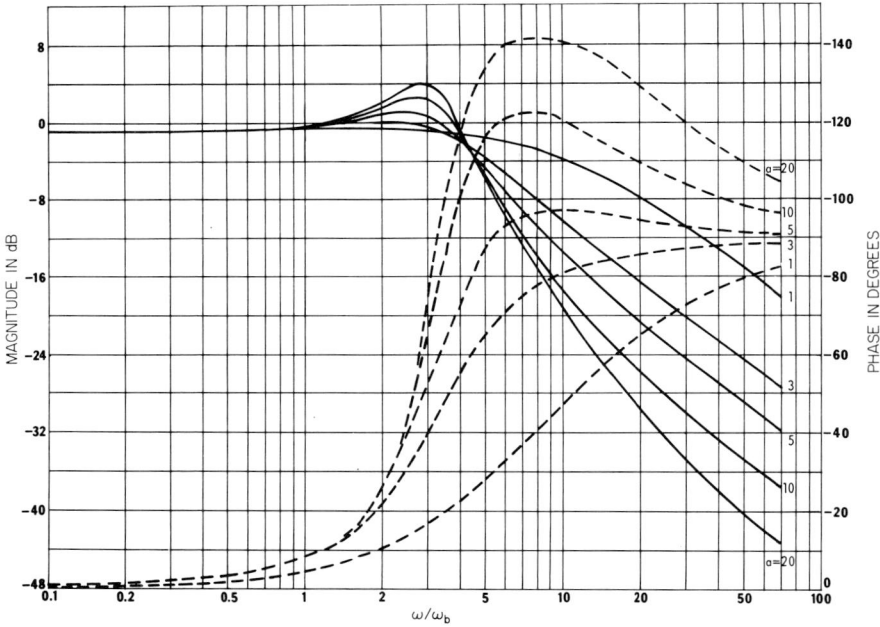

Fig. 4-4. FMFB closed-loop magnitude and phase response with zero constant a as a parameter (—: magnitude; - - -: phase). (Example: $K = 9$, $\xi = 2^{-1/2}$; Nominal design: $a = 5$.)

Starting with the closed-loop response, we may readily calculate the response to any point within the loop, viz.,

$$\frac{E_0}{\Phi_i}(s) = \frac{sH(s)}{K_3} \tag{4-57}$$

$$\frac{E_1}{\Phi_i}(s) = \frac{sH(s)}{K_2 K_3 H_2(s)} \tag{4-58}$$

4.3. Linear Operation of the Loop

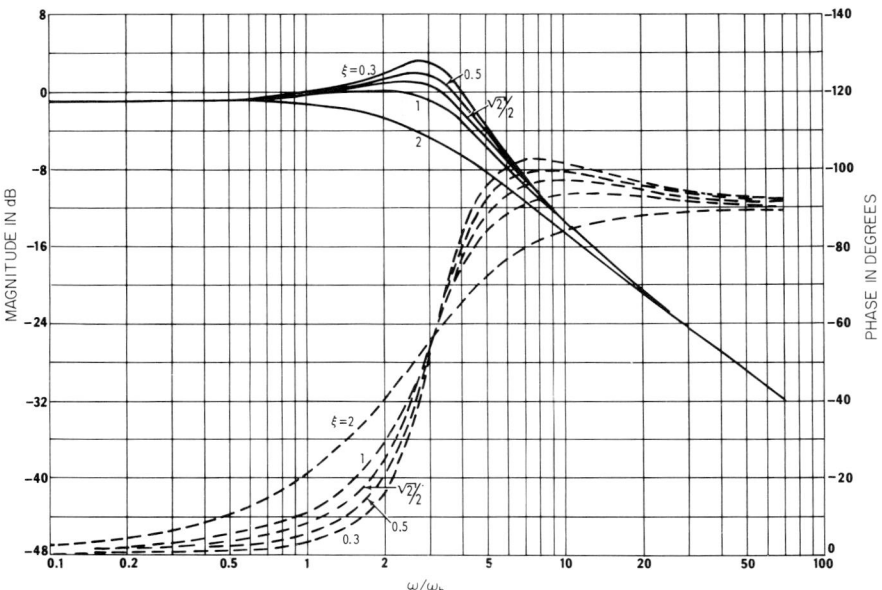

Fig. 4-5. FMFB closed-loop magnitude and phase response with damping factor ξ as a parameter (—: magnitude; ---: phase). (Example: $K = 9$, $a = 5$; Nominal design: $\xi = 2^{-1/2}$.)

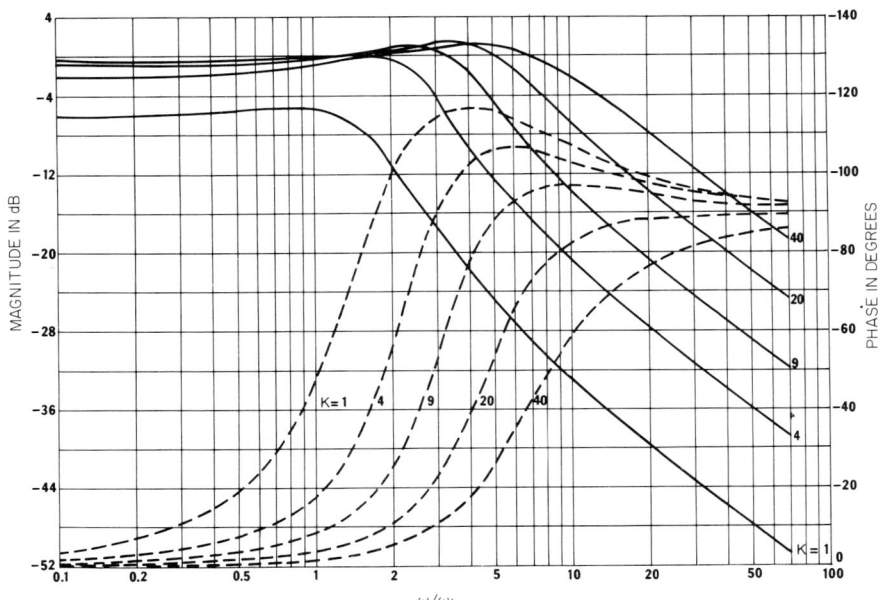

Fig. 4-6. FMFB closed-loop magnitude and phase response with gain K as a parameter (—: magnitude; ---: phase). (Example: $a = 5$, $\xi = 2^{-1/2}$; Nominal design: $K = 9$.)

$$\frac{\hat{\Phi}_e}{\Phi_i}(s) = \frac{H(s)}{K_1 K_2 K_3 H_2(s)} \qquad (4\text{-}59)$$

and

$$\frac{\Phi_e}{\Phi_i}(s) = \frac{H(s)}{K_1 K_2 K_3 H_{b1}(s, 0) H_2(s)} \qquad (4\text{-}60)$$

Previously, in the closed-loop transfer function calculation, we have considered only the case where the input signal is centered with respect to the loop. Off-tuning at the input will result, as discussed previously, in a frequency error within the loop IF and a modification of the equivalent modulation transfer response for the IF stage. In general, then, the closed-loop transfer function and the loop response to any point are a function of this off-tuning. As an illustrative example we consider a particular case that is representative of practice.

If the internal IF filter is a single-tuned circuit† of 3-dB bandwidth $2b\omega_b$, its equivalent modulation response to an on-tune carrier is

$$H_{L1}(s, 0) = 1/(s/b\omega_b + 1) \qquad (4\text{-}61)$$

with ω_b as the baseband normalization frequency and b as a proportionality factor; therefore, the baseband loop filter required to realize the specified open-loop transfer function $G(s)$, Eq. (4-50), is

$$H_2(s) = \left(\frac{s}{a\omega_b} + 1\right)\left(\frac{s}{b\omega_b} + 1\right) \bigg/ \left[\left(\frac{s}{\omega_b}\right)^2 + 2\xi\left(\frac{s}{\omega_b}\right) + 1\right] \qquad (4\text{-}62)$$

Note that $H_2(s)$ contains a zero designed to cancel $H_{L1}(s, 0)$ and realize the specified $G(s)$.

Consider the effects of off-tuning. The $H_{b1}(s, \Delta\omega_e)$ for the single-pole IF is no longer a simple real pole and is described (see Appendix E) as

$$H_{b1}(s, \Delta\omega_e) = \left[\frac{s}{b\omega_b \gamma^2} + 1\right] \bigg/ \left[\left(\frac{s}{b\omega_b \gamma}\right)^2 + 2\left(\frac{s}{b\omega_b \gamma^2}\right) + 1\right] \qquad (4\text{-}63)$$

with

$$\gamma = [1 + (\Delta\omega_e/b\omega_b)^2]^{1/2} \qquad (4\text{-}64)$$

The magnitude and phase of $H_{b1}(s, \Delta\omega_e)$ is plotted in Fig. 4-7, with the normalized magnitude of off-tuning as a parameter. Note that the response

---

† While a tuned circuit has in the frequency domain a pair of complex poles, its lowpass equivalent and its effect on a double sideband modulation is that of a single pole. Therefore, the single tuned circuit is often referred to in the literature and in this text as a *single pole*.

4.3. Linear Operation of the Loop

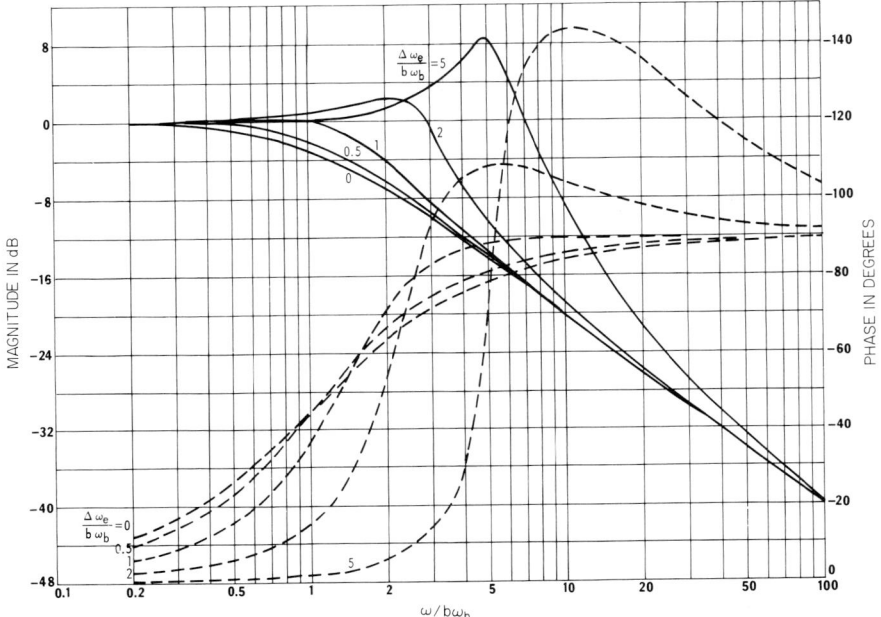

Fig. 4-7. Equivalent baseband response of single-pole filter, with signal carrier off-tuning as a parameter (—; magnitude; ---: phase).

may change considerably with a substantial off-tuning. Thus, the actual open-loop response in the presence of off-tuning is

$$G(s, \Delta\omega_e) = \frac{H_{b1}(s, \Delta\omega_e) G(s)}{H_{b1}(s, 0)}$$

$$= \frac{K\left(\dfrac{s}{a\omega_b}+1\right)\left(\dfrac{s}{b\omega_b\gamma^2}+1\right)\left(\dfrac{s}{b\omega_b}+1\right)}{\left[\left(\dfrac{s}{\omega_b}\right)^2+2\xi\left(\dfrac{s}{\omega_b}\right)+1\right]\left[\left(\dfrac{s}{b\omega_b\gamma}\right)^2+2\left(\dfrac{s}{b\omega_b\gamma^2}\right)+1\right]} \quad (4\text{-}65)$$

with the closed-loop response being

$$\frac{\Phi_r}{\Phi_i}(s) = \frac{H_{b1}(s, \Delta\omega_e)G(s)/H_{b1}(s, 0)}{1 + H_{b1}(s, \Delta\omega_e)G(s)/H_{b1}(s, 0)} \quad (4\text{-}66)$$

As an illustrative case, Fig. 4-8 contains the magnitude and phase of the closed-loop response function $H(j\omega, \Delta\omega_e)$ with $\Delta\omega_e$ as a parameter and the following set of loop constants:

$$a = 5, \quad b = 1.6, \quad K = 9, \quad \xi = 2^{1/2}/2 \quad (4\text{-}67)$$

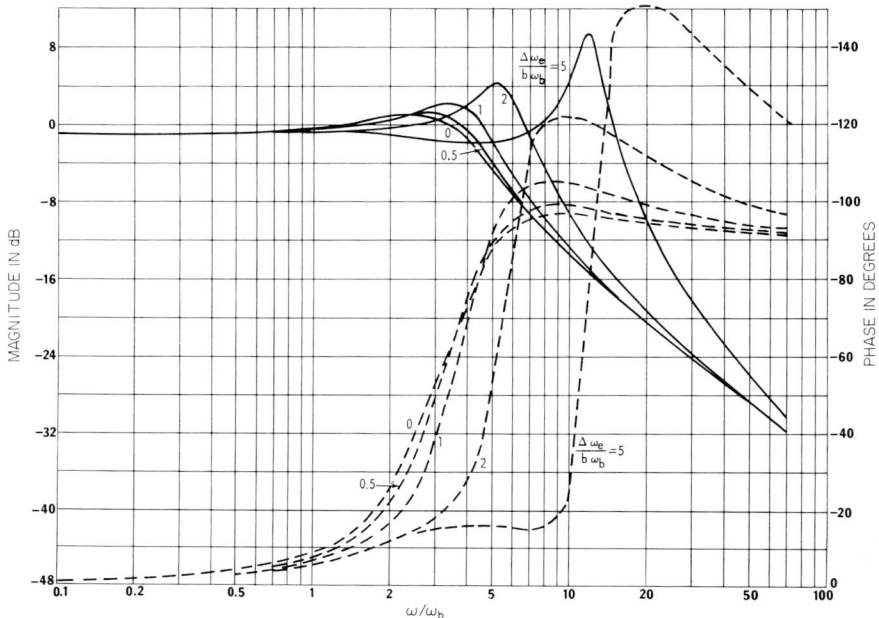

Fig. 4-8. FMFB closed-loop magnitude and phase response with input signal carrier off-tuning as a parameter (—; magnitude; – – –: phase).

A conclusion that may be reached is that for a moderate amount of off-tuning, with $\Delta\omega_e$ restricted to lie between the IF filter 3-dB points, the closed-loop response does not vary appreciably between 0 and ω_b. If we consider that the signal modulation components lie in this range (representative of a modulation tracking loop), the off-tuning does not affect the signal response. The closed-loop response beyond ω_b is progressively modified as off-tuning increases, and, as shall be found subsequently, has a distinct effect on the noise performance of the loop.

C. FM Demodulation

A main application for the FMFB loop is to perform modulation tracking and serve as a demodulator for FM signals. Under these conditions, the demodulated signal is available at the VCO input terminals, and Eq. (4-57) describes the loop response to the input modulation, expressed in terms of the phase modulation, $\phi_i(t)$. In the previous discussion, $\phi_i(t)$ was described as a low-level test signal that satisfies the linear model characterization of the internal IF filter. Note from Eq. (4-60) that the phase error is reduced by the gain factor $K_1 K_2 K_3$ and, hence, may be designed to be small for a given input phase deviation, thus satisfying the linearization requirements of the internal IF filter and validating the linear model approach. Low-frequency components in the input modulation that do not result in a small phase error can be included in the loop response by a quasi-stationary approach that associates these terms with the frequency-offset term $\Delta\omega_i t$ and, therefore, can be considered to be governed by the loop DC response. Note that the DC response corresponds to the magnitude of the phase modulation response, Eq. (4-60), at low frequencies, permitting Eq. (4-60) to fully describe the loop output. Therefore, the signal output from the loop is

$$E_o(s) = \frac{1}{K_3} sH(s)\Phi_i(s) \quad \text{or} \quad e_o(t) = \frac{1}{K_3} h(t) \circledast \frac{d}{dt} \phi_i(t) \quad (4\text{-}68)$$

Under conditions when the function $H(j\omega)$ is flat and has a linear phase response over the baseband frequency range, Eq. (4-68) reduces to

$$e_o(t) \cong \frac{1}{K_3} \frac{d}{dt} \phi_i(t) \quad (4\text{-}69)$$

The noise output from the loop, also calculable from the linear model, is of similar form

$$n_o(t) = \frac{1}{K_3} h(t) \circledast \frac{d}{dt} n(t) \quad (4\text{-}70)$$

where $h(t)$ is the impulse response of the closed-loop transfer function $H(s)$, and $n(t)$ is the equivalent-noise input to the loop.

These results are the same as those derived in Section 3.2 for the LD operating above its noise threshold and, therefore, indicate that above threshold the FMFB loop yields equivalent performance. The departure of the FMFB performance from that predicted by the linear model is brought on by factors that essentially invalidate the linear model. These factors are discussed in Section 4.4, and lead to a condition where Eqs. (4-69) and (4-70) can no longer be considered applicable. This departure from linear model performance brings forth the noise threshold of the loop.

4.4. Nonlinear Operation

We now turn to a discussion and evaluation of the factors that result in nonlinear loop performance. By this we mean performance that is not evident from the linear model, and is a result of the breakdown of, or noncompliance with, the requirements that form the linear model. These characteristics can be divided into two general areas: those brought about by inherent nonlinear characteristics exhibited by ideal systems, and those due to practical limitations of actual loop components. For example, in Section 4.3 a loop consisting of ideal elements, each characterized by an ideal mathematical operation, was found to contain a basic nonlinearity stemming from the FM filtering characteristics of the internal IF filter. However, when one considers the limitations found in the elements themselves, a distinct source of nonlinearity is found. For example, there is a limited range over which the VCO output frequency deviation is linearly related to the input baseband signal.

A. Element Limitations

Element limitations (VCO, discriminator, etc.) manifest themselves in their inability to function in an ideal manner. For the input mixer, we note that it is ideally characterized as a multiplier. Linear multiplication, however, is required only with respect to the signal input port, because the VCO output amplitude is constant. With this condition, linear multiplication (noninteraction between multiple signal inputs), or mixing, is achieved by specifying that the input signal level be significantly less than that supplied by the VCO. In usual applications the FMFB will be designed to discriminate between the signal and either noise or another interfering signal, with the primary selectivity determined by the internal IF characteristic. Hence, mixer saturation due to the presence of input interference producing nonlinear multiplication can adversely affect the loop performance. Mixer operating point, dynamic range, and pre-FMFB filtering characteristics are all to be considered in assuring linear loop operation. Mixer saturation is similar in effect to prelimiting ahead of the loop, and under certain circumstances is desirable, especially under large signal fluctuation conditions. This will be discussed further in Chapter 7. However, more often than not, the desired operating goal is linear mixing.

Considering next the IF filter, we see that one of the performance requirements is the rejection of the unwanted sideband emanating from the input mixer. Residue from this filtered sideband will subsequently be detected by the LD and appear in the loop output with particularly troublesome results in low-output-noise systems. The problem is alleviated by a judicious choice of the internal IF operating frequency, in conjunction with the internal IF bandwidth.

Two characteristics in the LD result in nonlinear performance. They are the failure of the limiter to clip arbitrarily close to zero, and the finite-detection bandwidth of the discriminator. Figure 4-9 illustrates both of these characteristics. As the signal component off-tunes in the IF, it is progressively attenuated by the IF response, and we depend upon the limiter to restore the signal level prior to detection. At the point where the limiter fails, the

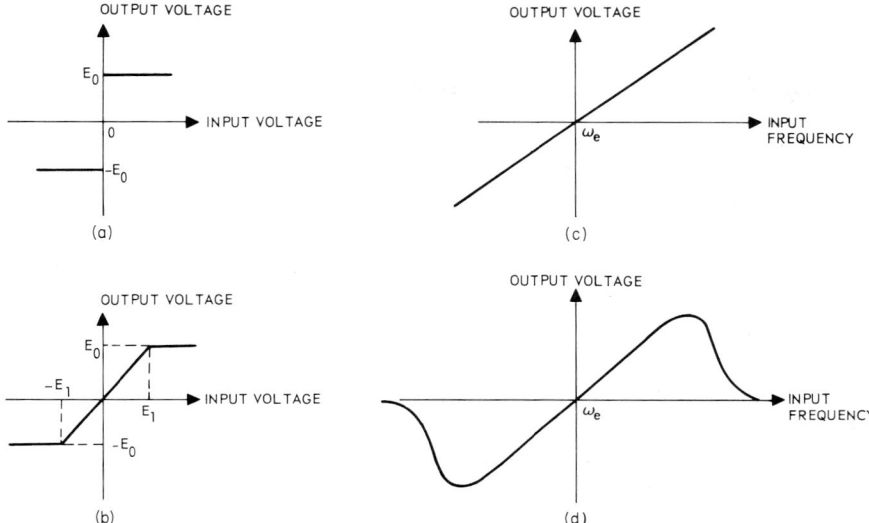

Fig. 4-9. (a) Ideal and (b) actual limiter; (c) ideal and (d) actual discriminator responses.

discriminator output becomes sensitive to the IF amplitude level, resulting in reduced detector sensitivity and reduced overall loop gain. Further IF off-tuning results in a reduction of loop gain to unity and less. The VCO now fails to track input frequency variations, resulting in a "runaway" condition. The loop unlocks, invalidating the tracking equations and the linear model representation. In addition, with insufficient limiting and/or signal level, the loop will not pull in and track the input signal. This problem manifests itself mainly in AFC systems, where highly selective internal IF filters are employed, and initial IF signal off-tuning beyond the filter passband precludes pull-in.

The second condition is the "turnabout" in the discriminator response for frequencies beyond the response peak. If the IF signal is off-tuned to the point where it passes the discriminator peak and falls on the discriminator skirt, the loop will no longer be stable, since the feedback slope is opposite to that required for tracking. The loop will effectively "runaway" and unlock. Thus, the frequency separation between the discriminator peaks defines the pull-in range of the FMFB system, assuming that one of the other factors under discussion does not predominate.

Finally, baseband dynamic range and VCO deviation range present practical limits on tracking and pull-in capability. It is evident that if the VCO cannot deviate to the maximum offset, or if the baseband circuitry cannot supply the necessary deviation voltage to the VCO, pull-in and tracking are not possible.

B. Fundamental Limitations

We now consider the nonlinear behavior exhibited by the ideal loop, being mindful of the practical limitations imposed by the system elements. Turning first to AFC applications, the stability of the loop is influenced by the off-tuning within the loop IF. The magnitude and phase of the equivalent FM response to a single IF pole as a function of off-tuning is given in Fig. 4-7. For a more general filter, the expression for the equivalent baseband response for the IF is given by Eqs. (4-17), (4-18), and (4-19). By this equivalent-response technique it is possible to calculate and ensure loop stability over the full pull-in range.

Consideration of the requirements in a modulation tracking loop results in an internal IF filter that is a single pole. This choice facilitates the realization of loop stability, and permits the application of simple compensation techniques. For instance, with an on-tune carrier, the IF equivalent-modulation response (as discussed in Section 4.3) is a real pole and, therefore, can be canceled from the overall loop response by placing a real zero in the baseband filter $H_2(s)$. The effectiveness of such compensation, in the presence of off-tuning, is demonstrated in Fig. 4-10, where the results of Fig. 4-7 (modulation response of the IF filter) are modified by the addition of a real zero at the

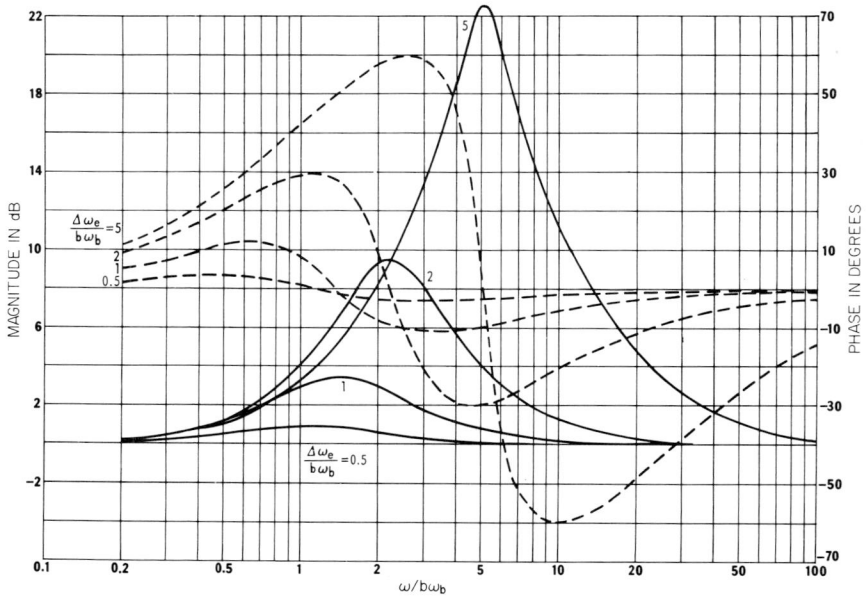

Fig. 4-10. Compensated response with signal carrier off-tuning as a parameter (—: magnitude; – – –: phase).

equivalent IF pole frequency (with carrier on-tune). The plot is for

$$T(s, \Delta\omega_e) = \left(\frac{s}{b\omega_b} + 1\right)\left(\frac{s}{b\omega_b\gamma^2} + 1\right) \bigg/ \left[\left(\frac{s}{b\omega_b\gamma}\right)^2 + 2\left(\frac{s}{b\omega_b\gamma^2}\right) + 1\right]$$

(4-71)

where

$$\gamma = \left[1 + \left(\frac{\Delta\omega_e}{b\omega_b}\right)^2\right]^{1/2}$$

(4-72)

Note that

$$T(s, 0) = 1$$

(4-73)

With off-tuning within the IF 3-dB bandwidth, reasonable compensation is achieved; however, with large amounts of off-tuning, the results indicate that placing the compensating zero at a progressively higher frequency is a more desirable choice. Exact compensation requires a complex-zero real-pole combination [refer to Eqs. (4-63) and (4-64)], the parameters of which are a function of the off-tuning. If it is recalled that the input off-tuning is actually an artifact to represent low-frequency modulation and noise components, a time-varying compensating network is implied for exact compensation. Such complexity is generally not required, and the real-zero compensation is adequate, especially if the off-tuning remains within the IF 3-dB bandwidth.

Two sources of nonlinearity and distortion stem from the IF filter. One source is the nonlinear modulation response exhibited by the IF filter, and is due mainly to phase nonlinearity. (This was discussed in Section 3.4.) The second source stems from inadequate equivalent IF response compensation (previously discussed) that results in a loop transfer function that varies quasi-statically with IF frequency off-tuning. This results in a form of intermodulation between high and low frequencies in a seemingly linear system.

C. Nonlinear Noise Performance (Threshold)

The noise analysis carried forth in Section 4.2 indicates the presence of two second-order noise components that are not linearly related to input noise. Recall

$$d(t) = \dot{\theta}_e(t) - \frac{1}{A}\frac{d}{dt}\left[\frac{\hat{\phi}_{es}(t) \cdot \hat{x}_1(t)}{E'(t)}\right] - \frac{1}{A}\frac{d}{dt}\left[\frac{\hat{\phi}_{en}(t) \cdot \hat{x}_1(t)}{E'(t)}\right]$$

(4-39′)

In order to establish a linear model representation, these second-order components, $(1/A)\,d/dt[\hat{\phi}_{es}(t) \cdot \hat{x}_1(t)/E'(t)]$ and $(1/A)\,d/dt[\hat{\phi}_{en}(t) \cdot \hat{x}_1(t)/E'(t)]$, are neglected. Their presence, however, indicates a nonlinear noise behavior and a threshold condition. The linear component, as discussed earlier, results in a loop response equivalent to that of the conventional LD operating above its noise threshold. The increase of these additional noise components (they are inversely related to the input signal level A) to a significant level, so as to degrade the output SNR below that predicted by the linear model, signals the onset of the loop noise threshold. For convenience, an output SNR degradation factor of 1 dB is taken as the threshold condition, as was done for the LD (refer to Section 3.2). Strictly speaking, the linear model does not predict this degradation; however, the model, in conjunction with the characteristics of the second-order noise components generally determined experimentally, can be used to predict and minimize the threshold condition.

The second term in Eq. (4-39′) may be neglected if one considers the case of no modulation. The third term indicates two separate threshold mechanisms or nonlinear noise-generating sources. In the absence of feedback, $\hat{\phi}_{en}(t)$ is essentially $\hat{\beta}(t)$, resulting in a term $(1/A)d/dt[\hat{\beta}(t) \cdot \hat{x}_1(t)/E'(t)]$ that is also found in conventional LD analyses [see Eqs. (4-26) and (3-32)]. This is indicative of the conventional threshold effect experienced by the internal demodulator and is, therefore, minimized by conventional FM demodulator design considerations. The introduction of feedback modifies $\hat{\phi}_{en}(t)$ to include the fed-back noise $\hat{\phi}_{rn}(t)$, i.e. $\hat{\phi}_{en}(t) = \overline{\hat{\beta}(t) - \phi_{rn}(t)}$ [see Eq. (4-29)]. This results in second-order noise terms that are a unique characteristic of the frequency feedback loop and represent intermodulation between the fed-back noise (the noise phase modulation appearing in the VCO) and the input noise, producing an additional detected noise voltage. The additional noise introduced by these two sources may collectively or individually bring forth the loop threshold condition. It is, therefore, convenient to identify threshold effects by their source and, thus, we will refer to loop noise degradation in terms of open-loop threshold and feedback threshold, the designations indicating additional noise in terms of that produced by the detector and that produced by closed-loop dynamics, respectively.

How is it that the threshold of an FMFB is lower than that of the LD? The answer resides in the following: The open-loop threshold is effectively reduced by the narrower IF passband permitted by the compressed signal deviation. The feedback threshold is prevented from dominating by appropriate design of the closed-loop response, as discussed next and in Chapter 7. In this manner, the combined effect of open-loop and feedback thresholds still results in a lower demodulator threshold for the FMFB.

4.4. Nonlinear Operation

The feedback threshold noise is controlled by fed-back noise $\phi_{rn}(t)$ and therefore, is controlled by the overall loop response. Enloe[1] empirically determined that the threshold effects attributed to this component are uniquely related to the mean square value of $\phi_{rn}(t)$ and, in turn, to the noise bandwidth B_n of the loop. We have then

$$\sigma^2 = \overline{\phi_{rn}^2(t)} = (2\eta/A^2) \int_0^\infty |H(j\omega)|^2 \, df = (2\eta/A^2) \, B_n |H(0)|^2 \quad (4\text{-}74)$$

and

$$B_n \equiv \int_0^\infty |H(j\omega)|^2 \, df / |H(0)|^2 \quad (4\text{-}75)$$

with the result that feedback threshold minimization requires minimization of the loop noise bandwidth.

If the second-order loop response characteristic is [see Eq. (4-50) with $\xi = 2^{-1/2}$]

$$G(s) = K\left(\frac{s}{a\omega_b} + 1\right) \Big/ \left[\left(\frac{s}{\omega_b}\right)^2 + 2^{1/2}\left(\frac{s}{\omega_b}\right) + 1\right] \quad (4\text{-}76)$$

then the closed-loop noise bandwidth, given a specific open-loop gain constant K, can be minimized by proper choice of the zero constant a. The result is that the minimum closed-loop noise bandwidth B_{nm} is a function of K alone. This is derived by Enloe,[2] and is presented in Fig. 4-11 in terms of the *feedback factor* $F = 1 + K$. In addition to the basic curve relating B_{nm} to F (indicated by $\phi_b = 0$), additional minimum noise bandwidth curves are presented which account for the effects of excess loop delay. The basis for these curves is discussed in Section 4.5.

This section is concluded by considering the effects of input off-tuning. Being mindful of the fact that off-tuning modifies the closed-loop characteristic, we also see that it modifies the closed-loop noise bandwidth and, hence, the feedback noise threshold. The open-loop transfer function is given by Eq. (4-65), from which the closed-loop response and the closed-loop noise bandwidth may be obtained. Taking as a representative condition the loop parameters encountered in the previous examples,

$$K = 9.0, \quad a = 5.0, \quad b = 1.6, \quad \xi = 2^{1/2}/2$$

we present the relative closed-loop noise bandwidth, i.e., $10 \log_{10}[B_n(\Delta\omega_e)/B_n(0)]$, as a function of off-tuning in Fig. 4-12. Note that IF off-tuning $\Delta\omega_e$, restricted to within the IF 3-dB bandwidth, results in a rather small increase in closed-loop noise bandwidth.

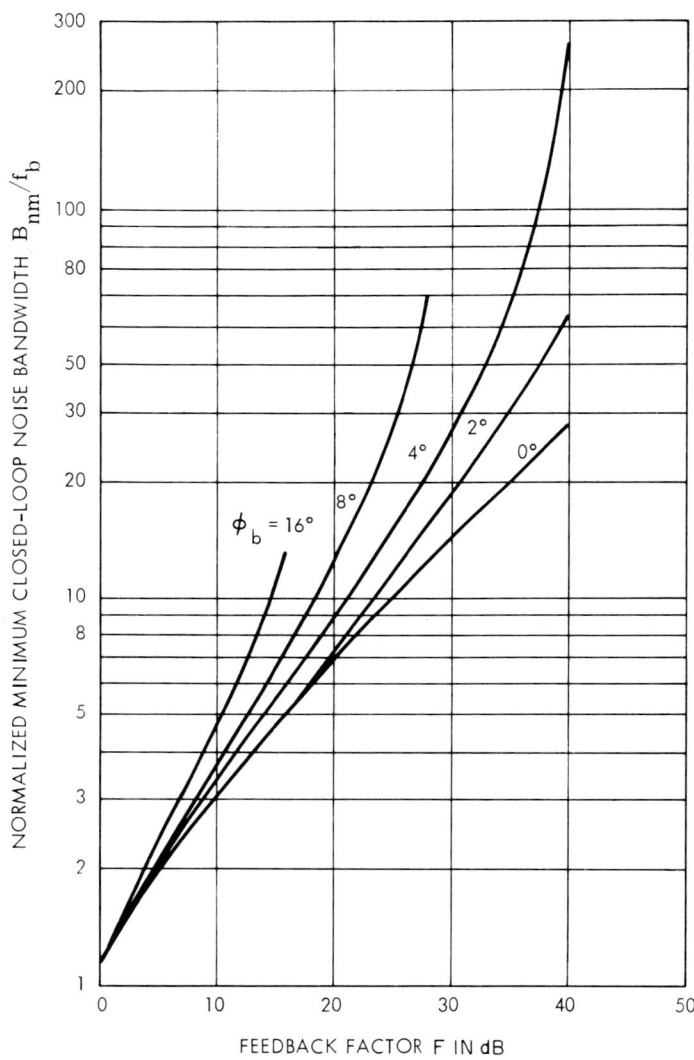

Fig. 4-11. Minimum closed-loop noise bandwidth versus feedback factor, with ϕ_b as a parameter (from Enloe[2]).

4.5 Effect of Excess Delay

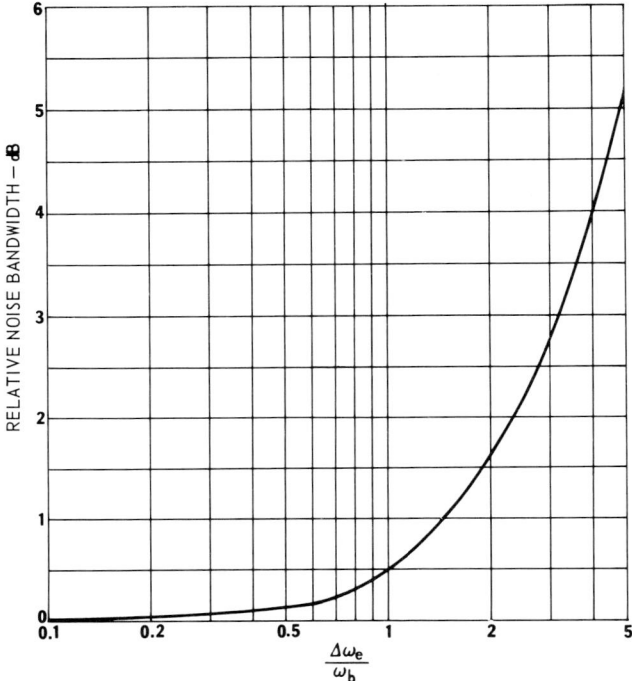

Fig. 4-12. FMFB relative noise bandwidth in decibels {i.e. $10 \log_{10}[B_n(\Delta\omega_e)/B_n(0)]$} as a function of signal carrier off-tuning. (Example: $a = 5$, $b = 1.6$, $K = 9$, and $\xi = 2^{-1/2}$.)

4.5. Effect of Excess Delay

In the implementation of FMFB systems, the loop inadvertently experiences excess open-loop phase shift, or delay. The sources of this delay are the physical electrical length of the signal path and the presence of high-frequency poles. Both factors are a function of circuit design and can be restricted to tolerable limits; however, they preclude the realization of an arbitrarily wideband system. The more difficult fact to cope with is the high-frequency poles, which are due partly to the presence of parasitic and stray energy storage elements. However, the major source of these poles resides in basic bandwidth characteristics associated with the loop elements. Examination of the LD (Section 3.2) indicates a basic lowpass frequency response characteristic related to the peak-to-peak response separation (see Fig. 4-9). Greater bandwidth, required to reduce low-frequency excess phase shift (and delay), necessitates greater discriminator peak-to-peak separation, and results in low-detection sensitivity. To realize an open-loop gain constant K compensation for loss in detection sensitivity may be achieved by an increase in baseband gain, with an associated increase in baseband delay. We can associate with the discriminator a performance index equal to the product of detection sensitivity (volt-seconds per radian) and bandwidth (radians per second) with the desire to maximize the index. Similarly, a performance index exists for the VCO which is dependent on circuit type and design. These indexes, together with the gain-bandwidth limitations of the baseband amplifying stages, form an overall gain-bandwidth product for the loop. The result, for a desired open-loop gain constant K, is the minimum unavoidable excess phase shift associated with the loop gain-bandwidth product. This excess phase shift will manifest itself as an apparent delay at low frequencies and, together with the transport delay, forms the overall signal delay within the loop.

Assuming that delay adequately represents the excess phase shift in the critical region of the open-loop response [where $G(j\omega) = 1$], the open-loop transfer function with the desired second-order response is

$$G(s) = K\left(\frac{s}{a\omega_b} + 1\right)e^{-s\tau} \bigg/ \left[\left(\frac{s}{\omega_b}\right)^2 + 2^{1/2}\left(\frac{s}{\omega_b}\right) + 1\right], \qquad \tau = \frac{\phi_b}{\omega_b} \qquad (4\text{-}77)$$

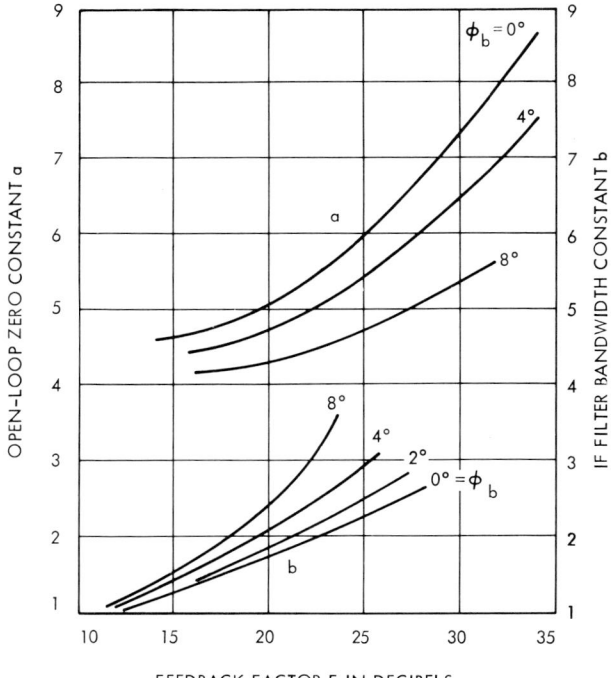

Fig. 4-13. Constants a and b versus feedback factor, with ϕ_b as a parameter (from Enloe[2]).

where ϕ_b is the excess phase shift, in radians, at baseband frequency ω_b. The closed-loop response $H(s)$, and subsequently the closed-loop noise bandwidth B_n, become a function of the delay (or phase shift ϕ_b), with the result that B_n minimization requires a choice for a the zero constant that is determined by both the gain K and the excess phase shift ϕ_b. Enloe[2] has carried forth this optimization with the minimum noise bandwidth (see Fig. 4-11) and the zero constant a (see Fig. 4-13) established as a function of feedback factor F and excess phase ϕ_b. The use of the optimized IF filter bandwidth constant b also shown in Fig. 4-13 will be discussed in Chapter 7. In extremely wideband systems, the excess phase in the critical region of the delay-loop response $[G(j\omega) = 1]$ cannot be adequately represented as delay. In those cases, the specific phase-frequency characteristics must be incorporated into the closed-loop calculation and loop optimization.

References

1. L. H. Enloe, Decreasing the threshold in FM by frequency feedback. *Proc. IRE* **50**, No. 1, 18–30 (1962).
2. L. H. Enloe, The synthesis of frequency feedback demodulators. *Proc. Nat. Electron. Conf.* **18**, 477–497 (1962).

CHAPTER

5

Phase-Locked Loop Principles

In this chapter we present, in a manner similar to that followed in Chapter 4, a qualitative description of, and the basic equations governing, the phase-locked loop (PLL). We begin by noting that the PLL is derivable from the FMFB loop (discussed in Chapter 4), and, therefore, find that certain describing characteristics are similar in the two systems. The PLL, however, has unique characteristics, particularly with respect to stability and acquisition, that must be considered in addition to those discussed for the FMFB. The results presented in this chapter serve as the basis for the design procedures found in Chapter 6.

5.1. Operational Principles

The essential elements that comprise the PLL are the *input phase detector*, *loop filter and amplifier*, and the *VCO*, as shown in Fig. 5-1. We note that this configuration is derivable from the FMFB loop, shown in Fig. 4-1, if the FMFB VCO center frequency is tuned to the input center frequency, the IF and LD are removed, and DC coupling is employed between the mixer output, through the baseband filter, to the VCO input. Note further that the phase detector function can be realized by use of a mixer with DC output coupling. The similarity between the loops results in similar describing characteristics; however, the basic mode of operation, which is different for the PLL, results in unique characteristics not paralleled in the FMFB.

5.1. Operational Principles

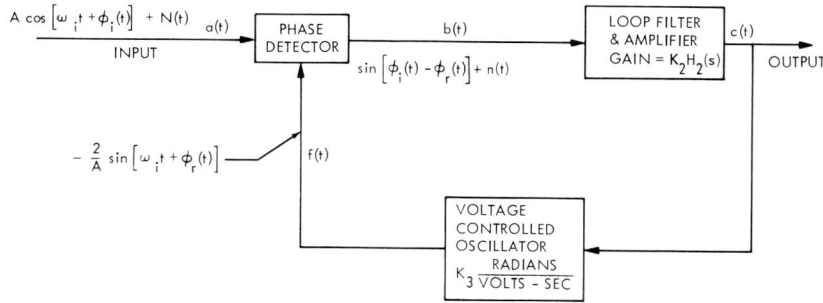

Fig. 5-1. Block diagram of basic PLL in synchronized mode.

The basic operational requirement for the loop is to track the phase of the input signal. This is accomplished by forming in the phase detector a voltage proportional to the phase difference between the input signal and VCO signal. This error signal is then amplified, filtered to remove excess noise and other undesirable components, and applied to the VCO to effect input signal phase tracking. By tracking signal phase, *frequency synchronism* and *frequency tracking* are achieved. Note that the difference between the FMFB and the PLL techniques is in the formation of the error signal; the FMFB tracks by forming a frequency error from which the loop control voltage is derived, while the PLL tracks by detecting the phase error between the signal input and the VCO output.

One of two tracking objectives may be desired. The first is related to AFC, where the PLL will lock (or synchronize) to a continuous spectral component (such as an unsuppressed carrier) appearing in the input signal, and be relatively insensitive to sideband or modulation components. The second objective is to closely track the angle modulation components appearing in the input signal, thus causing the VCO control (or input) signal to be a replica of the input signal modulation. Tracking fidelity and interference (or noise) rejection are both controlled by the loop parameters, and generally present conflicting design requirements, as discussed further in Chapter 6. This design results in a low-threshold FM demodulator for a large range of signal types and parameters. Depending upon the application, the loop output may be taken at either the VCO input (for demodulation applications) or the VCO output (for synchronization applications), with response characteristics being a function of the output point.

We continue by developing the describing equations for a PLL with idealized components. Consider an input signal of form

$$a(t) = A \cos[\omega_i t + \phi_i(t)] \qquad (5\text{-}1)$$

where A is peak signal amplitude (a constant), ω_i is input signal center frequency, and $\phi_i(t)$ is input signal phase modulation. Assume that the VCO output signal is of the form

$$f(t) = -(2/A) \sin[\omega_i t + \phi_r(t)] \qquad (5\text{-}2)$$

where $\phi_r(t)$ is VCO phase modulation, with the subscript r indicating a return signal. A particular form of VCO signal is chosen, with the main feature being the quadrature relationship between the input and the VCO output. A secondary feature is the $2/A$ amplitude, chosen solely as a matter of convenience. Both characteristics stem from the multiplier form of phase detector, as will be apparent presently. Further, the static frequency error between the input and the VCO signal is assumed to be zero, indicating the synchronized (or locked) condition; this is a basic operational requirement for all known PLL applications. Synchronization is discussed further in Section 5.4; we continue here by assuming that conditions necessary for synchronization have been satisfied.

Let the phase detector be characterized as an ideal multiplier which, in turn, is an idealization of the doubly-balanced mixer often used as a phase detector in practice. The result is an output of

$$b(t) = a(t)f(t) = -\sin[2\omega_i t + \phi_i(t) + \phi_r(t)] + \sin[\phi_i(t) - \phi_r(t)] \qquad (5\text{-}3)$$

If we neglect the upper sideband component [the first term in Eq. (5-3)], or assume that it is subsequently rejected by the loop filter, the filter output becomes

$$c(t) = K_2 h_2(t) \circledast \sin \phi_e(t) \qquad (5\text{-}4)$$

where K_2 is baseband amplifier gain, $h_2(t)$ is impulsive response of the filter,

$$\phi_e(t) = \phi_i(t) - \phi_r(t) \qquad (5\text{-}5)$$

and $\circledast$ denotes convolution. This baseband voltage is applied to the VCO, and results in an FM signal whose frequency deviation is directly proportional to the control voltage,

$$d\phi_r(t)/dt = K h_2(t) \circledast \sin \phi_e(t) \qquad (5\text{-}6)$$

with $K = K_2 K_3$, and K_3 being VCO sensitivity.

Equation (5-6) is a nonlinear differential equation relating the loop response $\phi_r(t)$ to the error signal $\phi_e(t)$. A general solution of this equation is not available; however, by restricting $|\phi_e(t)| < \pi/2$, some important loop characteristics may be derived.

We continue by considering an input signal of the form

$$\phi_i(t) = \Delta\omega_i t + \phi_i'(t) \qquad (5\text{-}7)$$

where $\Delta\omega_i$ represents a static input frequency offset from the center frequency

5.1. Operational Principles

ω_i and $\phi_i'(t)$ a dynamic signal modulation. Assuming that synchronization is maintained (i.e., no static frequency error is present) the response of the loop will take the form

$$\phi_r(t) = \Delta\omega_i t + \phi_r'(t) + \phi_0 \tag{5-8}$$

where ϕ_0 is a constant phase error due to $\Delta\omega_i t$, and $\phi_r'(t)$ is the dynamic signal modulation response. Substitution of Eqs. (5-7) and (5-8) into Eq. (5-6) results in

$$\Delta\omega_i + d\phi_r'(t)/dt = Kh_2(t) \circledast \sin[\phi_e'(t) - \phi_0] \tag{5-9}$$

where $\phi_e'(t) = \phi_i'(t) - \phi_r'(t)$, the dynamic phase error.

If we design the loop so that the dynamic phase error $\phi_e'(t) \ll 1$ rad, then the sine function can be approximated by

$$\sin[\phi_e'(t) - \phi_0] \cong \phi_e'(t)\cos\phi_0 - \sin\phi_0 \tag{5-10}$$

and Eq. (5-9) may be written in two parts

$$\Delta\omega_i = -Kh_2(t) \circledast \sin\phi_0 \tag{5-11}$$

and

$$d\phi_r'(t)/dt = K\cos\phi_0 h_2(t) \circledast [\phi_i'(t) - \phi_r'(t)] \tag{5-12}$$

where Eqs. (5-11) and (5-12) represent the static (or DC) describing equation and the linearized dynamic tracking equation, respectively.

By taking Laplace transforms of both equations, we may derive the transfer characteristic equations as

$$\sin\phi_0/\Delta\omega_i = -1/KH_2(0) \tag{5-13}$$

and

$$s\Phi_r'(s) = K\cos\phi_0 H_2(s)[\Phi_i'(s) - \Phi_r'(s)] \tag{5-14}$$

or

$$\frac{\Phi_r'}{\Phi_i'}(s) = \frac{K(\cos\phi_0)H_2(s)}{s + K(\cos\phi_0)H_2(s)} \tag{5-15}$$

where $H_2(s) = \mathscr{L}[h_2(t)]$. Note that the response to a static input frequency offset manifests itself as a gain reduction (by $\cos\phi_0$) in the linear model. Static offset is usually due to drift in transmitter and receiver oscillators, and doppler shift experienced by the signal. It also represents the drift of the VCO in the PLL. Normally, its effect is reduced by loop design, so that it may be neglected in its influence on dynamic response. Further discussion of the static response and loop performance is reserved for Section 5.3.

The linear describing equation is based on a small dynamic phase error. Conditions which cause the loop to depart from this restriction result in loop performance breakdown. With a departure from a small phase error, signal distortion becomes prominent; and with a phase error magnitude increase beyond $\pi/2$ radians, synchronism is lost. Note that as the input frequency increases, the phase error increases, forcing the VCO frequency to follow. When the phase error exceeds $\pi/2$ rad, the loop error signal (phase detector output) begins to decrease, not permitting the VCO frequency to follow the input frequency further, and resulting in a loss of synchronism.

As an aside, we note that the basic nonlinearity of the loop is due to the choice of a phase detector with a sinusoidal characteristic and one may wonder why a phase detector that results more readily in a linear describing equation is not used instead. The answer lies in a number of factors, including the "geneology" of the technique, the simplicity of this phase detector and its wideband characteristic, and the "linear translation characteristic" in frequency exhibited by the ideal multiplier to additive input interference. However, more complex phase detection schemes that result in a more linear characteristic and improved noise performance have been evaluated, and are discussed in Chapter 8.

We conclude by considering the effects of signal amplitude variations. This condition is included in the loop describing equations by modifying the input, Eq. (5-1), to be

$$a(t) = A(t) \cos[\omega_i t + \phi_i(t)] \tag{5-16}$$

resulting in a $c(t)$ (assuming multiplier-type of phase detector)

$$c(t) = K_2 h_2(t) \circledast [(A(t)/A) \sin \phi_e(t)] \tag{5-17}$$

Therefore, the effect of input signal level variation manifests itself as a varying loop gain component that might be associated with the loop amplifier.

Variations in loop gain about a quiescent value modify the tracking and noise immunity characteristics of the loop, as discussed in subsequent paragraphs. Different types of phase detectors, as well as nonlinear circuits employed before the loop, modify these results. However, the net effect in many cases can be represented by appropriately translating input signal amplitude variations to loop amplifier gain variations. It is also apparent that input signal variations may be compensated for by controlling loop amplifier gain.

5.2. The Linear Equivalent Circuit

A. The Linear Model[1-4]

We shall now derive equivalent circuit representations for the loop from the operational equations given in Section 5.1. Before considering the linear model, let us first present a model based on the synchronized condition alone. This quasi-linear model is helpful in understanding nonlinear loop behavior

5.2. The Linear Equivalent Circuit

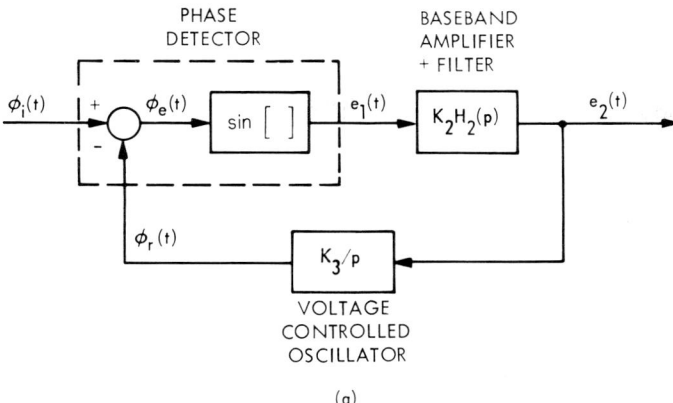

(a)

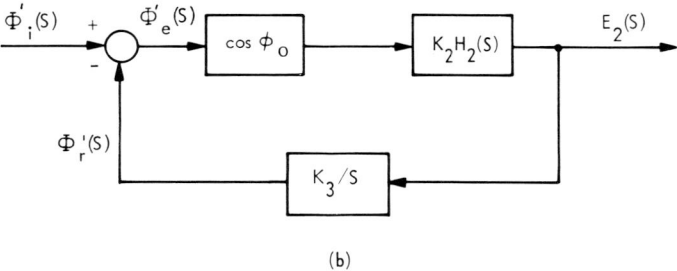

(b)

Fig. 5-2. PLL models. (Actual signals—Input: $A\cos[\omega_i t + \phi_i(t)]$ with $\phi_i(t) = \phi_i'(t) + \Delta\omega_i t$, $\phi_e(t) = \phi_e'(t) - \phi_0$; VCO output: $-(2/A)\sin[\omega_i t + \phi_r(t)]$ with $\phi_r(t) = \phi_r'(t) + \Delta\omega_i t + \phi_0$, $\phi_0 = -\sin^{-1}[\Delta\omega_i/K_2 K_3 H_2(0)]$. (a) Synchronous model $[|\phi_e(t)| < \pi/2]$. (b) Linear model ($|\phi_e'| \ll 1$, $|\phi_e'(t) - \phi_0| < \pi/2$).

discussed later in the chapter, and is a logical stepping stone in arriving at the linear model.

The equivalent circuit shown in (a) of Fig. 5-2 is the synchronous model. It is a representation of Eq. (5-6), and is derived in the following manner: The phase detector, being the ideal multiplier discussed earlier, forms the sine of the phase error and is, therefore, represented by the input-differencing network followed by a sine operation. The baseband filter and amplifier are transcribed directly from the basic block diagram (shown in Fig. 5-1), and the VCO is represented as an integrator. The condition $|\phi_e(t)| < \pi/2$ is required to assure synchronism.† The operational‡ notation $p = d/dt$ is used to describe the block's operation in the time domain.

† Figure 5-2a without the limitation $|\phi_e(t)| < \pi/2$ is an exact representation of the PLL, including the nonsynchronous condition.

‡ For the meaning of the operator p, see Section 2.1. Note: p is simply replaced by s for the Laplace domain.

The linear model shown in (b) of Fig. 5-2 evolves from (a) by considering the dynamic phase error to be small, i.e., $|\phi_e'(t)| \ll 1$ rad, linearizing the sine operation. The model follows from the development of Eq. (5-14) and can be transcribed directly from it.

These representations are in terms of ideal loop elements; methods for extending the model to include effects due to nonideal elements will be indicated later. The method for including input noise in these models is considered next.

B. Effects Due to Input Noise

In normal application, the PLL is subjected to a signal with additive input noise interference. A representative case, the same as that considered in Chapter 4, is a signal corrupted by band-limited Gaussian noise. We find that the loop exhibits a response to the input noise, resulting in a noise component of phase modulation in the VCO output. The phase detector output will then consist of a signal component that contains both signal and noise terms, and an additive noise component. We begin by formulating loop noise performance with the objective of reducing the results to a linear model representation to facilitate further noise calculations.

Let the input signal and noise be represented as

$$a(t) = A \cos[\omega_i t + \phi_i(t)] + N(t) \tag{5-18}$$

The VCO output contains a signal term and a noise term, $\phi_{rs}(t)$ and $\phi_{rn}(t)$, respectively, so that

$$f(t) = -(2/A) \sin[\omega_i t + \phi_r(t)] \tag{5-19}$$

where

$$\phi_r(t) = \phi_{rs}(t) + \phi_{rn}(t). \tag{5-20}$$

Finally, the phase detector output signal is of the form

$$b(t) = \sin[\phi_{es}(t) - \phi_{rn}(t)] + n(t) \tag{5-21}$$

where

$$n(t) = -(2N(t)/A) \sin[\omega_i t + \phi_r(t)]$$

and
$$\tag{5-22}$$

$$\phi_{es}(t) = \phi_i(t) - \phi_{rs}(t)$$

This result can be incorporated directly into the synchronous model to fully account for the input noise. The model, as shown in (a) of Fig. 5-3, is the same as that shown in (a) of Fig. 5-2, with the addition of the equivalent

5.2. The Linear Equivalent Circuit

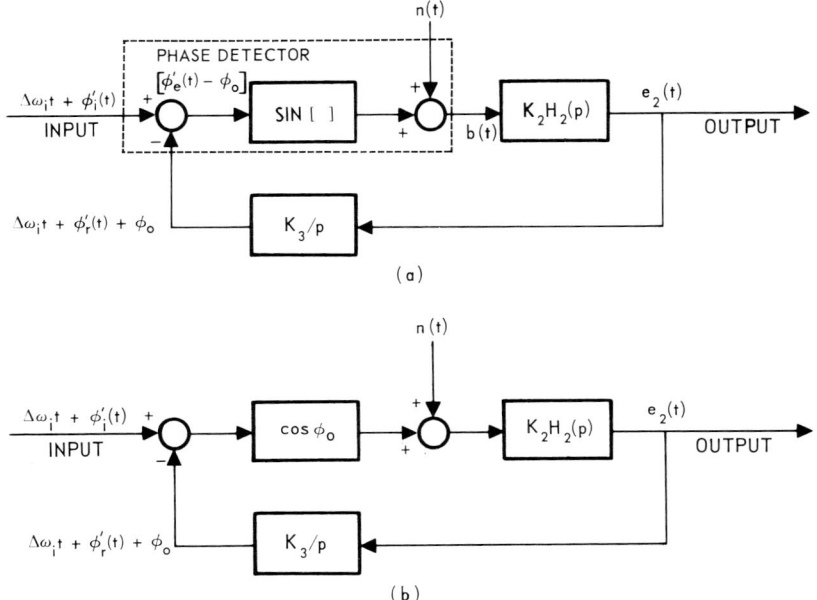

Fig. 5-3. PLL model with equivalent noise input. Actual signals—Input: $A \cos[(\omega_i + \Delta\omega_i)t + \phi_i'(t)] + N(t)$, with $N(t) = x(t) \cos \omega_i t - y(t) \sin \omega_i t$; VCO output: $-(2/A) \sin[(\omega_i + \Delta\omega_i)t + \phi_r'(t) + \phi_0]$. (a) Synchronous model. (b) Linear model.

input noise $n(t)$, and forms an exact representation of the loop equations. The linear model, as shown in (b) of Fig. 5-3 is again derived by requiring the dynamic phase error component to be small, linearizing the sine function.

We proceed by evaluating the equivalent input noise term $n(t)$. The input noise may be represented as

$$N(t) = x(t) \cos \omega_i t - y(t) \sin \omega_i t \qquad (5\text{-}23)$$

and from Eq. (5-22)

$$n(t) = -(2/A)[x(t) \cos \omega_i t - y(t) \sin \omega_i t] \sin[\omega_i t + \phi_r(t)] \qquad (5\text{-}24)$$

Carrying forth the indicated multiplication, combining terms, and neglecting terms containing $2\omega_i$ (since they are normally rejected by the loop), we have

$$n(t) = -[x(t)/A] \sin \phi_r(t) + [y(t)/A] \cos \phi_r(t) \qquad (5\text{-}25)$$

which is similar to the equivalent noise input found in the FMFB (see Fig. 4-3).

In order to utilize our model in the calculations, we need to know the power spectral density of $n(t)$. This is, in general, a complex problem, since $\phi_r(t)$ is in part derived from $n(t)$ and in part determines $n(t)$. However, it has been shown[3] that for the case where $N(t)$ has a symmetrical bandpass spectral density which is much wider than the bandwidth of $\phi_r(t)$, then as in the FMFB, $n(t)$ is essentially the lowpass analog of $N(t)$, appropriately scaled. Thus, if the input noise is white with density η, then the power spectral density of $n(t)$ is white with value $2\eta/A^2$.

The conclusion drawn is that for small dynamic phase error, induced by both modulation and noise, the PLL may be represented by a linear model with an equivalent noise input. It will be shown later that to realize small noise-induced phase errors, a high input CNR in terms of the loop bandwidth rather than input bandwidth is required. This permits linear loop operation to relatively low input CNR's, resulting in a sensitive signal detection system.

5.3. Linear Operation

We proceed here to calculate loop characteristics necessary to formulate and carry forth the loop design. These calculations are based on the linear model developed earlier. We recall that the loop gain, in the linear model, was found to be a function of the static tracking characteristic; therefore, we begin by reviewing and discussing further the static tracking characteristic.

A. DC Tracking

The DC tracking analysis is important in synchronization applications as well as dynamic tracking. The DC analysis pertains to the loop response to the $\Delta\omega_i t$ term in the input signal [refer to Eq. (5-7)] and, thus, represents the steady-state response to an input frequency offset $\Delta\omega_i$ from the loop center frequency. As shown in Eqs. (5-8)–(5-13) the loop responds with a phase tracking error ϕ_0 between the input signal and the VCO output

$$\sin \phi_0 / \Delta\omega_i = -1/[KH_2(0)] \qquad (5\text{-}13) \quad (5\text{-}26)$$

which may be written as

$$\phi_0 = -\sin^{-1}[\Delta\omega_i/KH_2(0)] \qquad (5\text{-}27)$$

where $H_2(0)$ is the DC gain of the loop filter. For small tracking errors, $|\Delta\omega_i/KH_2(0)| \ll 1$ rad, Eq. (5-27) can be approximated as

$$\phi_0 \cong -\Delta\omega_i/KH_2(0) \qquad (5\text{-}28)$$

Equation (5-27) indicates the static tracking range of the loop. The argument of the arcsin function lies between ± 1, and the loop, therefore, cannot produce a static response to frequency offsets greater than $KH_2(0)$.

5.3. Linear Operation

The static tracking range is, therefore, given by

$$|\Delta\omega_i|_{TR} \leq KH_2(0) \quad \text{rad/sec} \tag{5-29}$$

An increase in this range can be achieved by increasing the loop gain K. Note that the effective loop gain in terms of dynamic tracking is reduced by a factor of $\cos\phi_0$, as shown in (b) of Fig. 5-2. Two important characteristics of the PLL, namely, the *hold-in* and *pull-in* ranges are now defined.

Definitions

"*Hold-in*" *range*: Suppose the PLL is in a synchronized condition with $\omega_i = \omega_{r0}$, where ω_{r0} is the free-running VCO frequency. Now vary ω_i slowly and ω_r will follow. The "hold-in" range is the value of $\omega_i - \omega_{r0}$ for which ω_r fails to follow ω_i, resulting in a loss of synchronism. The hold-in range represents the maximum static tracking range and is given by the equality sign in Eq. (5-29).

The "*pull-in*" range is the maximum initial frequency difference between the input and the VCO, i.e., $|\omega_i - \omega_{r0}|$, for which the loop will eventually lock into synchronism. The pull-in range is (except for the first-order PLL) not easily calculable. It is related to the dynamic tracking characteristics of the loop which we consider next.

B. Open- and Closed-Loop Response

The synthesis of a stable closed-loop system, and the conditions necessary to maintain stability, depend upon the establishment of a suitable open-loop response. We begin by considering a modulated input signal with a small phase deviation at the loop center frequency, i.e., $|\phi_i(t)| \ll 1$ rad, and $\Delta\omega_i = 0$, $\phi_0 = 0$. The open-loop response to this "test signal," taken by figuratively breaking the loop at the VCO output, is [see linear model of Fig. 5-2(b), and note that with these restrictions $\phi_i(t) = \phi_i'(t)$]

$$\left.\frac{\Phi_r}{\Phi_i}(s)\right|_{\text{open loop}} = G(s) = \frac{KH_2(s)}{s} \tag{5-30}$$

where $K = K_2 K_3$.

The characteristic of the open-loop transfer function provides a classification that conveniently categorizes various PLL implementations. Following a convention established for servomechanisms,† we specify the *loop type*

---

† Often the PLL literature intermixes "type" and "order." The convention here is adapted from H. Chestnut and R. W. Mayer, "Servomechanisms and Regulating System Design," Wiley, New York, 1951. The second-order "type-one" loop is sometimes called "second-order loop with imperfect integrator."

by the number of *poles at the origin* in $G(s)$, and the *loop order* by the *total number of poles* in $G(s)$. For example, a $G(s)$ that contains one pole at the origin (contributed by the integration effect of the VCO) and a single real pole, results in a second-order type-one PLL. Within this framework, the FMFB is considered a "type-zero" loop in the family of phase tracking loops.

The closed-loop transfer function is, then,

$$\left.\frac{\Phi_r}{\Phi_i}(s)\right|_{\text{closed loop}} = H(s) = \frac{G(s)}{1+G(s)} = \frac{KH_2(s)}{s+KH_2(s)} \quad (5\text{-}31)$$

Examination of this transfer function provides the criteria for closed-loop stability, namely, the denominator $[s + KH_2(s)]$ must not contain right half s-plane zeros. It is presently shown, by root-locus examination of this denominator factor, that low-order systems are unconditionally stable, whereas third- and higher-order systems are conditionally stable. Note that the lowest loop order is limited to the loop type, and in order to realize loop types greater than 1, ideal integrators are required in the synthesis of $H_2(s)$. Normally this is not feasible, but under most practical conditions, not strictly required, either.

We continue by considering various forms of $H_2(s)$ and their influence on the closed-loop characteristic. The first case is that of no loop filter, i.e., $H_2(s) = 1$.

Case I: $H_2(s) = 1$, *first-order, type-one loop.*

Here the result is a closed-loop transfer function of the form

$$H(s) = \frac{\Phi_r}{\Phi_i}(s) = \frac{K}{s+K} = \frac{1}{s/K + 1} \quad (5\text{-}32)$$

which is that of a simple real pole at frequency $-K$ rads/sec. Note that the bandwidth increases linearly with K. Referring to Eq. (5-29), we see that K is also the hold-in range (in radians per second). Thus, the bandwidth and gain at any other frequency are predetermined, and may not be of independent design in the first-order loop. First-order loop response is shown in Fig. 5-4 by the curve marked $b/\omega_n = 1$. (Observe in part (b) of the figure that $b = \omega_n$ represents a first-order PLL response.)

The next case is representative of a second-order loop, which is formed by implementing $H_2(s)$ as $(s/a + 1)/s$.

Case II: $H_2(s) = (s/a + 1)/s$, *second-order, type-two loop.*

The closed-loop transfer function is of the form

$$H(s) = \frac{\Phi_r}{\Phi_i}(s) = \frac{K(s/a + 1)}{s^2 + (K/a)s + K} = \frac{(s/a + 1)}{(s^2/K) + (s/a) + 1} \quad (5\text{-}33)$$

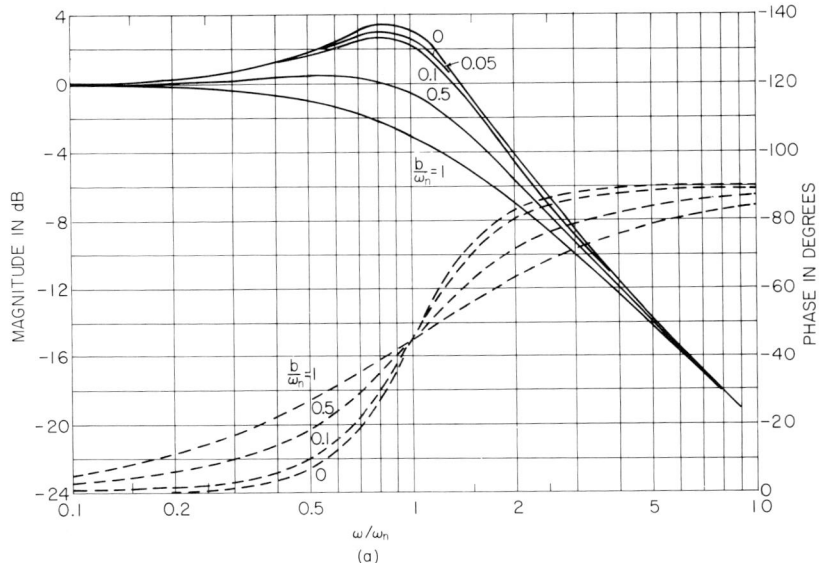

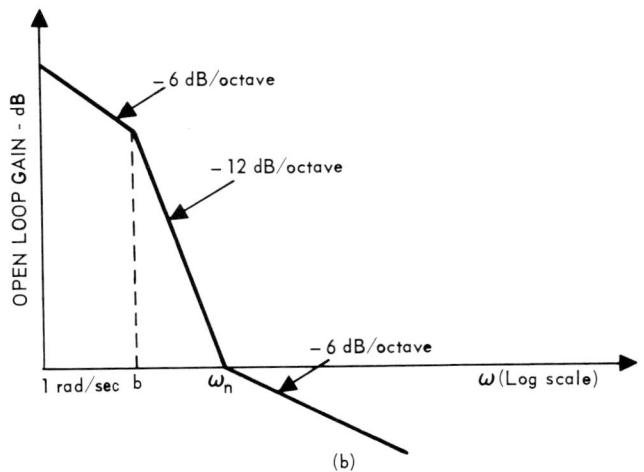

Fig. 5-4. Responses of first and second-order, type-one PLL. (a) Closed-loop response with the pole frequency as a parameter. (Example: $\omega_n = a = (Kb)^{1/2}$; Nominal design: $b = 0$; $b/\omega_n = 1$ is first-order loop, b/ω_n is less than unity for second-order loops.—: magnitude; – – –: phase.) (b) Open-loop amplitude response (asymptotic).

which is that of a pole pair and a real zero. In this case, the hold-in range is unbounded [due to the ideal integration introduced by $H_2(s)$], while the loop bandwidth characteristics are controlled both by the loop gain K and the "zero constant" a. The gain K is chosen (this is discussed in detail in Chapter 6) to achieve a desired open-loop gain characteristic, and is primarily determined by the received signal modulation. Once a K is determined, a may be adjusted to minimize the bandwidth of $H(s)$ (this is discussed further in Section 5.5). The second-order loop response is seen to be equivalent to the normalized response discussed earlier in Chapter 4, and can be re-expressed in the form

$$H(s) = \frac{s/(a_0 \omega_n) + 1}{(s/\omega_n)^2 + 2\xi(s/\omega_n) + 1} \quad (5\text{-}34)$$

with

$$\omega_n = K^{1/2}, \quad \xi = K^{1/2}/2a, \quad a_0 = a/K^{1/2} \quad (5\text{-}35)$$

The variation of the magnitude and phase of $H(j\omega)$ is illustrated in Figs. 5-5 and 5-6 as a function of a and K, respectively. This form of second-order loop

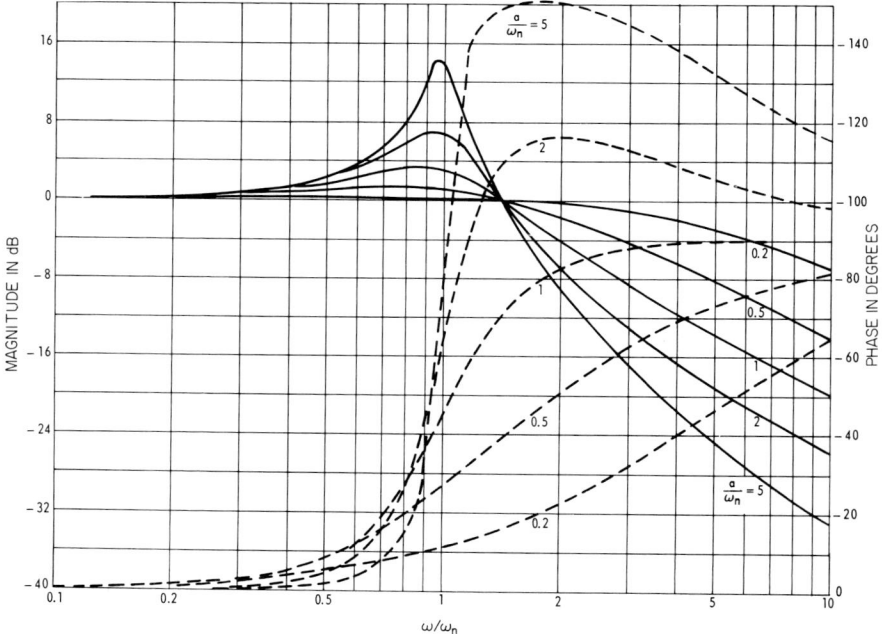

Fig. 5-5. Closed-loop magnitude and phase response of second-order, type-two PLL, with zero frequency as a parameter (—: magnitude; - - -: phase). (Example: $\omega_n = K^{1/2}$; Nominal design: $a/\omega_n = 1$; For comparison open-loop response see Fig. 6-16.)

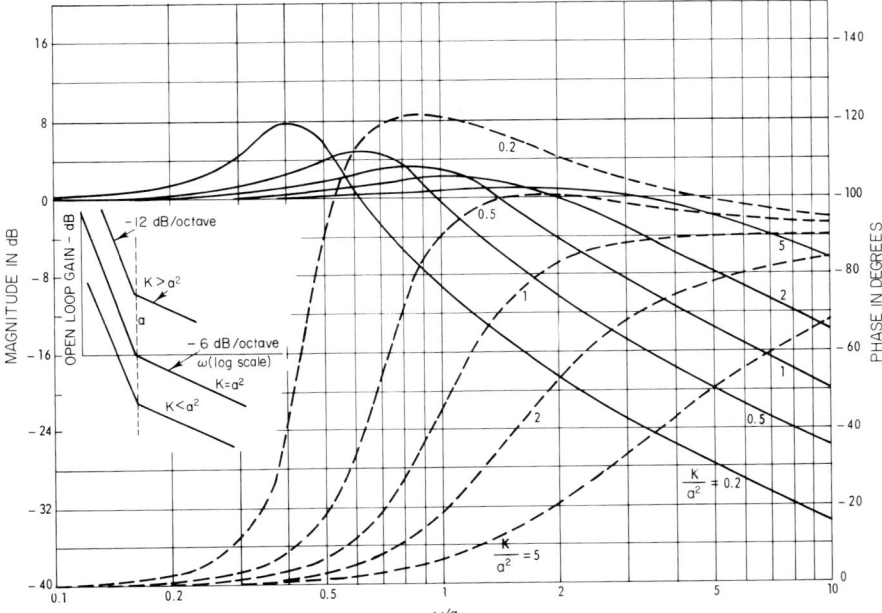

Fig. 5-6. Closed-loop magnitude and phase response of second-order, type-two PLL, with gain K as a parameter (—: magnitude; ---: phase; nominal design: $K^{1/2}/a = 1$).

has limited practicality in that an ideal integration is required in $H_2(s)$. In practice, an approximation is made by implementing the loop filter $H_2(s)$ as $(s/a + 1)/(s/b + 1)$ leading to a second-order, type-one loop.

Case III: $H_2(s) = (s/a + 1)/(s/b + 1)$, *second-order, type-one loop.*

The closed-loop transfer function is now of the form

$$H(s) = \frac{\Phi_r}{\Phi_i}(s) = \frac{\left(\dfrac{s}{a} + 1\right)}{\dfrac{s^2}{Kb} + \left(\dfrac{1}{K} + \dfrac{1}{a}\right)s + 1} \tag{5-36}$$

which is similar to the complex pole pair and real zero found for Case II. The result may be normalized by setting

$$\omega_n = (Kb)^{1/2}, \quad \xi = \tfrac{1}{2}(b/K)^{1/2} + (Kb)^{1/2}/2a, \quad a_0 = a/(K)b^{1/2} \tag{5-37}$$

This gives the normalized response described by Eq. (5-34). The influence of b on the closed-loop transfer function becomes minimal as b approaches zero. Note that K must be increased to compensate for decreasing b, so as to maintain a nominal loop response. The Kb product plays the same role here as K alone in the type-two loop discussed previously.

The closed-loop transfer function dependence on b is illustrated in Fig. 5-4a, where the compensating zero a and the square root of the Kb product are both held fixed at the loop natural frequency. The asymptotic open-loop response characteristic in (b) of the figure illustrates the meaning of the parameters. For b a decade or more below a, the loop response is seen to be primarily determined by a and the Kb product, i.e., the response closely approaches that of the second-order type-two loop.

The second-order loop provides the flexibility necessary to design hold-in range and loop bandwidth independently, a feature not exhibited by the first-order loop. This permits the realization of a narrow-band tracking filter that has high noise immunity and a wide tracking range. For a given loop gain K, greater selectivity over the second-order loop is achievable with third- and higher-order loops. We consider next the third-order type-three loop.

Case IV: $H_2(s) = 1/a + 1/bs + 1/s^2 = (s^2/a + s/b + 1)/s^2$, *third-order, type-three loop*

For this case the closed-loop transfer function becomes

$$H(s) = \frac{\Phi_r}{\Phi_i}(s) = \frac{K\left(\frac{s^2}{a} + \frac{s}{b} + 1\right)}{s^3 + \frac{K}{a}s^2 + \frac{K}{b}s + K} = \frac{\left(\frac{s^2}{a} + \frac{s}{b} + 1\right)}{\left(\frac{s^3}{K} + \frac{s^2}{a} + \frac{s}{b} + 1\right)} \qquad (5\text{-}38)$$

We note that if K, a, and b are all positive, the response consists of a pair of complex zeros, a pair of complex poles, and an additional real pole. The locus of the complex poles, as K is increased from zero, traverses a portion of the right-half s-plane, indicating that the loop is unstable for low gain. The root loci of the $H(s)$ denominators for the various loops discussed are given in Fig. 5-7; they indicate that type-one and type-two loops are unconditionally stable, whereas the type-three loop requires a gain constant of $K \geq ab$ to realize stability. This requirement may be understood by observing that in the type-three loop the third-order pole [two due to $H_2(s)$, and one due to the VCO transfer characteristic] in $G(s)$ produces a phase lag of 270°. For a stable closed-loop system this must be reduced to at least 180°, by the complex zero pair, at the $G(s)$ unity gain point. The complex zero pair produces 90° phase advance at $s = ja^{1/2}$, providing the required phase lag reduction in $G(s)$. The resulting loop gain at this frequency is

$$|G(ja)| = K(a^{1/2}/b)/a(a)^{1/2} = K/ab \qquad (5\text{-}39)$$

By using a gain constant greater than ab, the unity gain point is advanced to a frequency above $ja^{1/2}$, producing greater phase lead, increasing the stability margin, and leading to the stated gain requirement.

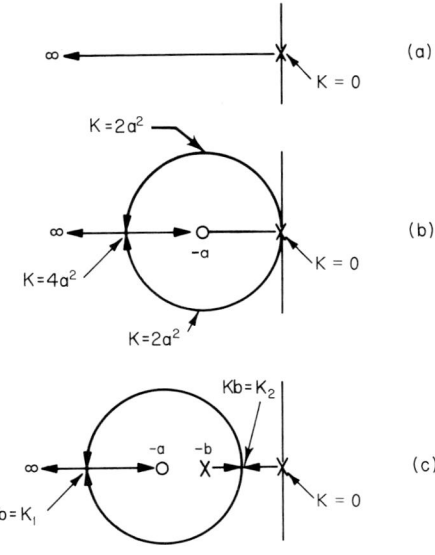

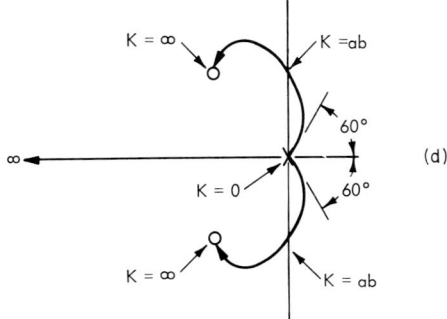

Fig. 5-7. Root loci for PLL transfer functions. (a) First-order. (b) Second-order type-two. (c) Second-order type-one ($K_1 = a^2[1 + (1 - b/a)^{1/2}]^2$, $K_2 = a^2[1 - (1 - b/a)^{1/2}]^2$). (d) Third-order type-three. (From "Principles of Coherent Communication" by A. J. Viterbi. Copyright 1966 by McGraw-Hill, Inc. Used with permission of McGraw-Hill Book Co.)

C. Transient Response

The PLL is often required to track transient phase modulation components appearing in the input signal. These transients may occur through medium characteristics (such as doppler shift, fading and multipath, etc.) or may be simply the nature of the intended modulation [such as TV baseband, frequency shift keying (FSK), etc.].

The basic requirements for transient tracking are dictated by the acquisition properties of the loop—a subject that is treated in Section 5.4. The conclusions drawn there are that the extremes in tracking without loss of synchronism, are achieved in the presence of large phase errors that prohibit linear model analysis. In the presence of noise (covered in Chapter 6), however, we find that the signal phase error will be restricted, by design, to be a fraction of $\pi/2$ rad, and, therefore, loop transient tracking characteristics, for small phase error, are directly applicable to this discussion. The linear model derived on the small phase error assumption provides the linear loop transfer functions derived in Section 5.3. The loop transient response may be derived easily from the loop transfer function by application of linear network theory. In particular, the reader is referred to the normalized transient characteristics for step, ramp, and ramp-integral excitation of the normalized second-order system in Gardner,[2] and for step responses in Chapter 9. These responses apply to phase step, frequency step, and frequency ramp excitation of the second-order type-two loop, and approximate the response of the second-order type-one loop.

An indication of successful transient tracking is the final value of the phase error as the transient progresses in time. This evaluation is carried out by application of the final value theorem for linear systems in the Laplace domain (see Table I in Chapter 2):

$$\phi_e(\infty) = \lim_{s \to 0} s\Phi_i(s)[1 - H(s)] \qquad (5\text{-}40)$$

where

$\phi_i(t)$ = signal phase modulation (transient)

and $\quad \phi_e(\infty)$ = loop phase error final value $\quad (5\text{-}41)$

The phase error final value is summarized in Table II in terms of excitation and loop characteristics. The conclusion that may be drawn is that for excitations of the form

$$\phi_i(t) = At^{n-1}u(t) \qquad (5\text{-}42)$$

where $u(t)$ is a step at $t = 0$, a type-n loop establishes a zero final value, a type-$(n - 1)$ loop results in a finite final value, and a type-$(n - 2)$ loop results in an unbounded final value, for $\phi_e(t)$. Therefore, the second-order type-one loop will eventually unlock (or lose synchronism) when tracking a frequency ramp ($n = 3$, which is characteristic of doppler acceleration). The tracking duration in the type-one loop depends on how closely the type-two loop can be approximated or, in other words, how low in frequency can the lag-filter pole "b" be placed [refer to Eq. (5-36)]. Taking into account the fact that transients do not exist for all time in physical systems, we find that the

TABLE II

Basic Characteristics of First-, Second-, and Third-Order Phase-Locked Loops

Order and type	Closed-loop transfer function $H(s)$	Loop noise bandwidth B_n (Hz)	Filter transfer function $H_2(s)$	Received-signal phase modulation $\phi_i(t)$	Steady-state error $\lim_{t\to\infty} \phi_e(t)$
First, one	$\dfrac{1}{s/K+1}$	$\dfrac{K}{4}$	1	Frequency step $\Delta\omega_i t + \phi_{i0}$	$\dfrac{\Delta\omega_i}{K}$
Second, two	$\dfrac{(s/a+1)}{s^2/K + s/a + 1}$	$\dfrac{(K/a)+a}{4}$	$\dfrac{1}{a} + \dfrac{1}{s}$	Frequency step $\Delta\omega_i t + \phi_{i0}$	0
Second, two	$\dfrac{(s/a+1)}{s^2/K + s/a + 1}$	$\dfrac{(K/a)+a}{4}$	$\dfrac{1}{a} + \dfrac{1}{s}$	Frequency ramp $\tfrac{1}{2}Rt^2 + \Delta\omega_i t + \phi_{i0}$	$\dfrac{R}{K}$
Second, one	$\dfrac{(s/a+1)}{s^2/(Kb) + (1/K + 1/a)s + 1}$	$\dfrac{Kb\left(\dfrac{Kb}{a}+a\right)}{4a\left(\dfrac{Kb}{a}+b\right)}$	$\dfrac{s/a+1}{s/b+1}$	Frequency step $\Delta\omega_i t + \phi_{i0}$	$\dfrac{\Delta\omega_i}{K}$
Third, three	$\dfrac{s^2/a + s/b + 1}{s^3/K + s^2/a + s/b + 1}$	$\dfrac{K\left[\dfrac{K}{a}\dfrac{b}+\left(\dfrac{a}{b}\right)^2\right]-a}{4\left(\dfrac{K}{b}-a\right)}$	$\dfrac{1}{a} + \dfrac{1}{bs} + \dfrac{1}{s^2}$	Frequency ramp $\tfrac{1}{2}Rt^2 + \Delta\omega_i t + \phi_{i0}$	0

second-order type-one loop may be applied, by careful design, to the tracking and demodulation of signals undergoing limited-duration doppler acceleration.

We conclude the transient tracking discussion by noting that experimental evaluation of the second-order type-one loop by Viterbi[3] indicates a maximum frequency step that may be tolerated before the loop loses synchronism. Refer to Section 5.4 for PLL acquisition characteristics and further discussion.

D. FM Demodulation

The main application for the PLL discussed in this book is in modulation tracking, i.e., as a demodulator for FM signals, with the demodulated signal available at the VCO input terminals. Equation (5-31) and the VCO transfer characteristic can be combined to describe the loop response, in terms of the input modulation $\phi_i(t)$. The model from which the response characteristics are derived depends upon maintaining a small signal- and noise-induced phase error. With this condition, the loop response to both signal and noise is additive, as indicated in the linear model shown in (b) of Fig. 5-3, and may be calculated as follows: The signal output from the loop is

$$E_2(s) = sH(s)\Phi_i(s)/K_3 \tag{5-43}$$

or

$$e_2(t) = [h(t) \circledast d\phi_i(t)/dt]/K_3 \tag{5-44}$$

The noise output from the loop, also calculable from the linear model, is of similar form

$$n_0(t) = \frac{1}{K_3} h(t) \circledast \frac{d}{dt} n(t) \tag{5-45}$$

where $h(t)$ is the impulse response of the closed-loop transfer function $H(s)$, and $n(t)$ is the equivalent noise input to the loop.

Now let us examine $n(t)$. From Eq. (5-25) we find

$$n(t) = -\frac{x(t)}{A} \sin \phi_r(t) + \frac{y(t)}{A} \cos \phi_r(t) \tag{5-46}$$

and under high input CNR conditions $\phi_r(t) \cong \phi_i(t)$, i.e., the noise component is negligible with respect to the signal component in the VCO phase modulation. This results in an output noise very similar to that experienced in the conventional LD above threshold [Section 3.2., Eq. (3-13)]. Note that both signal and noise components are effectively filtered by $H(s)$; hence, the PLL gives the same output signal and noise performance at high CNR as that exhibited by an LD followed by a filter $H(s)$. The effects of $H(s)$, if desired, may be compensated for at the loop output by a post-PLL filter having factors of $1/H(s)$. Demodulation at lower CNR is discussed in Section 5.4.

5.4. Nonlinear Operation

We now turn to a discussion and evaluation of the factors that result in nonlinear loop performance. By this we mean performance not evident from the linear model and which is a result of the breakdown of, or noncompliance with, the requirements that form the linear model. These characteristics are divided into two general areas: those brought about by inherent nonlinear characteristics exhibited by ideal systems, and those due to practical limitations of actual loop components. For example, we found in Section 5.1 that a loop consisting of ideal elements, each characterized by an ideal mathematical operation, is found to contain a basic nonlinearity. This nonlinearity stems from the multiplier form of phase detector which produces a loop control voltage proportional to the sine of the phase error. Further, when one considers the limitation found in the elements themselves, a distinct source of nonlinearity is found.

A. Element Limitations

The PLL is subject to the same element characteristics that inhibit FMFB performance discussed previously. These characteristics are phase detector saturation (analogous to the input mixer saturation in the FMFB), baseband dynamic range, and VCO deviation range saturation. Saturation occurs in the double-balanced mixer type of phase detector when the input signal port level approaches that of the VCO port. In the absence of input interfering components, this form of saturation has the effect of "linearizing" the phase error detection characteristic, as was discussed in Section 3.4. It may be beneficial in some applications, especially where the objective is to phase lock to a "clean" signal. In the case of FM demodulation, under additive interference, generally speaking, linear operation (no prelimiting or phase detector saturation) is the desired operating goal.

The same comments on baseband dynamic range and VCO deviation range, as made for the FMFB, apply to the PLL. If the VCO cannot deviate to the maximum offset, or if the baseband circuitry cannot supply the necessary deviation voltage to the VCO, pull-in and tracking are not possible to the theoretical limits based on ideal loop elements.

B. Pull-in and Acquisition Characteristics

Until this point we have assumed that the PLL is synchronized to the signal, and have proceeded to investigate loop characteristics based on the synchronized condition. We now turn to a discussion of loop characteristics exhibited prior to and during the pull-in or acquisition phase.

Consider a PLL with no input, and the VCO in a quiescent free-running state described as

$$f(t) = -(2/A)\sin[\omega_i t + \phi_r(t)] \tag{5-47}$$

where

$$\phi_r(t) = 0, \quad t < 0 \tag{5-48}$$

and ω_i is VCO free-running frequency. At a specific instant, which we designate as time $t = 0$, a signal is applied to the PLL input which we may characterize as

$$a(t) = A\cos[\omega_i t + \Delta\omega_i t + \phi_{i0}] \tag{5-49}$$

where $\Delta\omega_i$ is input signal frequency offset. Therefore, an error-control voltage is created at the VCO input

$$c(t) = K_2 h_2(t) \circledast \sin[\Delta\omega_i t + \phi_{i0} - \phi_r(t)] = K_2 H_2(p)\sin\phi_e(t) \tag{5-50}$$

Also, $c(t)$ is proportional to the frequency deviation of the VCO

$$c(t) = \dot{\phi}_r(t)/K_3 = [\Delta\omega_i - \dot{\phi}_e(t)]/K_3 \tag{5-51}$$

Equating (5-50) and (5-51), we have the describing equation for the loop in terms of ϕ_e, $\dot{\phi}_e$, and $\Delta\omega_i$

$$\Delta\omega_i - \dot{\phi}_e(t) = KH_2(p)\sin\phi_e(t) \qquad K = K_2 K_3 \tag{5-52}$$

This is a nonlinear differential equation to which a general analytic solution is not available. However, some important properties and specific responses of the loop may be obtained from a *phase plane* analysis which is essentially a graphic solution of the equation. Principles of phase plane analysis are discussed in texts on control systems,† and we briefly summarize one procedure in the following steps:

(1) As shown in Fig. 5-8, the vertical axis of the phase plane is calibrated in terms of $\dot{\phi}_e$ (rad/sec), and the horizontal axis in terms of ϕ_e (rad).

(2) The trajectory describing the response of the system is started by

---

† See, for example, J. G. Truxal, "Automatic Feedback Control Systems Synthesis," McGraw-Hill, New York, 1955, or W. W. Seifert and C. W. Steeg, Jr., "Control Systems Engineering," McGraw-Hill, New York, 1960.

5.4. Nonlinear Operation

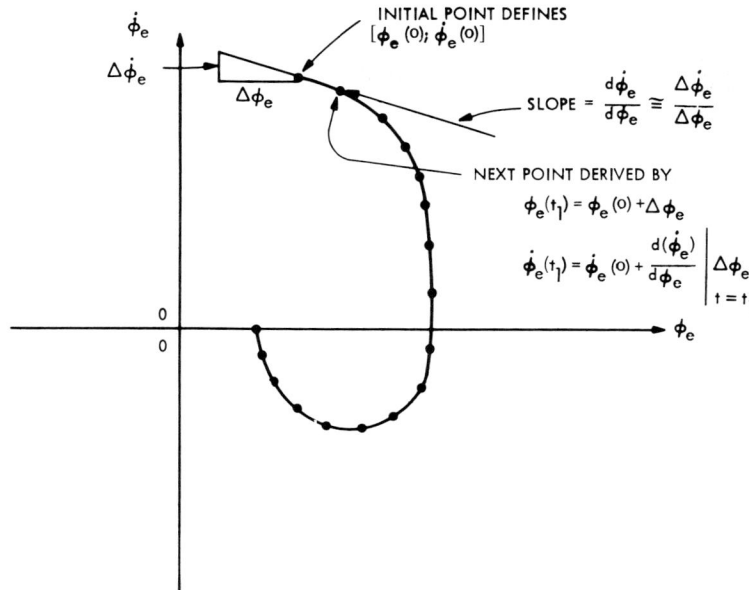

Fig. 5-8. Graphical construction of phase-plane portrait.

evaluating $\dot{\phi}_e(t)$ at $t = 0$ in terms of the known initial condition $\phi_e(0)$, which in this case is

$$\phi_e(0) = \phi_{i0} \qquad (5\text{-}53)$$

We now have the point $[\dot{\phi}_e(0), \phi_e(0)]$.

(3) The trajectory is formed by noting that its slope at the point $[\dot{\phi}_e(0), \phi_e(0)]$ is

$$d\dot{\phi}_e/d\phi_e|_{t=0} \cong \Delta\dot{\phi}_e/\Delta\phi_e|_{t=0} \qquad (5\text{-}54)$$

and, therefore, the next point in the trajectory is approximately

$$\phi_e(t_1) = \phi_e(0) + \Delta\phi_e \qquad (5\text{-}55)$$

$$\dot{\phi}_e(t_1) \cong d\dot{\phi}_e/d\phi_e|_{t=0} \cdot \Delta\phi_e + \dot{\phi}_e(0) \qquad (5\text{-}56)$$

where a new slope is calculated and the trajectory continued. A small $\Delta\phi_e$ is used for accuracy.

(4) The time interval between trajectory points is calculated by noting that

$$\Delta\phi_e/\Delta t \cong \dot{\phi}_e(t_1) \qquad (5\text{-}57)$$

thus

$$\Delta t = \Delta\phi_e/\dot{\phi}_e(t_1) \qquad (5\text{-}58)$$

and

$$t_1 = \Delta t \tag{5-59}$$

permitting the calculation of $\dot{\phi}_e(t)$ and $\phi_e(t)$ in terms of time.

(5) Finally, we may compute the time necessary to traverse a path on the phase plane by summing the individual time elements Δt.

Needless to say, this method of analysis is tedious and time-consuming; however, it is a basic means of evaluating the pull-in or acquisition properties of the PLL.

We proceed by examining phase-plane characteristics for various orders and types of the PLL.[3,5]

Case I: Loop Filter $H_2(s) = 1$, *first-order type-one loop*

Here the describing equation is simply [from Eq. (5-52)]

$$\dot{\phi}_e(t) = -K \sin \phi_e(t) + \Delta \omega_i \tag{5-60}$$

with the phase-plane solution shown in Fig. 5-9. Whenever $\dot{\phi}_e(t)$ is positive, $\phi_e(t)$ must necessarily increase with time. Similarly, a negative $\dot{\phi}_e(t)$ implies a decreasing $\phi_e(t)$ with time. This gives the directions of the arrows in the plot. Note that if $|\Delta \omega_i| \leq K$, there are points on the curve at which $\dot{\phi}_e(t) = 0$.

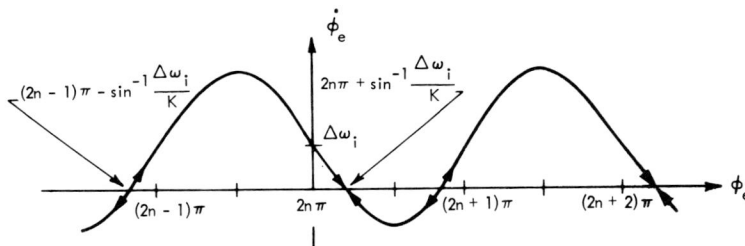

Fig. 5-9. First-order PLL phase-plane characteristic.

Suppose $\phi_e(0) = 2n\pi$, where n is an integer, then the system tends to move along the sinusoidal trajectory until it reaches the ϕ_e-axis and remains there at steady state

$$\phi_e(\infty) = 2n\pi + \sin^{-1}(\Delta \omega_i / K) \tag{5-61}$$

This is a stable point; $\dot{\phi}_e(t)$ cannot become negative because $\phi_e(t)$ would then tend to decrease and return the system to the $\phi_e(t)$-axis. If $\phi_e(0) = (2n+1)\pi$, the system would go through a larger part of the sinusoidal trajectory until it reached a stable point at

$$\phi_e(\infty) = (2n+2)\pi + \sin^{-1}(\Delta \omega_i / K) \tag{5-62}$$

5.4. Nonlinear Operation

If $|\Delta\omega_i| > K$, however, the trajectory never crosses the $\phi_e(t)$-axis, and phase lock is never achieved. The pull-in range of a first-order loop is, thus,

$$|\Delta\omega_i|_{\text{p-i}} = \text{one-sided pull-in range} = K \quad (5\text{-}63)$$

Whenever $|\Delta\omega_i| < K$, the loop ultimately locks with no frequency error, but with a phase error predicted by the DC static response [refer to Eq. (5-27)]. The first-order loop when synchronizing does not skip cycles, i.e., the phase error always traverses less than 2π rad during pull-in. The n multiples of π may be omitted, because the complete portrait of first-order loop pull-in is contained in the principal 2π region.

The pull-in time may be determined from Eq. (5-60) as follows

$$dt/d\phi_e(t) = 1/[\Delta\omega_i - K \sin \phi_e(t)] \quad (5\text{-}64)$$

$$\int_{t_1}^{t_2} dt = \int_{\phi_e(t_1)}^{\phi_e(t_2)} [1/(\Delta\omega_i - K \sin \phi_e)] d\phi_e \quad (5\text{-}65)$$

The indicated integrations may be carried out from some initial phase error $\phi_e(t_1)$ and frequency error $\Delta\omega_i$ to another phase error $\phi_e(t_2)$. Since the denominator in the above integral vanishes at the lock-in point, an infinite time is required before $\dot{\phi}_e(t) = 0$. However, if some small error δ is allowed between the phase error $\phi_e(t_2)$ and $\phi_e(\infty)$, i.e., $|\phi_e(t_2) - \phi_e(\infty)| < \delta$, then the maximum time $t_2 - t_1$, which we designate as the *acquisition time* τ, is finite and given approximately by[1]

$$\tau = \frac{2 \ln (2/\delta)}{K \cos \phi_e(\infty)} \quad (5\text{-}66)$$

We proceed now to higher-order loops.

Case II: Loop Filter $H_2(s) = (s/a + 1)/s$, second-order type-two loop

Here the describing equation is [from Eq. (5-52)]

$$a \, d^2\phi_e(t)/dt^2 + K \cos \phi_e(t) \, d\phi_e(t)/dt + aK \sin \phi_e(t) = 0 \quad (5\text{-}67)$$

which can be reduced to

$$a \, d\dot{\phi}_e/d\phi_e = -K \cos \phi_e - aK \sin \phi_e/\dot{\phi}_e \quad (5\text{-}68)$$

The phase portrait for this equation takes on the form shown in Fig. 5-10, where we see for a large initial frequency offset an almost sinusoidal trajectory. As the initial frequency error is reduced, there is a critical trajectory for which synchronism will occur without skipping cycles, i.e., the trajectory will not traverse to the next 2π region. The conditions for the critical trajectory, taken from Viterbi[3,5] are plotted in Fig. 5-11 in terms of its normalized frequency error $(a/K)\dot{\phi}_{ea}$ at $|\phi_e| = \pi$, as a function of a^2/K. For initial trajectory

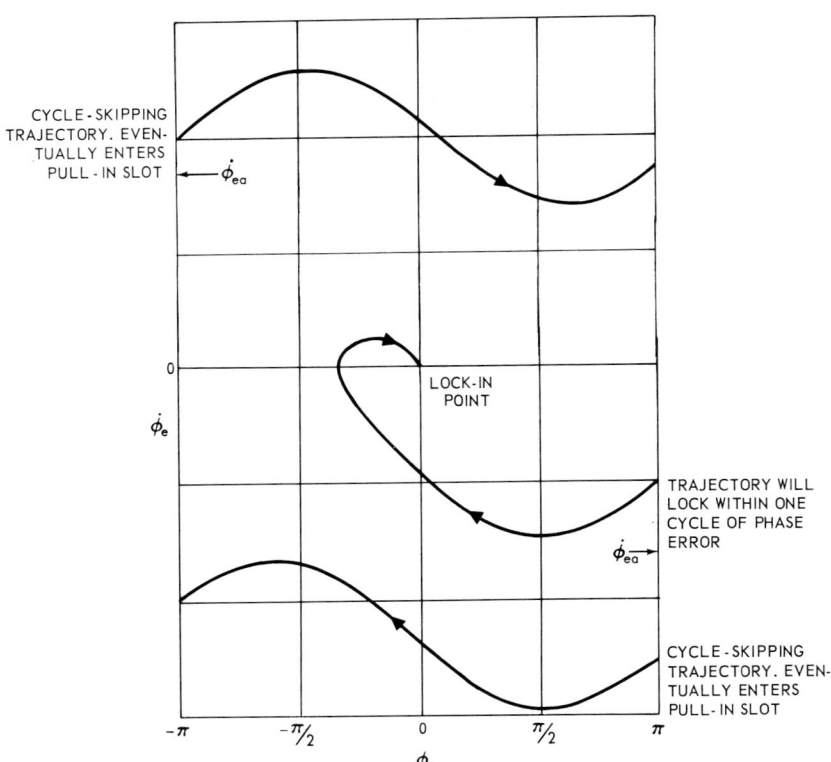

Fig. 5-10. Phase-plane trajectories for second-order type-two PLL.

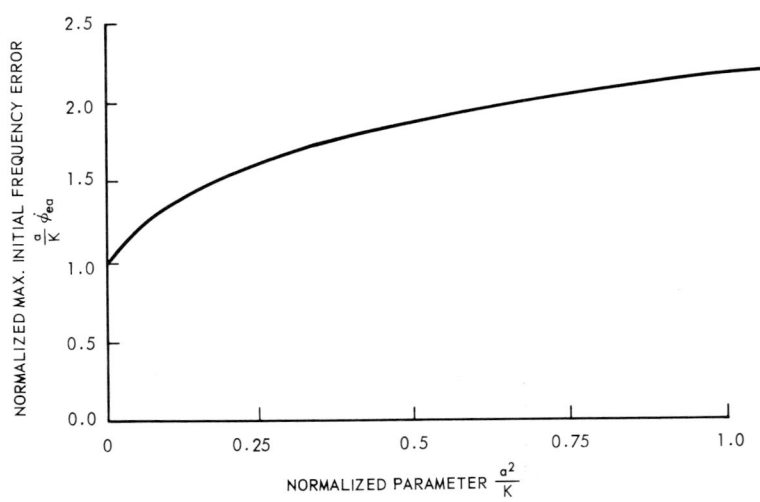

Fig. 5-11. Limit of frequency lock in second-order type-two PLL ($a^2/K = 0.25/\xi^2$). (From "Principles of Coherent Communication" by A. J. Viterbi. Copyright 1966 by McGraw-Hill, Inc. Used with permission of McGraw-Hill Book Co.)

5.4. Nonlinear Operation

frequency errors greater than $\dot{\phi}_{ea}$, the loop will follow one of the "sinusoidal" trajectories advancing (or retarding, depending on whether $\Delta\omega_i$ is positive or negative) through successive cycles of 2π phase error. This effect is called "cycle skipping," because the loop maintains a frequency error for a sufficient length of time to produce a beat type signal at the phase detector output, as shown in Fig. 5-12. This beat signal is also characteristic of the first-order loop when the initial frequency offset is outside the pull-in range [refer to Eq. (5-63)].

The beat signal characteristic is produced in the following way: Initially, the input offset frequency beats with the VCO signal in the phase detector to produce a waveform containing the difference frequency. This waveform,

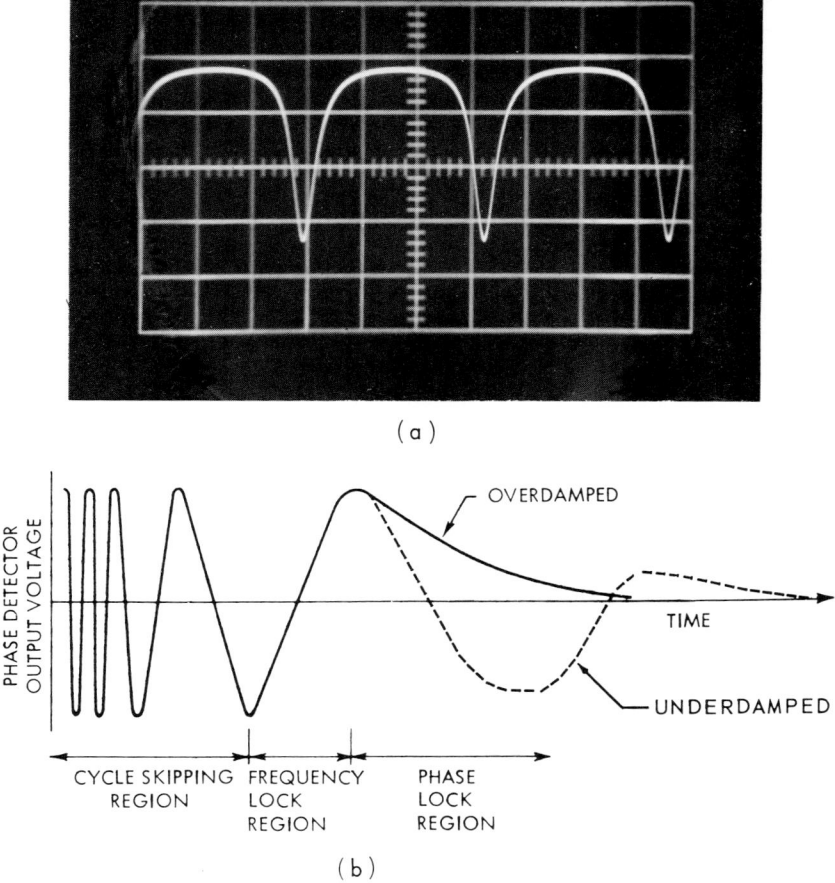

Fig. 5-12. Phase detector outputs in nonsynchronous mode. (a) First-order PLL outside pull-in range, $\Delta\omega_i/K = 1.1$. (b) Second-order type-two PLL during pull-in.

fed around the loop, frequency modulates the VCO. As the VCO frequency approaches that of the offset input signal, the beat frequency goes down, producing the shallow bottom in the beat waveform. As the VCO frequency reverses direction,† the beat frequency increases, producing the peaking characteristic in the beat waveform. This unsymmetrical waveform contains a DC component which deviates the VCO in the direction of the frequency offset. In the case of the first-order loop, frequency offsets past the pull-in range do not produce beat waveforms containing sufficient DC voltage to produce a pull-in. In the type-two loop, the ideal integration utilized provides sufficient amplification for whatever little DC voltage is produced in the beat waveform to effect pull-in. Hence, the type-two loop will always pull in; however, the pull-in time may be extraordinarily large. The pull-in time τ is given by Viterbi[3] as approximately

$$\tau \cong \frac{1}{a}\left(\frac{\Delta\omega_i}{K} - \phi_{i0}\right)^2 \tag{5-69}$$

The infinite pull-in range in the type-two loop is due to the ideal integration employed in the loop filter.

We consider next the second-order type-one loop, which has finite DC gain and, as we might expect, contains a compromise in characteristics between the first-order type-one and second-order type-two loops.

Case III: Loop filter $H_2(s) = (s/a + 1)/(s/b + 1)$, second-order type-one loop

Here the describing equation is

$$\ddot{\phi}_e + \dot{\phi}_e[b + (Kb/a)\cos\phi_e] + bK\sin\phi_e = b\,\Delta\omega_i \tag{5-70}$$

It may be written as

$$\frac{d\dot{\phi}_e}{d\phi_e} = -\left(b + \frac{Kb}{a}\cos\phi_e\right) + \left(\frac{b\,\Delta\omega_i - Kb\sin\phi_e}{\dot{\phi}_e}\right) \tag{5-71}$$

Examination of the phase-plane portrait (example given in Fig. 5-13) for this equation indicates the cycle-skipping phenomenon and the finite pull-in range exhibited by the second-order type-one loop. Viterbi[3,5] has examined the pull-in characteristic for this loop and concluded that the pull-in range is dependent upon the initial conditions, i.e., $\Delta\omega_{i0}$ and ϕ_{i0}. Further, for all values of ϕ_{i0} and $\Delta\omega_{i0}$ pull-in is guaranteed if

$$|\Delta\omega_i| < 2\left[(Kb)\left(1 + \frac{1}{2}\frac{K}{a}\right)\right]^{1/2} \tag{5-72}$$

† The direction reversal is basically due to the decrease in the output of the phase detector when the phase error increases beyond $\pi/2$.

5.4. Nonlinear Operation

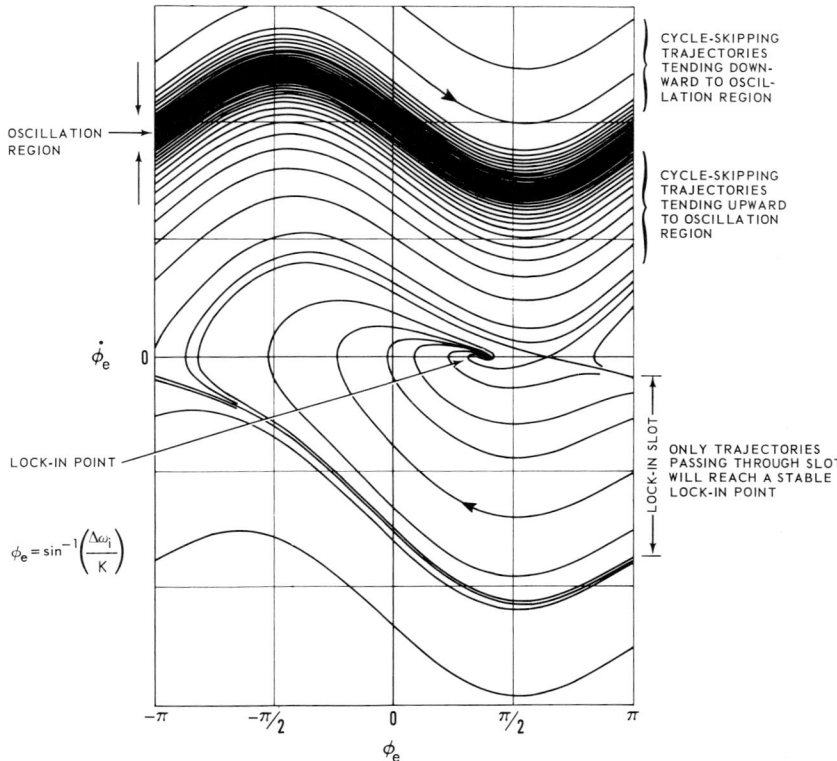

Fig. 5-13. Phase-plane trajectories for second-order type-one PLL. (From Tausworthe.[1] Provided through the courtesy of the Jet Propul. Lab., California Institute of Technology.)

Experimental evidence indicates that this is a fairly sharp bound[1]; that is, for $\Delta\omega_i$ very close to the bound above, lock always ultimately occurs, although as the bound is approached, longer and longer times are required to lock-in.

Acquisition time for this loop is difficult to compute. As an approximation, Viterbi has computed the time required before no more cycles are skipped; when b is small, it is[5]

$$\tau \cong a(\Delta\omega_i/Kb)^2 \quad \text{sec} \tag{5-73}$$

We conclude this section by noting that the presence of additive interference (or noise) accompanying the input signal modifies the pull-in or acquisition characteristics considerably. Experimental work carried out by Frazier and Page[6] on a PLL whose VCO was externally swept (in order to acquire signals in wide bandwidths), produced an empirical formula relating maximum sweep rate to loop parameters for a 90% probability of pull-in.

Their equation adapted by Gardner[2] is

$$(\dot{\Delta\omega})_{max} = [1 - \sigma]\omega_n^2/(1 + d) \tag{5-74}$$

which is maximum sweep frequency rate in radians per second squared, where

$$\sigma = [2\eta B_n/A^2]^{1/2} \tag{5-75}$$

is rms NCR (noise-to-carrier ratio) referred to the PLL noise bandwidth, ω_n is loop natural frequency, and d is a factor derived as

$$d = \begin{cases} \exp[-\xi\pi/(1-\xi^2)^{1/2}], & \xi < 1 \tag{5-76} \\ 0, & \xi \geq 1 \tag{5-77} \end{cases}$$

where ξ is the closed-loop response damping factor for the second-order PLL.

As expected, loop acquisition becomes more difficult with increasing input noise, since the sweep rate $\dot{\Delta\omega}$ must be reduced to preserve the 90% probability of acquisition.

C. Loss-of-Lock Phenomenon

We will now amplify on the meaning of synchronism and develop further the concept of loss-of-lock and cycle-skipping. We defined earlier the locked or synchronized condition as that condition where the static frequency error is zero. This means that there is an equal cycle-to-cycle count of the input and VCO frequencies. Phase difference (or error) between the input and VCO may vary, however, between $-\pi$ and π and, therefore, create a nonzero instantaneous phase-error derivative. This indicates an apparent frequency error that appears to run contrary to our notion of synchronism.

This is resolved by considering the following case. Consider that the loop is at rest at the [0, 0] point of the phase plane, and a sudden input perturbation disturbs the loop operating point. If the loop operating point in its subsequent trajectory stays within the boundary that defines the lock-in region without cycle-skipping, the loop will resettle on the $\dot{\phi}_e = 0$ axis within the central 2π region. If, however, the input perturbation is such as to propel the loop operating point outside the boundary, the loop will settle in a stable region 2π rad (or a multiple thereof) separated from the principal, or initial, region. A cycle-to-cycle count of the input and VCO frequencies will then indicate a skipped or added cycle in the VCO response. We conclude that instantaneous frequency errors with accompanying phase errors that are less than $\pm\pi$ rad are permissible within the framework of the definition of synchronism.

The question arises as to what is the maximum input frequency step that might be sustained by the loop maintaining synchronism. The bound on this condition is a maximum phase error approaching π rad at the zero-frequency error point in the phase-plane trajectory. Below this point the loop can reduce, without loss of synchronism, the phase error to that predicted by the DC tracking equation, Eq. (5-27). Those characteristics are not predictable from the linear model or the linear transient analysis discussed earlier, as the phase error traverses values greater than those permitted by the linear model approximation. Gardner,[2] has determined from Viterbi's experiments[5] the maximum permissible frequency step amplitude for the second-order type-two loop. This is presented in Fig. 5-14a, where the normalized frequency step is related to the loop damping factor. It is to be noted that this maximum frequency step is not the same as the static hold-in range derived earlier, and should not be confused with that characteristic. For a frequency step of greater amplitude than the maximum tolerable step, the loop may or may not pull in, depending on the loop type. For both type-one and type-two loops, synchronism may be lost, and the phase-error and loop response will exhibit an oscillatory condition. Eventually, if the signal is within the loop pull-in range, the loop will reacquire the signal. Note that the second-order type-one loop has a finite pull-in range, whereas the type-two loop will always reacquire synchronism. In PLL literature, we often find that $\pi/2$ is given as the maximum phase error before losing lock. This is strictly correct for the first-order loop. However, even in the second-order PLL, the difference between the $\pi/2$ and the π phase-error bounds is not substantial, as is illustrated in Fig. 5-14b.

D. Nonlinear Noise Performance (Threshold)

The noise analysis carried forth in Section 5.2, and the discussion on loss-of-lock and cycle-skipping, indicate potential sources for nonlinear noise performance. The nonlinear noise performance produces additional loop output noise that tends to degrade the output SNR below that predicted by the linear model. This degradation marks the onset of threshold in FM receiving systems, and as discussed previously, is characteristic of all FM demodulators. By judicious loop design (covered in Chapter 6), the factors that bring forth the threshold may be restricted, so that the PLL serves as a low-threshold FM demodulator of performance comparable to that of the FMFB discussed earlier.

Examination of Eq. (5-21), which describes the loop response to input signal and noise, indicates the basic nonlinear noise mechanism. We have

$$b(t) = \sin[\phi_{es}(t) - \phi_{rn}(t)] + n(t) \tag{5-21}$$

while with respect to the loop output

$$\dot{\phi}_r(t) = KH_2(p)\{\sin[\phi_{es}(t) - \phi_{rn}(t)] + n(t)\} \tag{5-78}$$

and with respect to the phase-error components $\phi_{es}(t)$ and $\phi_{rn}(t)$,

$$\dot{\phi}_r(t) = \dot{\phi}_i(t) - \dot{\phi}_e(t) = \dot{\phi}_i(t) - \dot{\phi}_{es}(t) + \dot{\phi}_{rn}(t) \tag{5-79}$$

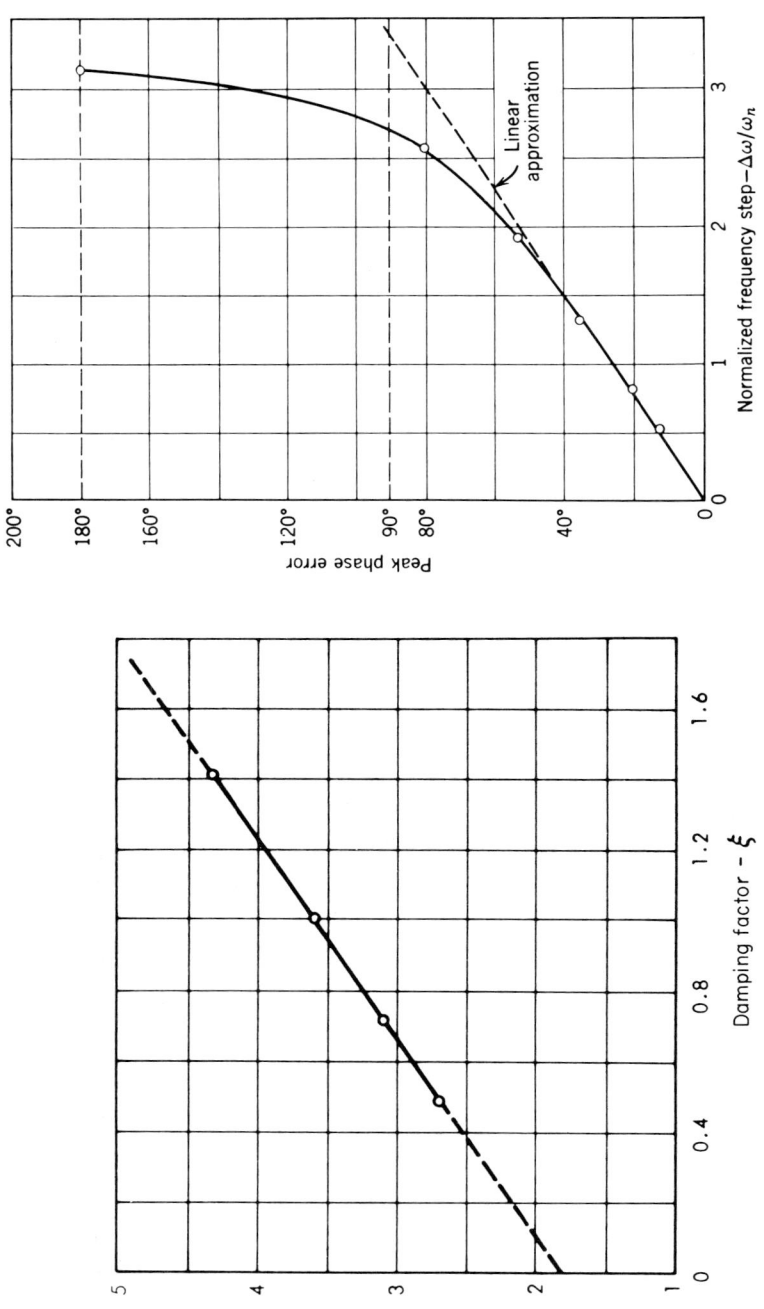

Fig. 5-14. Pull-out frequency and phase error due to frequency step in second-order PLL (from Gardner[2]). (a) Pull-out frequency. (b) Peak phase error ($\xi = 0.707$).

5.4. Nonlinear Operation

Finally, we have

$$\dot{\phi}_{es}(t) - \dot{\phi}_{rn}(t) = -KH_2(p)\sin[\phi_{es}(t) - \phi_{rn}(t)] - KH_2(p)n(t) + \dot{\phi}_i(t) \quad (5\text{-}80)$$

Now consider that the noise response $\phi_{rn}(t)$ is no longer small, as was assumed in formulating the linear model representation. The first-order effect is that the sine term no longer renders the signal and noise components as additive terms in the loop response, for we may expand the sine term as

$$\sin[\phi_{es}(t) - \phi_{rn}(t)] = \phi_{es}(t)\cos\phi_{rn}(t) - \sin\phi_{rn}(t) \quad (5\text{-}81)$$

if $\phi_{es}(t) \ll 1$; and by expanding $\cos\phi_{rn}(t)$ and $\sin\phi_{rn}(t)$ in power series,† we have

$$\begin{aligned}
-[\dot{\phi}_{es}(t) - \dot{\phi}_{rn}(t)] \\
= KH_2(p)\Bigg[\phi_{es}(t)\bigg(1 - \frac{\phi_{rn}^2(t)}{2!} + \frac{\phi_{rn}^4(t)}{4!} - \cdots\bigg) \\
- \bigg(\phi_{rn}(t) - \frac{\phi_{rn}^3(t)}{3!} + \frac{\phi_{rn}^5(t)}{5!} - \cdots\bigg)\Bigg] + KH_2(p)n(t) - \dot{\phi}_i(t)
\end{aligned} \quad (5\text{-}82)$$

which shows explicitly some higher-order noise and intermodulation (between signal and noise) components ignored in formulating the linear model. In addition, $n(t)$ is affected as given by Eq. (5-25).

An exact solution to the PLL dynamics [Eq. (5-80)] is that based on Fokker–Planck techniques.[3,4] Complete results were obtained only for the following set of conditions: (1) a *first-order* loop [$H_2(s) = 1$], (2) white, stationary, Gaussian noise input (no filter preceding the loop), and (3) no modulation. Nevertheless, it is very useful, especially for evaluating approximate methods under the same conditions. The two salient quantities of interest about the noise at the output of the loop are (a) its mean-square value, and (b) its statistical distribution which is generally not Gaussian in view of the nonlinear process. The conclusion is that both are a function of the input noise-to-carrier ratio (NCR) referred to the closed-loop noise bandwidth B_n. Figure 5-15 presents Viterbi's calculations for the *exact* mean-square value of the VCO phase noise $\overline{\phi_{rn}^2(t)}$ [where $\phi_{rn}(t)$ is defined modulo 2π], with a comparison made against that predicted by the linear model. The linear model prediction may be derived from Fig. 5-3 simply as the product of the power spectral density of the equivalent noise input $2\eta/A^2$ and the closed-loop noise bandwidth B_n, i.e.,

$$\overline{[\phi_{rn}(t)]^2} = (2\eta/A^2)B_n \quad (5\text{-}83)$$

---

† See, for example, B. O. Peirce, "A Short Table of Integrals." Ginn, Waltham, Massachusetts, 1929, p. 91.

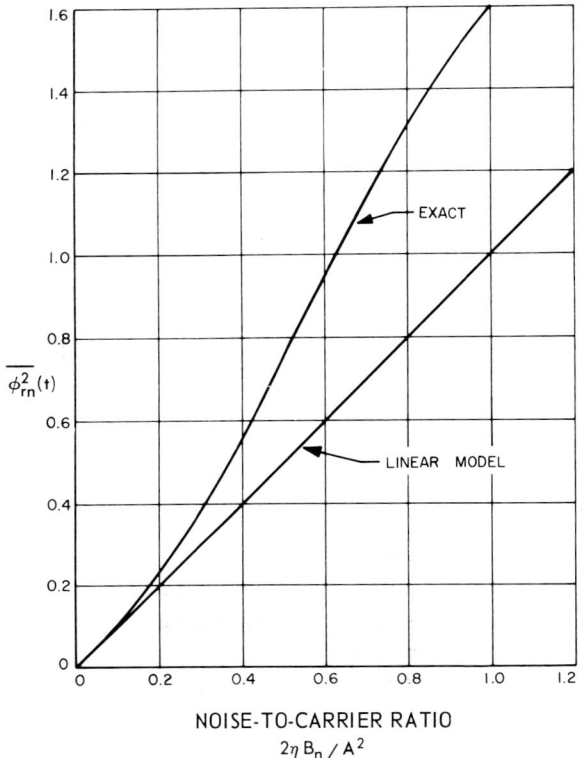

Fig. 5-15. Mean-square VCO phase noise of first-order PLL. (From "Principles of Coherent Communication" by A. J. Viterbi. Copyright 1966 by McGraw-Hill, Inc. Used with permission of McGraw-Hill Book Co.)

Note that at NCR = 0.25 (referred to the loop noise bandwidth, B_n), $\overline{[\phi_{rn}(t)]^2}$ has increased by 1-dB over that predicted by the linear model. This is an indication that the nonlinear noise is becoming significant in the loop response.

Let us return to Eq. (5-80), which relates the total frequency error to the total phase error. The operating point of the loop (in terms of phase and frequency error) will migrate in the central region of the phase plane as the input signal modulation and input noise components fluctuate. Depending upon the loop realization, there is a boundary within which the loop operating point will always tend toward a minimum phase and frequency error. If the operating point of the loop should cross the boundary, the dynamics of the loop, portrayed by the phase-plane trajectories, will tend to move the loop operating point toward another stable operating region 2π (or a multiple thereof) removed from the central (or initial) region. A phase-error migration of 2π is a skipped (or added) cycle which appears as a rapid phase step of 2π in the

5.4. Nonlinear Operation

VCO output. As the PLL output is proportional to frequency, the phase step is differentiated and appears as a sharp impulsive transient, or spike. These transients are called *loss-of-lock impulses* (LLI), and are similar in effect to the threshold impulses ThI which are characteristic of the conventional LD operating in the threshold region (Section 3.2).

Thus in the PLL, we have two sources of noise spikes:

(1) When the resultant phasor of carrier and noise at the loop input encircles the origin, a threshold impulse (ThI) exists (see Section 3.2). Depending upon the parameters of the loop and the noise, the ThI may or may not be tracked by the loop. If the ThI is tracked, then it is reflected as a spike at the PLL output. However, due to the compressive and filtering actions of the loop, the greater number of the ThI are generally not tracked. It should be clear that a PLL of infinite bandwidth will track all ThI and, therefore, have a noise performance identical to that of the LD. Approximately, an order of magnitude reduction of ThI is equivalent to a 1-dB increase in CNR.

(2) When the noise, signal error, or both, force the phase error beyond $\pi/2$, the PLL enters a regenerative mode, because of the change in the sign of the slope of the phase detector characteristic. This generally results in the shifting of the stable operating point by 2π or a multiple thereof, producing LLI.† Again, a PLL of infinite bandwidth does not produce LLI since its phase error is negligible at all times.

A basic aim in the design of a PLL is to minimize the combined effect of the ThI and the LLI, i.e., the total spike rate. The fact that a PLL is capable of a lower FM threshold than the LD is an indication that the total spike rate in a PLL may, by proper design, be made less than that of the ThI alone in an LD. Calculation of the spike rate N for the first-order loop (without signal modulation) has been made by Viterbi[3,4] in terms of the PLL parameters. Viterbi's results are

$$N = \frac{2B_n}{\pi^2 \alpha I_0^2(\alpha)} \cong \frac{4B_n}{\pi} \exp(-2\alpha) \quad \text{for large } \alpha \quad (5\text{-}84)$$

where $I_0(\alpha)$ is the modified Bessel function of the first-kind, zero-order, and argument α, and α is the input CNR referred to B_n. Exact values for N are available from Viterbi's results only for the first-order loop, no modulation, and white-noise interference. The influence on the calculations due to the presence of a pre-PLL filter cannot be incorporated directly.‡

---

† In the second-order PLL, it is possible to have a phase error approaching π without losing lock. However, phase errors beyond $\pi/2$ will mostly result in LLI [see Fig. 5-14b].

‡ For a discussion on the required bandwidth over which the spectrum must be flat in order for the theory to hold, see R. E. Langseth and R. F. Lambert, "Influence of bandwidth on some nonlinear transformations of a Gaussian process," *IEEE Trans.* **IT-14**, (1), 88–93 (1968).

Experimental results are available for the second-order type-one loop ($\xi = 2^{-1/2}$) which indicate a cycle-skipping rate of the following form

$$N \cong c\omega_n \exp(-D\alpha) \qquad (5\text{-}85)$$

with reported experimental constants of

$$\begin{aligned}
C &= 1.57; & D &= 1.64 \,(\text{Ref. 7}) \\
C &= 1.0; & D &= 1.6 \,(\text{Ref. 8}) \\
C &= 1.91; & D &= 1.89 \,(\text{Ref. 9})
\end{aligned} \qquad (5\text{-}86)$$

It is observed from Eqs. (5-84) and (5-85) that for an undeviated carrier, the spike rate decreases with decreasing loop bandwidth. While the spike rate is directly proportional to the loop bandwidth, the more important factor is generally the exponent $D\alpha$ which is also proportional to B_n. Why then do we not reduce the cycle slipping rate by reducing loop bandwidth ad infinitum? The answer is that we are generally interested in detecting a *modulated* input carrier and with modulation the situation is different. First, in order for the demodulated output to be undistorted, the loop must have a certain minimal bandwidth. More important to our present discussion, however, is the fact that with modulation, the rate of the LLI increases very rapidly when the loop bandwidth is decreased below a critical value. Thus there is an optimum loop bandwidth for which the combined effect of the LLI and ThI is minimum. The combined effect is also expressible in terms of the total phase error of the VCO, which contains the effects of the phase error due to the noise (which decreases with decreasing loop noise bandwidth), and the effect of the phase error due to imperfect tracking of the signal when the loop bandwidth is reduced. Both the mean-square-noise phase error and the cycle-skipping rate provide a means to evaluate and minimize PLL threshold. This is discussed further in Chapters 6 and 9, where design examples are treated.

5.5. Excess Delay and Minimum Noise Bandwidth

In the implementation of phase-locked systems the loop inadvertently experiences excess open-loop phase shift, or delay. The sources of this delay are the physical electrical length of the signal path and the presence of high-frequency poles. In certain PLL applications the electrical path lengths may be quite large. For example, interplanetary space probes are tracked in velocity and range by phase-locked techniques, where the round-trip delay forms part of the closed loop. In other applications, the demodulation of ultra-wideband FM signals is hampered by delay limitations that prevent the attainment of

5.5. Excess Delay and Minimum Noise Bandwidth

arbitrarily wideband systems. Element limitations (finite gain-bandwidth products), as well as stray energy storage elements, provide the major sources of excess phase shift. The finite response time, exhibited by both the VCO and phase detector, contribute to the problem, with the net result being an appearance of excess phase shift in the open-loop response.

Assuming that pure delay adequately represents the excess phase shift in the critical region of the open-loop response [where $|G(j\omega)| = 1$], the open-loop transfer function of the linear model for the second-order type-two loop that includes delay is

$$G(s) = K[s/a + 1]e^{-s\tau}/s^2 \quad (5\text{-}87)$$

where τ is the delay due to the excess phase shift. The closed-loop response $H(s)$ and, subsequently, the closed-loop noise bandwidth B_n become a function of the delay; however, B_n may be minimized by a proper choice for a, the zero constant. This optimization is carried out for a fixed gain constant K, in order to maintain the necessary compressive feedback.

For the case of no delay, the noise bandwidth of $H(s)$ is minimized by (Section 6.3)

$$a = K^{1/2} = \omega_n \quad \text{rad/sec} \quad (5\text{-}88)$$

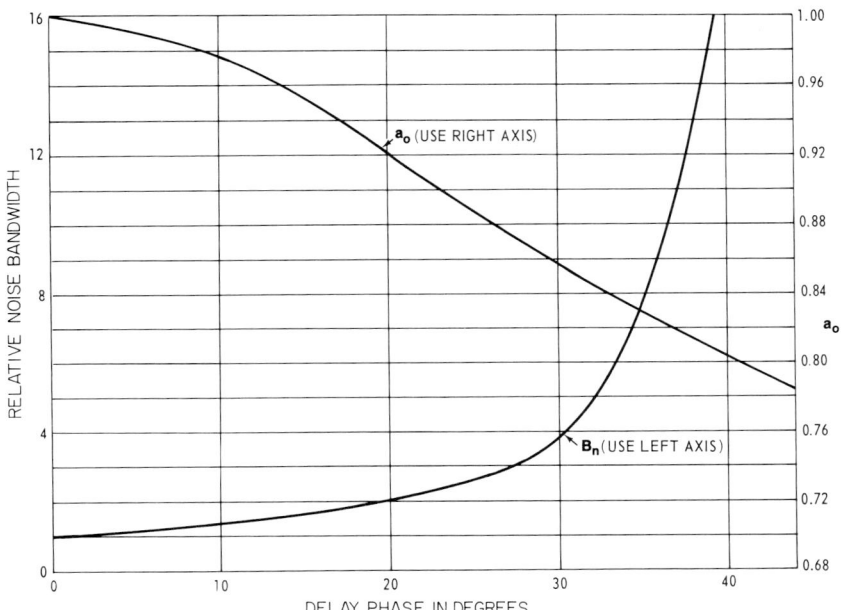

Fig. 5-16. Minimum relative noise bandwidth of second-order type-two PLL, i.e., $B_n(\phi)/B_n(0)$, as a function of delay phase $\phi = \tau\omega_n$.

resulting in

$$B_n = \int_0^\infty |H(j\omega)|^2 \, df = \tfrac{1}{2} K^{1/2} \quad \text{Hz} \qquad (5\text{-}89)$$

By fixing K we may normalize a as

$$a = a_0 K^{1/2} \quad \text{rad/sec} \qquad (5\text{-}35') \quad (5\text{-}90)$$

Figure 5-16 presents the normalized minimum noise bandwidth of the second-order type-two loop, as well as the normalized zero constant a_0 necessary to minimize the noise bandwidth in the presence of delay. Noise bandwidth minimization plays an important role in the attainment of low-threshold FM demodulators, which are discussed further in Chapter 6.

References

1. R. C. Tausworthe, Theory and practical design of phase-locked receivers. Tech. Rep. 32-819, Vol. I. Jet Propul. Lab., California Inst. of Technol., Pasadena, Calif., February 15, 1966.
2. F. M. Gardner, "Phaselock Techniques." Wiley, New York, 1966.
3. A. J. Viterbi, "*Principles of Coherent Communication.*" McGraw-Hill, New York, 1966.
4. A. J. Viterbi, Phase-locked loop dynamics in the presence of noise by Fokker-Planck techniques. *Proc. IEEE* **51**, 1737–1753 (1963).
5. A. J. Viterbi, Acquisition and tracking behavior of phase-locked loops. *External Publ. No. 673*. Jet Propul. Lab., California Inst. of Technol., Pasadena, Calif., July 14, 1959; *Proc. Symp. Active Networks and Feedback Systems*, Polytech. Inst. of Brooklyn, New York, *April* 19–21, 1960, **10**, pp. 583–619.
6. J. P. Frazier and J. Page, Phase-lock loop frequency acquisition study. *IRE Trans.* **SET-8**, 210–227 (1962).
7. B. M. Smith, A semi empirical approach to the PLL threshold. Letters to the Editor. *IEEE Trans.* **AES-2**, 463 (1966).
8. E. J. Baghdady, Advanced threshold reduction techniques study. 1st Quart. Rep., NASA Rep. N65-30842. Adcom Inc., Cambridge, Mass. 1964.
9. B. M. Smith, The phase-lock loop with filter: Frequency of skipping cycles. Letters to the Editor. *Proc. IEEE* **54**, 296 (1966).

CHAPTER

Design of Phase-Locked Loops for FM Demodulation

This chapter develops the design concepts supplemented by the necessary design curves to carry forth PLL design for FM demodulation. A variety of signal modulations with their associated performance indexes (signal-to-noise or noise-power ratio, for example) are considered, with design examples illustrating each concept. The chapter concludes with inclusive examples that incorporate all the considerations in low-threshold PLL design.

We begin by considering the performance characteristics of the receiving system depicted in Fig. 6-1, with particular attention to the interplay between system performance and PLL design. Similar to the case of the conventional FM detector (Section 3.2), the system will display the distinctive performance curve shown in Fig. 6-2, which illustrates the three basic regions of operation we wish to control by proper PLL design.

6.1. FM Improvement Region

Consider first the central region of the curve in Fig. 6-2; by its very nature it may simply be called the *linear SNR-CNR region*. It is basically derived from the above-threshold detection characteristic of the PLL, which was seen, in Chapter 5, to be equivalent to that of an ideal frequency detector (operating above its threshold) followed by a filter that represents the PLL closed-loop response $H(j\omega)$. By requiring that the postdetection filter compensate for $H(j\omega)$, in addition to supplying the ideal sharp cutoff characteristic, we may

6. Design of PLL for FM Demodulation

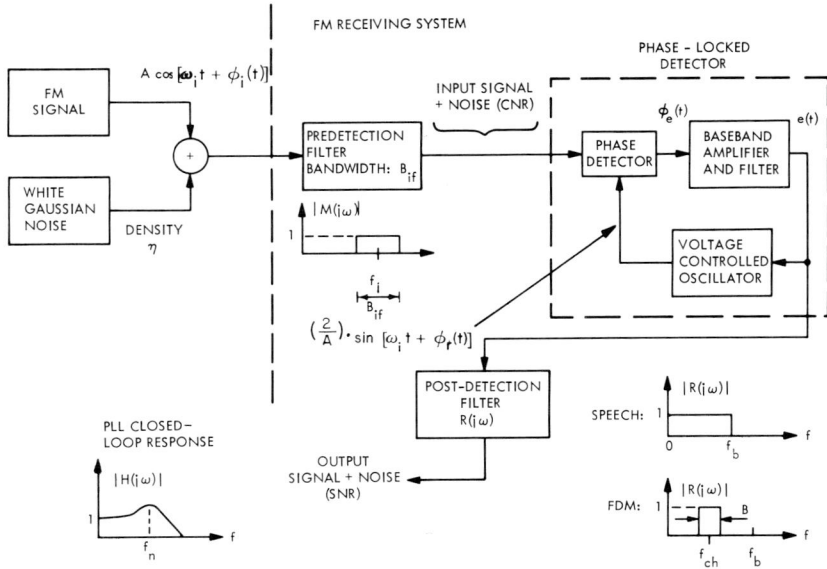

Fig 6-1. FM receiving system employing phase-locked demodulation.

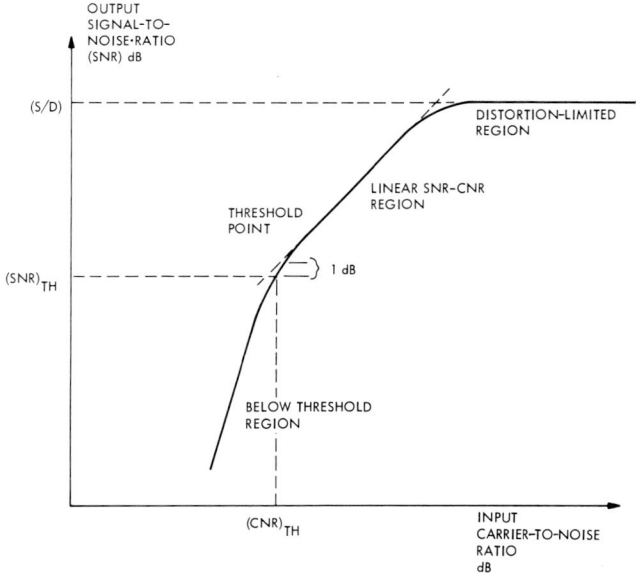

Fig. 6-2. Typical performance characteristic for FM receiving system.

6.1. FM Improvement Region

apply directly the "above-threshold" FM detection equations. These have already been noted as Eqs. (3-15)–(3-24), but are repeated below to facilitate discussion.

A. Single Channel Speech (FM and PM)

$\text{SNR}_{\text{SP}} = 6\sigma^2(\text{CNR})_{\text{AM}}$ (FM) (3-19) (6-1)

$\text{SNR}_{\text{SP}} = 2(\omega_b/\omega_a)\sigma^2(\text{CNR})_{\text{AM}}$ (PM) (3-20) (6-2)

$\text{SNR}_{\text{TT}} = 3m_p^2(\text{CNR})_{\text{AM}}$ (FM) (3-15) (6-3)

$\text{SNR}_{\text{TT}} = m_p^2(\omega_b/\omega_T)^2(\text{CNR})_{\text{AM}}$ (PM) (3-21) (6-4)

B. Frequency Division Multiplex† (FM and PM)

$\text{NPR} = 2(\omega_b/\omega_{\text{CH}})^2\sigma^2(\text{CNR})_{\text{AM}}/[1 - \omega_a/\omega_b]$ (FM) (3-22) (6-5)

$\text{NPR} = 6\sigma^2(\text{CNR})_{\text{AM}}/[1 - (\omega_a/\omega_b)^3]$ (PM) (3-23) (6-6)

$\text{SNR}_{\text{TT}} = (f_b/B)(\omega_b/\omega_{\text{CH}})^2 m_p^2(\text{CNR})_{\text{AM}}$ (FM or PM) (3-24) (6-7)

with the symbols defined as follows:

- SNR = output SNR for speech waveform (SP) or test-tone (TT)
- NPR = noise power ratio
- $(\text{CNR})_{\text{AM}}$ = input CNR referred to twice base-bandwidth ($2f_b$)
- σ = rms modulation index = $\dfrac{\text{rms carrier frequency deviation}}{\text{top baseband frequency}} = \dfrac{\Delta\omega_{\text{rms}}}{\omega_b}$
- m_p = peak modulation index = $\dfrac{\text{peak carrier frequency deviation}}{\text{top baseband frequency}} = \dfrac{\Delta\omega_p}{\omega_b}$
- ω_a = bottom frequency of speech spectrum model (nominally $2\pi \times 10^3$ radians per second) or bottom frequency of frequency division multiplex (FDM) baseband (radians per second)
- ω_b = top baseband frequency of transmission (radians per second)
- ω_{ch} = channel center frequency (radians per second)
- ω_T = test-tone frequency (radians per second, nominally $2\pi \times 10^3$ radians per second)
- B = channel noise bandwidth (FDM) in hertz

Several simplifying assumptions underlie the development of these equations. To begin with, the above-threshold output noise is calculated in the absence of modulation, with the signal residing in the center of the predetection filter. Then, the predetection filter and noise are assumed to be flat over

† High-Q channels are assumed ($2\pi B \ll \omega_{\text{CH}}$).

a $2f_b$ band about the center frequency, and the postdetection filter and demodulator responses are assumed to be flat from 0 to f_b Hz with sharp cutoff at f_b Hz (no response is exhibited beyond f_b Hz). For most applications, the equations may be applied usefully, even though strict adherence to the underlying assumptions is not the case.

The major error in applying the above equations to actual systems stems from the lack of ideal filtering. There is often inadequate compensation for the PLL closed-loop response $H(j\omega)$, and insufficiently rapid rolloff in the postdetection filter $R(j\omega)$. We can account for these nonideal characteristics by modifying the results as follows: The full expressions for the above-threshold output signal and noise powers† are

$$S_{po} = (1/\mu^2) \int_0^\infty S(\omega) |M_L(j\omega)|^2 |H(j\omega)|^2 |R(j\omega)|^2 \, df \qquad (6\text{-}8)$$

$$N_{po} = [4\pi^2(2\eta)/A^2\mu^2] \int_0^\infty f^2 |M_L(j\omega)|^2 |H(j\omega)|^2 |R(j\omega)|^2 \, df \qquad (6\text{-}9)$$

where

N_{po}	= output noise power from postdetection filter following the PLL (watts)
S_{po}	= output signal power from postdetection filter (watts)
$S(\omega)$	= power spectral density of transmitted baseband
$|M_L(j\omega)|$	= magnitude of response of predection filter (lowpass-equivalent, see pp. 10–11)
ω_i	= Signal center frequency (radians per second)
$|H(j\omega)|$	= magnitude of PLL closed-loop response
$|R(j\omega)|$	= magnitude of response of postdetection filter
μ	= PLL VCO sensitivity (radians per volt-second)
η	= noise power spectral density in predetection filter (watts per hertz)
A	= Signal Amplitude (volts)

The ratio of these two factors is the output SNR,

$$\text{SNR} = S_{po}/N_{po} \qquad (6\text{-}10)$$

To serve as an example of those corrections that may be applied to the detection equations we consider the FM single-channel case in detail.

Example

Consider first the output noise component over a range of system filter characteristics. For the FM single-channel case, the following characteristics are representative of actual practice.

† Unmodulated signal carrier at predetection filter center frequency is assumed, with an arithmetically symmetric predetection filter.

6.1. FM Improvement Region

$$H(s) = \frac{(s/\omega_n + 1)}{(s/\omega_n)^2 + (s/\omega_n) + 1}$$

= second-order PLL closed-loop response; (6-11)

$$\text{damping factor } \zeta = \frac{1}{2}$$

$$|R(j\omega)|^2 = \frac{1}{1 + (\omega/\omega_b)^{2n}}$$

= post-PLL filter, nth order Butterworth, with cutoff frequency ω_b (6-12)

$$|M_L(j\omega)|^2 = \begin{cases} 1, & 0 \le f \le B_{IF}/2 \\ 0, & f > B_{IF}/2 \end{cases}$$

= predetection filter (crystal) lowpass equivalent (6-13)

From Eq. (6-9) the output noise power is

$$N_{po} = \frac{4\pi^2(2\eta)f_b^3}{A^2\mu^2} \int_0^{b'} \frac{[(x/a')^2 + 1]}{[1 - (x/a')^2]^2 + (x/a')^2} \cdot \frac{x^2}{1 + x^{2n}} dx \quad (6\text{-}14)$$

where

$x = \omega/\omega_b$

$a' = \omega_n/\omega_b$ = ratio of loop natural frequency to base bandwidth (6-15)

$b' = B_{IF}/2f_b$ = ratio of predetection semibandwidth to base bandwidth. (6-16)

Normalization of this result to the quantity† $4\pi^2(2\eta)f_b^3/3A^2\mu^2$ provides a direct measure of the noise penalty (or increase) due to nonideal filtering, and results in the penalty factor P

$$P = 3 \int_0^{b'} \frac{[(x/a')^2 + 1]}{[1 - (x/a')^2]^2 + (x/a')^2} \cdot \frac{x^2}{1 + x^{2n}} dx \quad (6\text{-}17)$$

By evaluating P over a range of n, the influence on output noise due to postdemodulator selectivity (selectivity increases for increasing n) is determined, whereas evaluation over a' illustrates the influence of the PLL closed-loop response. Figures 6-3 and 6-4 present P in this manner, with b' chosen

---

† This is the above-threshold output noise in a system with perfect filtering (see Chapter 3), where we have set $\mu = 1/\lambda$ (λ is ideal frequency detector sensitivity in volt-seconds per radian).

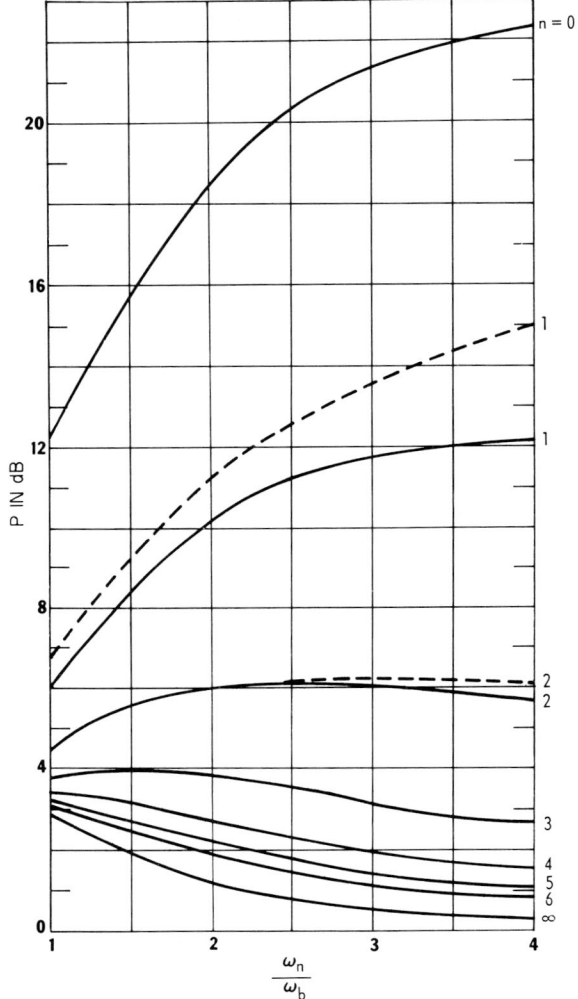

Fig. 6-3. Correction factor for PLL FM output noise (—: $b' = 4.5$; – –: $b' = \infty$). (No compensation for PLL closed-loop response.)

as 4.5 (representative of an FM transceiver) and infinity (representative of white input noise to the PLL). Several observations and comments are made on the behavior of P:

(1) Significant noise penalty exists for moderate PLL bandwidths when compensation for $H(j\omega)$ is not utilized. For example, a typical design (covered more fully in Section 6.4) has $a' = 1.4$, $b' = 4.5$; with $n = 6$, this yields

6.1. FM Improvement Region

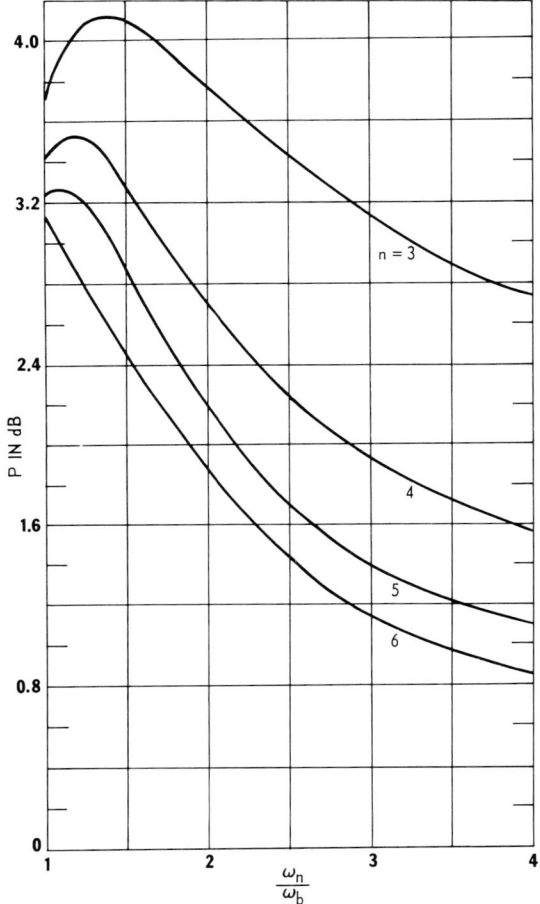

Fig. 6-4. Correction factor for PLL FM output noise ($b' = \infty$). (No compensation for PLL closed-loop response.)

$P = 2.6$ dB. If, however, suitable compensation is employed for $H(j\omega)$, we have for $n = 6$, $P = 0.8$ dB†, which indicates a potential 1.8-dB output noise penalty when failing to compensate for $H(j\omega)$.

(2) Fairly selective postdetection filters are required to restrict P to tolerable limits. By referring to the results in Figs. 6-3 and 6-4 we see that (except for cases where the predetection filter is relatively narrow, $b' = 1$ to $b' = 2$, which is characteristic of narrowband FM), postdetection filter selectivity plays the critical role in determining the output noise. Furthermore, with the use of highly selective postdetection filters, the predetection filter

† The result is arrived at by letting $a' \to \infty$.

bandwidth has negligible influence on the output noise in all cases. This rather strong dependence of output noise power on the high-frequency baseband filtering characteristics of an FM detection system stems from the parabolic spectral nature of the detected noise.

For PM transmission, an integrator is utilized after the PLL, in order to render the detected voltage proportional to phase modulation. This deemphasis of the high-frequency noise makes nonideal filtering a less critical factor in system performance. Analogous to the results presented for FM, a set of correction factor curves is derived for PM and presented in Figs. 6-5 and 6-6.

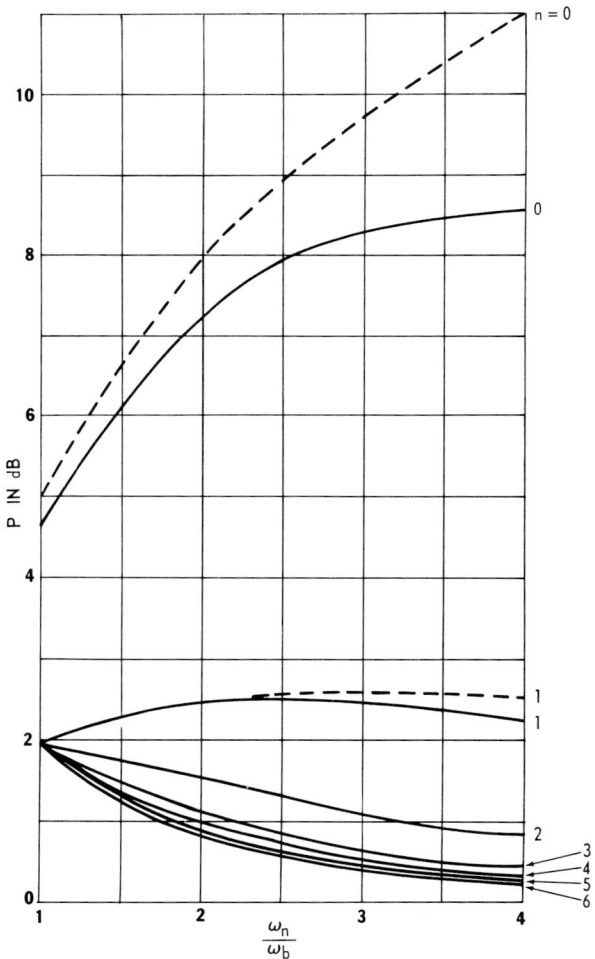

Fig. 6-5. Correction factor for PLL PM output noise (—: $b' = 4.5$; – –: $b' = \infty$). (No compensation for PLL closed-loop response.)

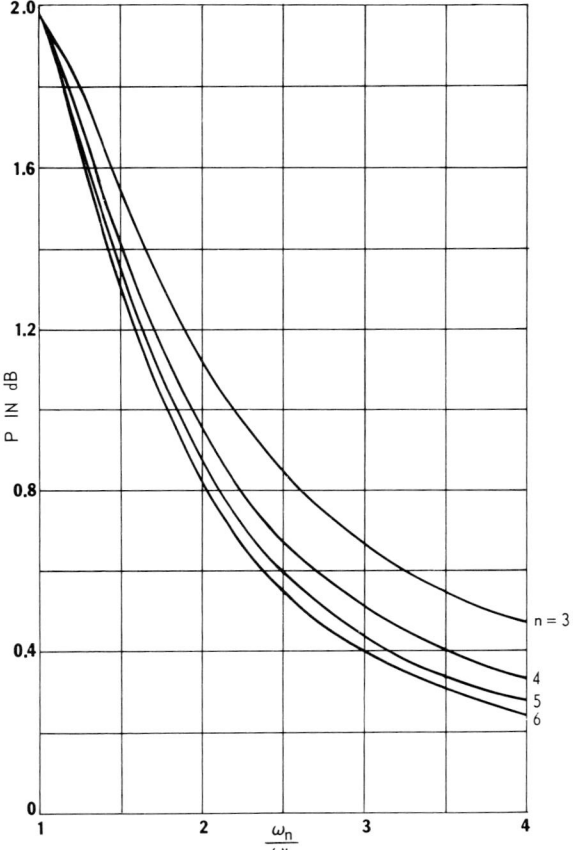

Fig. 6-6. Correction factor for PLL PM output noise ($b' = \infty$). (No compensation for PLL closed-loop response.)

The equations for the postdetection noise are

$$N_{po} = 2\eta f_b P/A^2\mu^2$$

with

$$P = \int_0^{b'} \frac{[(x/a')^2 + 1]}{[1-(x/a')^2]^2 + (x/a')^2} \cdot \frac{dx}{1+x^{2n}} \tag{6-17a}$$

To conclude the example, we consider the actual output-signal power incorporating the effects due to nonideal filtering. For TT modulation it is necessary only to evaluate the magnitude-squared frequency response $|H(j\omega)|^2$ at the TT frequency of the detection system. Normally, a single speech channel is tested at 1 kHz, whereas $f_b = 4$ kHz. Using the parameters employed in the previous example ($a' = 1.4$, $b' = 4.5$, $n = 6$) we have a 0.3-dB rise in signal transmission due to $H(j\omega)$ (see Fig. 5-4, Chapter 5, at $\omega/\omega_n = 0.18$ and $b/\omega_n = 0$), if we choose not to compensate $H(j\omega)$ in the design of $R(j\omega)$. Therefore, SNR_{TT} may be corrected by multiplying the ideally calculated SNR_{TT} by the factor $|H(j\omega_T)|^2/P$, i.e.,

$$\text{SNR}_{TT} = \frac{|H(j\omega_T)|^2 \, 3m_p^2}{P} (\text{CNR})_{AM} \qquad (6\text{-}18)$$

which in this case becomes $\text{SNR}_{TT} = 10 \log_{10}[3m_p^2 (\text{CNR})_{AM}] - 2.6 + 0.3$ dB. In the case of a compensated $H(j\omega)$, Eq. (6-18) applies with the proper choice for P found in Figs. 6-3 and 6-4 by letting $a' \to \infty$. For this case,

$$\text{SNR}_{TT} = 10 \log_{10}[3m_p^2 (\text{CNR})_{AM}] - 0.8 \quad \text{dB} \qquad (6\text{-}19)$$

For speech modulation, Eq. (6-8) must be applied directly to evaluate the modulation component of the SNR correction factor; however, under most circumstances (where $\omega_n/\omega_b > 1$) the modulation correction factor is negligible.

In conclusion, we consider the characteristics of FDM† and observe that $H(j\omega)$ does not enter the NPR or SNR_{TT} calculation. The reason for it is that $H(j\omega)$ modifies both the channel noise and channel signal in the same manner.

6.2. Distortion-Limited Region

With the reduction of the additive input noise to zero (or conversely, with the input signal becoming arbitrarily large), Eqs. (6-1)–(6-7) predict that the PLL output SNR grows without bound. Practical limitations in the receiving system, however, produce a residual noise that results in a maximum attainable output SNR, as indicated in the performance curve of Fig. 6-2. This residual noise stems from two major sources. They are: the idle noise which is inherent in the circuit elements; and intermodulation (and harmonic) distortion of the signal, which is characteristic of dynamic nonlinearities in the system elements. Idle noise is primarily oscillator frequency jitter, contributed mainly by the VCO's employed in the transmitter (FM modulator) and receiver (PLL). Discussion of VCO frequency jitter is reserved for Section 6.3. The discussion here will be concerned primarily with harmonic and intermodulation distortion noise.

For conditions of low distortion, i.e., less than 1% of the primary signal

† High-Q narrow-band channels are assumed.

component, the distortion introduced by an individual loop element may be included in the loop analysis and design by substituting for the nonlinear element a linear equivalent element that contains a dependent generator, as indicated in Fig. 6-7. The resulting output distortion is then calculated from the linear transfer characteristic of the PLL.

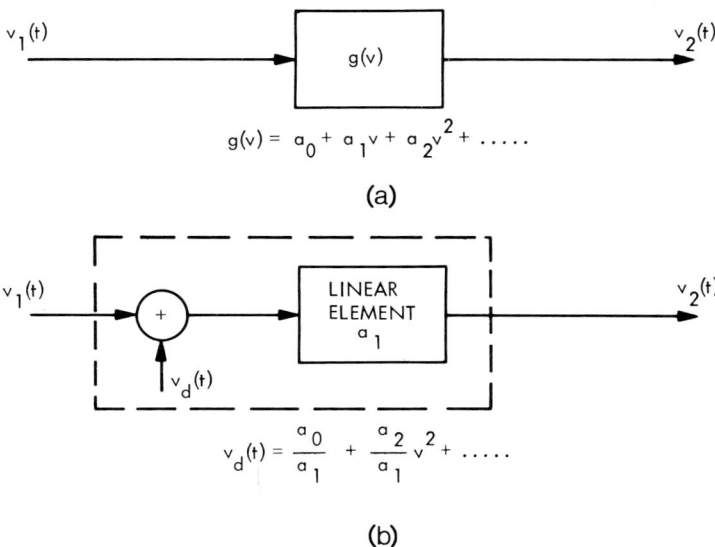

Fig. 6-7. Linear element representation for nonlinear loop element. (a) Nonlinear element (v is total voltage across nonlinear element). (b) Linear element with dependent generator representation.

The following examples clarify the technique:

Example: Phase Detector Distortion

The phase detector, when implemented as the product device discussed earlier, has an output voltage $b(t)$

$$b(t) = \sin \phi_e(t) = \left[\phi_e(t) - \frac{\phi_e^3(t)}{3!} + \frac{\phi_e^5(t)}{5!} - \cdots\right] \quad (6\text{-}20)$$

When the phase error is small, the principal distortion component is that due to the $\phi_e^3(t)$ term, which may be included in the linear model, as discussed above, and as indicated in Fig. 6-8. The distortion generator $\phi_d(t)$ is [see Eq. (2-24)]

$$\phi_d(t) = -\frac{\phi_e^3(t)}{6} = -\frac{1}{6}\left[\left(1 - \frac{\Phi_r}{\Phi_i}(p)\right)\phi_i(t)\right]^3 \quad (6\text{-}21)$$

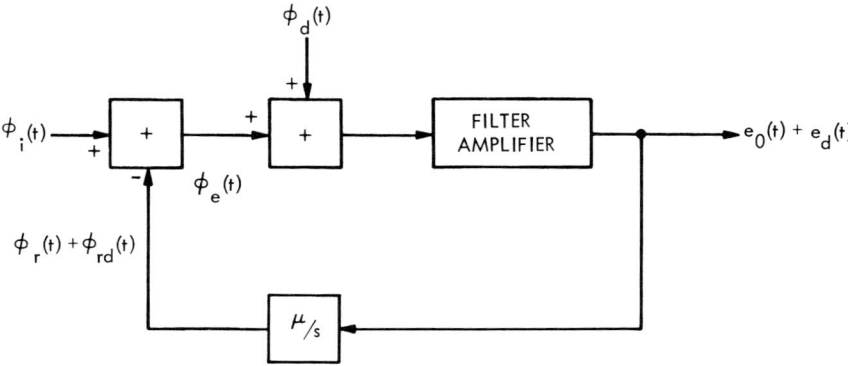

Fig. 6-8. Linear PLL model including equivalent distortion input. (Phase detector distortion.)

where $\Phi_r(s)/\Phi_i(s)$ is PLL closed-loop transfer function $H(s)$, $p = d/dt$ (see Section 2.1), and $\phi_i(t)$ is input signal phase modulation.

The loop response $\phi_{rd}(t)$ to this distortion term $\phi_d(t)$ is

$$\phi_{rd}(t) = -\frac{1}{6}\frac{\Phi_r}{\Phi_i}(p)\left[\left(1 - \frac{\Phi_r}{\Phi_i}(p)\right)\phi_i(t)\right]^3 \qquad (6\text{-}22)$$

and the loop output distortion voltage is

$$e_d = \frac{p}{\mu}\phi_{rd}(t) = -\frac{p}{6\mu}\frac{\Phi_r}{\Phi_i}(p)\left[\left(1 - \frac{\Phi_r}{\Phi_i}(p)\right)\phi_i(t)\right]^3 \qquad (6\text{-}23)$$

A representative test condition is two-tone modulation, for which the input modulation angle is written as

$$\phi_i(t) = (\Delta\omega_1/\omega_1)\cos\omega_1 t + (\Delta\omega_2/\omega_2)\cos\omega_2 t \qquad (6\text{-}24)$$

If we consider the second-order loop, the distortion term $\phi_d(t)$ arising from the input test tones can be evaluated by noting from Eq. (5-34) that

$$1 - \frac{\Phi_r}{\Phi_i}(s) = \frac{(s/\omega_n)^2}{(s/\omega_n)^2 + 2\xi(s/\omega_n) + 1} \qquad (6\text{-}25)$$

The magnitude can be approximated by

$$\left|1 - \frac{\Phi_r}{\Phi_i}(s)\right| \cong \left|\left(\frac{s}{\omega_n}\right)^2\right| \qquad (6\text{-}26)$$

for test tones at frequencies substantially less than ω_n. (Note that the magnitude of the denominator of Eq. (6-25) is relatively flat,[1] for ω over the range 0 to ω_n, and for damping factors $\xi \cong 0.5$). The distortion term $\phi_d(t)$ therefore, may be approximated by

6.2. Distortion-Limited Region

$$\phi_d(t) = -\frac{1}{6}\left[\frac{\omega_1\,\Delta\omega_1}{\omega_n^2}\cos(\omega_1 t + \psi) + \frac{\omega_2\,\Delta\omega_2}{\omega_n^2}\cos(\omega_2 t + \theta)\right]^3 \quad (6\text{-}27)$$

with $\omega_1, \omega_2 < \omega_n$. The phase angles ψ and θ are phase shifts imposed by the transfer function given in Eq. (6-25) for $s = j\omega_1$ and $s = j\omega_2$, respectively. Expanding Eq. (6-27) results in†

$$\begin{aligned}\phi_d(t) = -\frac{1}{6}\Bigg\{ &\left[\frac{3}{2}\left(\frac{\omega_1\,\Delta\omega_1}{\omega_n^2}\right)\left(\frac{\omega_2\,\Delta\omega_2}{\omega_n^2}\right)^2 + \frac{3}{4}\left(\frac{\omega_1\,\Delta\omega_1}{\omega_n^2}\right)^3\right]\cos(\omega_1 t + \psi) \\ &+ \left[\frac{3}{2}\left(\frac{\omega_1\,\Delta\omega_1}{\omega_n^2}\right)^2\left(\frac{\omega_2\,\Delta\omega_2}{\omega_n^2}\right) + \frac{3}{4}\left(\frac{\omega_2\,\Delta\omega_2}{\omega_n^2}\right)^3\right]\cos(\omega_2 t + \theta) \\ &+ \left[\frac{3}{4}\left(\frac{\omega_1\,\Delta\omega_1}{\omega_n^2}\right)\left(\frac{\omega_2\,\Delta\omega_2}{\omega_n^2}\right)^2\right]\cos[(2\omega_2 \mp \omega_1)t + 2\theta \mp \psi] \\ &+ \left[\frac{3}{4}\left(\frac{\omega_1\,\Delta\omega_1}{\omega_n^2}\right)^2\left(\frac{\omega_2\,\Delta\omega_2}{\omega_n^2}\right)\right]\cos[(2\omega_1 \mp \omega_2)t + 2\psi \mp \theta] \\ &+ \left[\frac{1}{4}\left(\frac{\omega_1\,\Delta\omega_1}{\omega_n^2}\right)^3\right]\cos(3\omega_1 t + 3\psi) \\ &+ \frac{1}{4}\left(\frac{\omega_2\,\Delta\omega_2}{\omega_n^2}\right)^3\cos(3\omega_2 t + 3\theta)\Bigg]\Bigg\} \quad (6\text{-}28)\end{aligned}$$

Several types of distortion are evident, and may be classified by referring to the arguments of the individual cosine functions:

(1) self-distortion—components at frequencies ω_1 and ω_2;
(2) intermodulation distortion—components at frequencies $2\omega_1 \pm \omega_2$ and $2\omega_2 \pm \omega_1$;
(3) harmonic distortion—components at frequencies $3\omega_1$ and $3\omega_2$.

For typical single-channel tests, test tones at 1 and 4 kHz are selected, with the in-band intermodulation distortion component appearing at 2 kHz, and harmonic distortion component appearing at 3 kHz. Assuming that the PLL closed-loop transfer function is flat over the TT frequencies, as well as over the distortion components of interest, the output signal-(tone at 1 kHz)-to-distortion ratios are from Eqs. (6-23), (6-24), and (6-28) approximately:

A. SIGNAL-TO-INTERMODULATION-DISTORTION VOLTAGE RATIO

$$\omega_1 = 1 \text{ kHz}, \quad \omega_2 = 4 \text{ kHz}, \quad \omega_2 - 2\omega_1 = 2 \text{ kHz}$$

$$\left(\frac{S}{D}\right)_{\text{IM}} = \frac{(\Delta\omega_1/\omega_1)[\omega_1/(\omega_2 - 2\omega_1)]}{(\frac{1}{6})(\frac{3}{4})(\omega_1\,\Delta\omega_1/\omega_n^2)^2(\omega_2\,\Delta\omega_2/\omega_n^2)} = \frac{8\omega_n^6/(\omega_2 - 2\omega_1)}{(\Delta\omega_1)(\Delta\omega_2)\omega_1^2\,\omega_2} \quad (6\text{-}29)$$

† Writing $\cos(\alpha - \beta) + \cos(\alpha + \beta)$ as $\cos(\alpha \mp \beta)$.

B. SIGNAL-TO-HARMONIC-DISTORTION VOLTAGE RATIO

$$\left(\frac{S}{D}\right)_H = \frac{\Delta\omega_1}{(\frac{1}{6})(\frac{1}{4})(\omega_1 \Delta\omega_1/\omega_n^2)^3(3\omega_1)} = \frac{8\omega_n^6}{(\Delta\omega_1)^2(\omega_1)^4} \quad (6\text{-}30)$$

The ratio of intermodulation to harmonic distortion is for the same conditions

$$\frac{(S/D)_{IM}}{(S/D)_H} = \frac{\omega_1^2(\Delta\omega_1)}{(\Delta\omega_2)(\omega_2)(\omega_2 - 2\omega_1)} = \frac{1}{8}\left(\frac{\Delta\omega_1}{\Delta\omega_2}\right) \quad (6\text{-}31)$$

Since $\Delta\omega_2$ is usually taken equal to $\Delta\omega_1$, intermodulation distortion will generally dominate.

To illustrate the relationship between loop bandwidth and phase-detector distortion, Eq. (6-29) may be rewritten in terms of $m_p = \Delta\omega_p/\omega_b$ using the following conditions

$$\omega_2 = 4\omega_1 = \omega_b, \quad 2\Delta\omega_1 = 2\Delta\omega_2 = \Delta\omega_p,$$

where $\Delta\omega_p$ is the system peak frequency deviation in radians per second. The resulting equation is

$$(\omega_n/\omega_b)^6 = (S/D)_{IM}(m_p)^2/1024 \quad (6\text{-}32)$$

which gives rise to the constant distortion contours given in Fig. 6-9.

To illustrate the application of these contours, assume an 80-dB specification for high-carrier-level TT signal-to-distortion ratio, for a system with a peak modulation index of 2.5; from Fig. 6-9 we have that the ω_n/ω_b ratio must be 2 or greater to meet the specification. To conclude, we remark that these results are somewhat pessimistic, in that performance for a given ω_n/ω_b ratio can be improved by allowing the phase detector to saturate (both input levels become equal) at high input CNR's. With saturation, the phase detector exhibits a triangular (rather than sinusoidal) detection characteristic (refer to Chapter 3 for details) which provides greater linearity and, therefore, lower distortion.

We consider next an example of distortion contributed by the VCO employed within the PLL.

Example: VCO Distortion

A major source of distortion in the PLL is the nonlinear modulation characteristics of the VCO. In general, the dynamic modulation equation is of the form

$$\Delta\omega(t) = a_1 v(t) + a_2 [v(t)]^2 + a_3 [v(t)]^3 + \cdots \quad (6\text{-}33)$$

where $\Delta\omega(t)$ is VCO frequency deviation (radians per second) about quiescent center frequency, $v(t)$ is VCO input control voltage, and $a_1, a_2, \ldots$ are coefficients determined by the oscillator circuit characteristics. The distortion

6.2. Distortion-Limited Region

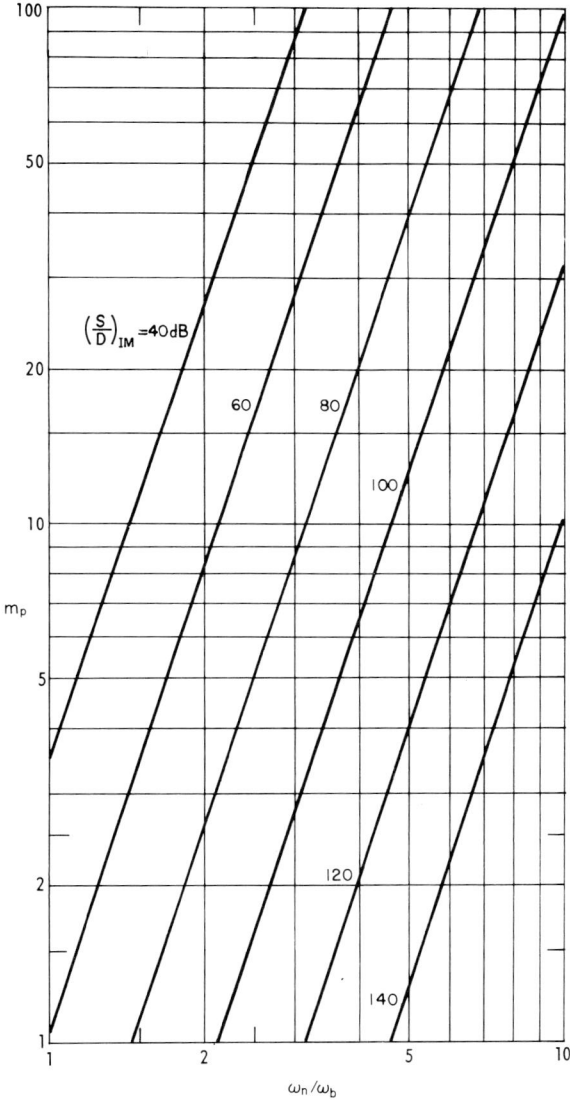

Fig. 6-9. Constant distortion contours for phase detector in PLL. (Test tone at $f_b/4$.)

coefficients are related to the parameters of the varactor of a varactor-controlled VCO in Appendix D.

A characteristic of the single-varactor-controlled VCO is that the distortion is principally due to the voltage-squared term, indicating a predominance of second-harmonic and first-order intermodulation distortion components. This would be fully evident by carrying out the two-tone analysis given for the phase detector; however, rather than repeating the previous type of analysis, we consider an exercise applicable to PLL design in FDM-FM systems (see also Section 7.2).

For low-distortion conditions, Eq. (6-33) may be reduced to

$$\Delta\omega(t) = a_1\{v(t) + (a_2/a_1)[v(t)]^2\} \tag{6-34}$$

from which the dependent distortion generator (as indicated in Fig. 6-10) can be derived as

$$v_d(t) = (a_2/a_1)[v(t)]^2 \tag{6-35}$$

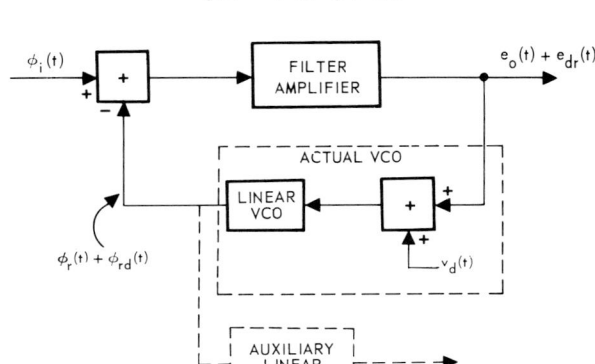

Fig. 6-10. Linear PLL model including equivalent distortion input. (VCO distortion.)

For the case of FDM, the modulating voltage $v(t)$ may be represented by Gaussian noise of two-sided power spectral density $W(f)$ which is of a constant value η_2 over the baseband range from frequency f_1 to f_2, and zero elsewhere.[2]

The power spectral density $W_d(f)$ of $v^2(t)$ is[3]

$$W_d(f) = 2\int_{-\infty}^{\infty} W(u)W(f-u)\,du + \left[\overline{v^2(t)}\right]^2 \delta(f) \tag{6-36}$$

where $\delta(f)$ is the dirac delta function. The second term in Eq. (6-36) represents a DC component which may be omitted in the remainder of the analysis. Thus the NPR is

$$\text{NPR} = (a_1/a_2)^2 W(f)/W_d(f) \tag{6-37}$$

Equation (6-37) assumes narrowband FDM channels over which both the signal and the distortion power spectral densities are essentially flat, and therefore, the noise power ratio in a channel is proportional to the ratio of the power spectral densities. The calculation of the power spectral density by the convolution integral (6-36) is illustrated graphically in Fig. 6-11.

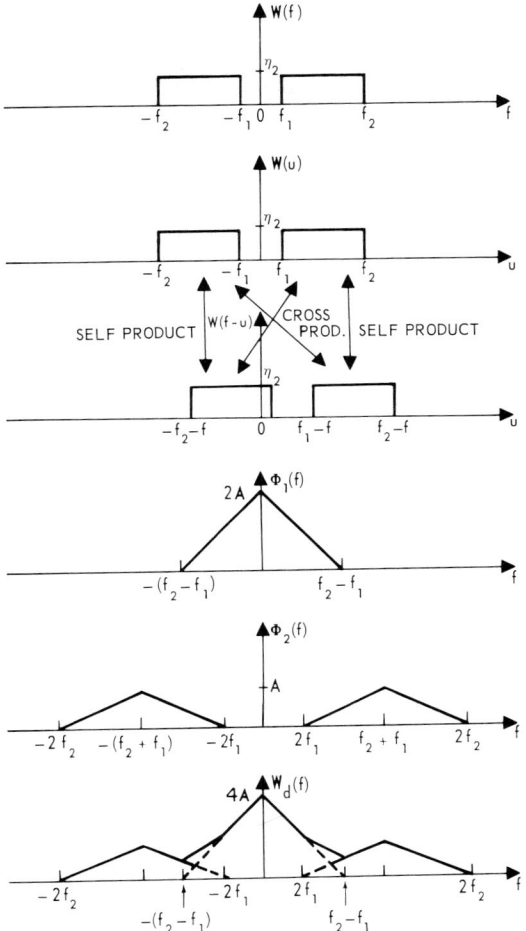

Fig. 6-11. Calculation of distortion spectrum by convolution. {$\Phi_1(f)$ is spectrum of self-product; $\Phi_2(f)$ = spectrum of cross-product; $A = \eta_2^2(f_2 - f_1)$; $W_d(f) = 2[\Phi_1(f) + \Phi_2(f)]$.}

It is convenient to define a normalized distortion spectrum

$$W_{dn}(f) = W_d(f)/[4\eta_2^2(f_2 - f_1)] \tag{6-38}$$

plotted in Fig. 6-12. Note that

$$(\Delta\omega_{rms})^2 = 2\eta_2(f_2 - f_1) = \text{mean square frequency deviation (radians per second)}^2 \tag{6-39}$$

(since both sides represent the total signal power in the baseband). Therefore, the NPR equation [Eq. (6-37)] reduces to

$$\text{NPR} = \left(\frac{a_1}{a_2}\right)^2 \frac{\eta_2}{4\eta_2{}^2(f_2 - f_1)W_{dn}(f)} = \left(\frac{a_1}{a_2}\right)^2 \frac{1}{2(\Delta\omega_{rms})^2 \, W_{dn}(f)} \tag{6-40}$$

The ratio (a_1/a_2) is related to the percentage linearity of the oscillator. If the oscillator is peak-deviated by V_p volts, the error in frequency at the deviation peak is $(a_2/a_1)V_p{}^2$ and the percentage error is

$$\Delta = \text{percentage error} = 100\,(a_2/a_1)V_p \tag{6-41}$$

or

$$\Delta^2 = 10^4 (a_2/a_1)^2 V_p{}^2 \tag{6-42}$$

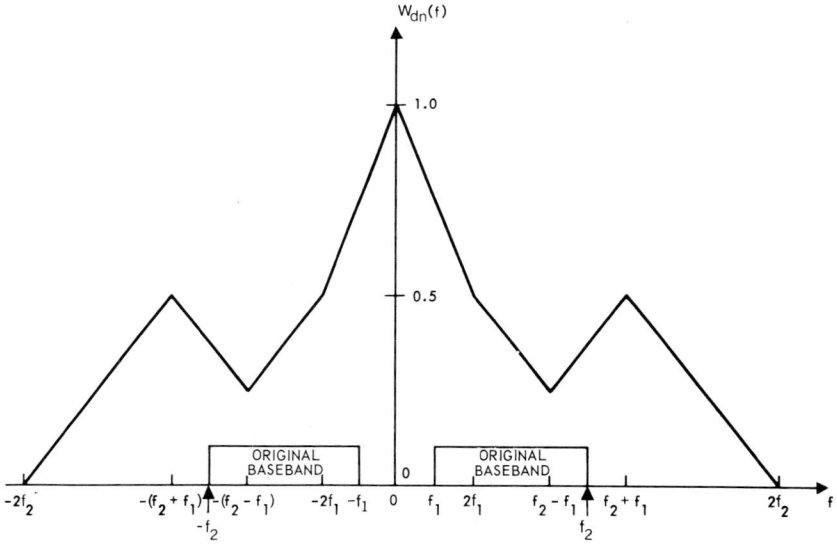

Fig. 6-12. Normalized spectrum for square-law distortion.

Assuming a peak-to-rms ratio of 10 dB for the input modulation, Eqs. (6-40) and (6-42) reduce to

$$\text{NPR} = 10^5/[2\Delta^2 W_{dn}(f)] \tag{6-43}$$

If we examine Fig. 6-12, we observe that the greatest distortion resides in the low-frequency portion of the baseband. Note from Eq. (6-43) that the lowest NPR, from a PLL which incorporates a VCO exhibiting a linearity of 1% over the full deviation range, is 47 dB.

A. Techniques to Reduce Distortion

Several methods may be employed to reduce signal distortion introduced by the PLL. Of first consideration is element design, which has the objective of realizing suitably linear loop components. The VCO can benefit greatly by attention to those factors that limit linearity. For example, the varactor-controlled VCO is seen to exhibit primarily square-law distortion, which may be eliminated by push–pull VCO design. The result, of course, is a VCO exhibiting primarily cubic distortion, however, at a greatly reduced level. Other types of VCO's, such as voltage-controlled multivibrators, afford superior linearity, but usually at a sacrifice of stability and internal noise, as discussed in Chapter 3.

The phase detector, on the other hand, contains a fundamental source of distortion (the sinusoidal detection characteristic) when implemented as a multiplying device. The multiplying action desirable from low-threshold design considerations is, however, not strictly required at high signal levels where threshold and the threshold mechanism do not predominate. Specifically, the doubly-balanced mixer, which approximates the multiplying device for low input-signal levels, exhibits substantially improved linearity over the sinusoidal characteristic when the input levels are equal for both input ports. Phase detector distortion may, therefore, be reduced by allowing the PLL input signal to rise, as the input SNR increases toward the distortion-limited region. Thus, reduction of phase detector distortion depends somewhat on the design of pre-PLL stages, specifically prelimiting and AGC.

Finally, output distortion is seen to be controlled by the choice of the PLL output point within the loop. For example, by taking the signal output at the output of the distorting element, the full loop gain effectively reduces the distortion with respect to the signal component. This may be understood by examining the case for VCO distortion. If the loop output is taken from the VCO output, through a linear discriminator, we have as the output distortion $e_d(t)$ (see Fig. 6-10; the discriminator's sensitivity is taken as unity without loss of generality)

$$e_d(t) = v_d(t)[1 - H(p)]\mu/p \qquad (6\text{-}44)$$

and the output signal as

$$e_o'(t) = e_o(t)\mu/p \qquad (6\text{-}45)$$

The signal-to-distortion ratio is, therefore,

$$e_o'(t)/e_d(t) = e_o(t)/\{v_d(t)[1 - H(p)]\} \qquad (6\text{-}46)$$

Taking the PLL output at the VCO input we have, similarly,

$$e_o(t) = \text{output signal} \qquad (6\text{-}47)$$

$$e_{dr}(t) = v_d(t)H(p) = \text{distortion component} \qquad (6\text{-}48)$$

and

$$e_o(t)/e_{dr}(t) = e_o(t)/[v_d(t)H(p)] \qquad (6\text{-}49)$$

The increase in signal-to-distortion ratio by taking the PLL output after the VCO rather than before it, is

$$\frac{e_o'(t)/e_d(t)}{e_o(t)/e_{dr}(t)} = \frac{H(p)}{1 - H(p)} = G(p) \qquad (6\text{-}50)$$

where $G(s)$ is open-loop response characteristic of the PLL. For sinusoidal signals and distortion, Eq. (6-50) may be recast in terms of the power ratio

$$\rho_D = \frac{\overline{|e_o'(t)/e_d(t)|^2}}{\overline{|e_o(t)/e_{dr}(t)|^2}} = |G(j\omega_d)|^2 \qquad (6\text{-}51)$$

where ω_d is frequency of distortion component.

6.3. Threshold-Limited Region

A. Factors Affecting Threshold

Before considering the design approach applicable to low-threshold FM demodulation, a discussion of the factors that influence threshold is in order. The threshold itself is a function of the additional output noise introduced by the nonlinear detection characteristics of the PLL, as indicated in Chapter 5, and thus we begin by considering some of the characteristics of this additional, or threshold-producing noise.

The excess noise is generated primarily by the cycle-skipping phenomenon that results in an impulsive type of noise appearing at the PLL output. As described in Section 5.4, there are two distinct sources of noise spikes in a PLL: the ThI which are generated in the same manner as in the LD, and the LLI which are due to the periodic response of the phase detector. Of interest is the rate of the sum of these, to which we will refer as the spike or cycle skipping rate. The result is very similar to the threshold characteristic of the LD. Accordingly, we may develop a threshold noise model where the total PLL output noise is represented by the sum of noise calculable from the PLL linear model, and impulse noise with 2π intensity which is due to the 2π phase jumps experienced in cycle skipping.

For the condition of unmodulated input carrier with additive white Gaussian noise, the cycle-skipping rate N is known for the first-order loop,

6.3. Threshold-Limited Region

with some experimental documentation for the second-order loop. From Section 5.4 we have

$$N = c\omega_n \exp(-D\alpha) \tag{6-52}$$

where α is input CNR referred to PLL noise bandwidth B_n, ω_n is loop natural frequency (usually defined for second-order system), and constants of Table III.

TABLE III

PLL CYCLE-SKIPPING PARAMETERS

Case	Loop order	c	D	B_n
I	First (theoretical)	$1/\pi$	2.0	$\omega_n/4$
II	Second (experimental)	1.57	1.64	$0.53\omega_n{}^a$
III		1.0	1.6	$0.53\omega_n{}^a$
IV		1.91	1.89	$0.53\omega_n{}^a$

^{a} Experimental performance based on $\xi = 2^{1/2}/2$.

The spike noise power spectral density is essentially flat, except for the shaping by both the PLL closed-loop response $H(j\omega)$ and the postdetection filter $R(j\omega)$. The result is an output impulse noise spectral density of

$$W_I(f) = 8\pi^2 N |H(j\omega)|^2 |R(j\omega)|^2 \tag{6-53}$$

Neglecting for the moment nonideal filtering characteristics that may be contained in the $|H(j\omega)|^2 |R(j\omega)|^2$ product, we have for the FM system an output noise power density consisting of a parabolic component (predicted from above threshold characteristics) and a flat component (predicted from the spike noise model), as illustrated in Fig. 6-13. From this model the total output noise power for a single-channel baseband as well as for a narrow-band FDM channel may be calculated by Eqs. (6-9) and (6-53); we have

$$N_{po} = 4\pi^2(2\eta/A^2)(f_b{}^3/3) + 8\pi^2 N f_b \quad \text{(single channel)} \tag{6-54}$$

$$N_{po} = 4\pi^2(2\eta/A^2)f_{CH}^2 B + 8\pi^2 N B \quad \text{(FDM channel)} \tag{6-55}$$

The threshold condition is derived by following the convention established earlier, namely the threshold point is where the total output noise increases by 1 dB over that predicted by the linear model. For the cases under consideration we have, from Eq. (6-52) and with $\alpha = A^2/(2\eta B_n)$,

$$\alpha \exp(-D\alpha) = (0.26/6c)(f_b{}^2/\omega_n B_n) \quad \text{(single channel)} \tag{6-56}$$

$$\alpha \exp(-D\alpha) = (0.26/2c)(f_{CH}^2/f_b{}^2)(f_b{}^2/B_n \omega_n) \quad \text{(FDM channel)} \tag{6-57}$$

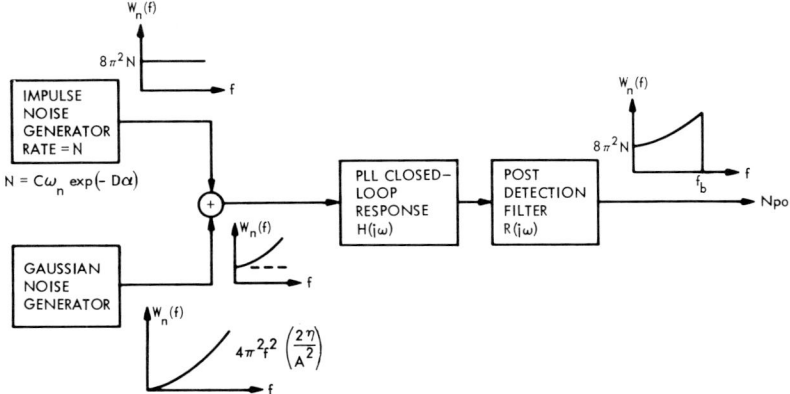

Fig. 6-13. Diagrammatical representation of spike-noise model to calculate PLL threshold [$W_n(f)$ is noise power density in watts per hertz].

Equation (6-56) is solved for the four sets of parameters given in Table III with the results given in Fig. 6-14. Note that the results are comparable, with the first-order loop (based on analytical data) exhibiting a somewhat lower threshold than the second-order loop (based on experimental data). On the other hand, the first-order PLL design generally results in a larger B_n/f_b. For subsequent calculations it is convenient to select an α representative of PLL

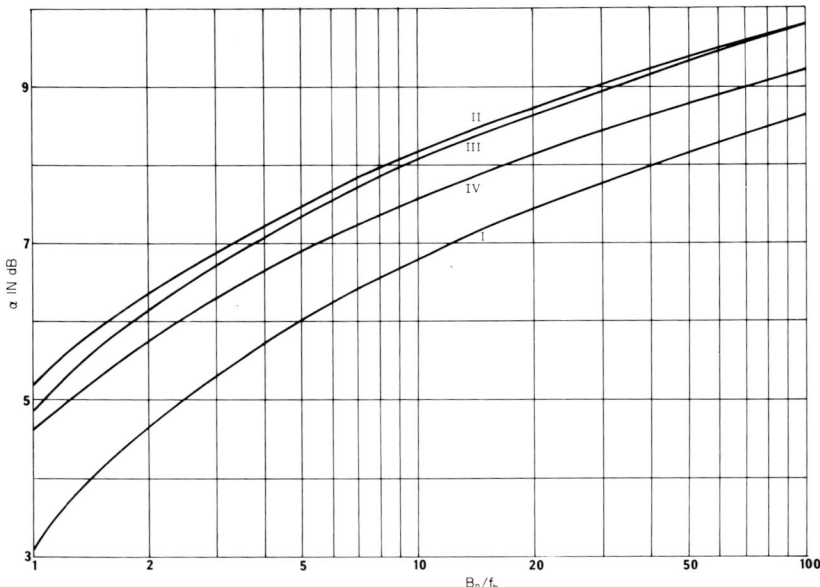

Fig. 6-14. Threshold CNR α in PLL noise bandwidth B_n. (Refer to Table III for curve identification.)

6.3. Threshold-Limited Region

threshold over a significant range of loop design parameters. In particular, for the typical $B_n/f_b = 5$ the results for the first-order loop indicate a threshold $\alpha = 4$ (6 dB) as being representative of the single-channel case. We will use this factor of $\alpha = 4$ also in the design of the second-order loop as well as for other B_n/f_b ratios as an approximation, although it appears to give a somewhat optimistic prediction of threshold. This value of α is widely used in the literature.[4,5]

Considering next the characteristics inherent in FDM, we note in particular that the threshold-inducing noise component has a flat spectral density (which increases with increasing N). This is in contrast to the parabolic distribution predicted by the linear model. It is, therefore, possible to deduce for a channelized baseband that, for a given α, the threshold effect will be distributed and will not occur simultaneously in each channel. This is called the "channel-threshold spread" and is derivable from the solution of Eq. (6-57) in terms of the (f_{CH}/f_b) ratio. It is desirable to normalize the channel-threshold spread to the threshold found for the entire baseband. This is achieved by plotting the solutions of Eq. (6-57) in terms of α_{CH}/α_b as indicated in Fig. 6-15; α_{CH} is channel threshold CNR, and α_b is entire baseband threshold CNR (taken from first-order loop curve, Fig. 6-14), both referred to B_n; First-order

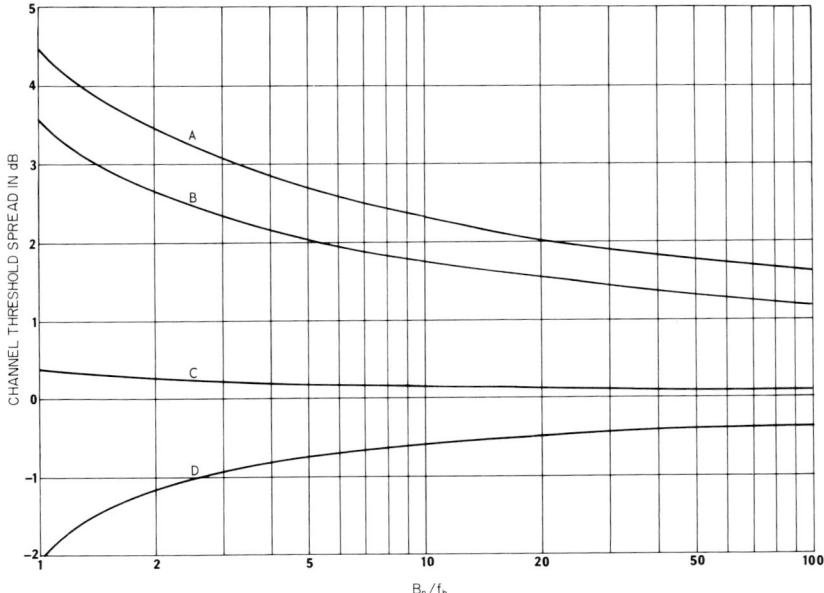

Fig. 6-15. Channel threshold spread α_{CH}/α_b for FDM-FM (or PM) demodulation by PLL. (0 dB corresponds to the threshold of the entire baseband; A: bottom channel with $f_{CH}/f_b = 1/40$, corresponding to 120- and 600-channel systems; B: bottom channel with $f_{CH}/f_b = 1/15$, corresponding to 240-channel systems; C: middle channel, $f_{CH}/f_b = 0.5$; D: top channel, $f_{CH}/f_b = 1$.)

loop cycle skipping parameters were used in Eq. (6-57). For the case of loop designs that fall significantly outside of this area, so that the $\alpha = 4$ threshold criterion is not justified, another choice, based on Fig. 6-14, for the threshold performance index may be made.

Modulation, or any other factor that changes the input or VCO frequency from the quiescent unmodulated center frequency condition, will tend to create a phase-error component which increases the total loop phase error in both the instantaneous and mean-square senses. In the linear model and with noise and signal statistically independent, $\phi_{es}(t)$ may be considered an independent additive term which increases the total mean-square phase error $\overline{\phi_e^2(t)}$ linearly. For the particular case of Gaussian noise modulation which is representative of FDM and single-channel speech, the modulation-induced phase error is Gaussian, as is to a first approximation the above-threshold noise-induced phase error. The result is a total phase error having essentially the same statistics with and without modulation, with the modulation basically representing a mean-square value increase. This strongly suggests that the threshold criterion for a PLL operating in the presence of Gaussian modulation may simply be adapted from the unmodulated case by stating that

$$\overline{\phi_{rn}^2(t)} + \overline{\phi_{es}^2(t)} = \overline{\phi_e^2(t)} \tag{6-58}$$

with $\overline{\phi_e^2(t)}$ equal to the mean-square phase error which is indicative of threshold in the absence of modulation, i.e.,

$$\overline{\phi_e^2(t)} = \tfrac{1}{4} \text{ rad}^2 \qquad \text{(Equivalent to } \alpha = 4\text{)}† \tag{6-59}$$

Modulation other than Gaussian should be treated on an individual basis. For example, in test-tone (TT) modulation we observe that the modulation-induced phase error, by being sinusoidal, resides near its peak value a significant portion of the time. At the point of peak modulation phase error, the loop has the least phase-error margin to unlock and, hence, is most prone to generating a loss-of-lock impulse. The means, then, of incorporating the effect of TT modulation is to reduce, by the peak modulation phase error, the effective peak phase error that is indicative of unlock. This is best understood by the following development.

Taking the threshold criterion $\overline{\phi_{rn}^2(t)} = \tfrac{1}{4} \text{ rad}^2$, and noting that the noise in the loop is essentially Gaussian in this operating region, the $\tfrac{1}{4} \text{ rad}^2$ mean-square value indicates that threshold occurs when the noise amplitude exceeds $\pi/2$ rad with 0.15% probability. Retaining this requirement in the presence of modulation, the new equivalent peak phase error ϕ_{rnp} that must be exceeded by the noise is

$$\phi_{rnp} = \frac{\pi}{2} - \frac{\Delta\omega_p \omega_T}{\omega_n^2} \tag{6-60}$$

† Note that in the linear model $\overline{\phi_{rn}^2(t)} = 1/\alpha$, see Eq. (5-83).

6.3. Threshold-Limited Region

where ω_T is the radian frequency of the test tone. To obtain Eq. (6-60), the following arguments are used:

(1) The loop is most vulnerable to loss-of-lock at the peak phase error, since this leaves least margin to unlock. For test-tone modulation, the phase error being sinusoidal, spends much of the time at its peak where the loss-of-lock impulses will generally occur. Since the peak of the input phase is the modulation index $\Delta\omega_p/\omega_T$, the peak absolute value of the modulation phase error is, from Eq. (6-26),

$$|\phi_{es}| = |1 - H(j\omega)|\frac{\Delta\omega_p}{\omega_T} \cong \left(\frac{\omega_T}{\omega_n}\right)^2 \frac{\Delta\omega_p}{\omega_T} = \frac{\Delta\omega_p\,\omega_T}{\omega_n^2} \qquad (6\text{-}61)$$

(2) An LLI occurs every time the magnitude of the total instantaneous phase error exceeds $\pi/2$, throwing the loop into the regenerative mode.

(3) The rate of the LLI is indicative of threshold. This is an approximation in that the effect on the rate of the ThI is ignored.

The above approximations are made to facilitate analytical design, and the results were found meaningful in practice. At threshold, ϕ_{rnp} will be exceeded with 0.15% probability when

$$\phi_{rnp}/[\overline{\phi_{rn}^2(t)}]^{1/2} \cong \pi \qquad (6\text{-}62)$$

(Note, $\alpha = 1/\overline{\phi_{rn}^2(t)}$), resulting in the threshold α of

$$\alpha = 1 \bigg/ \left[\frac{1}{2} - \frac{\Delta\omega_p\,\omega_T}{\pi\omega_n^2}\right]^2 \qquad (6\text{-}63)$$

The PLL is normally preceded by a predetection filter which rejects excess out-of-band noise. This filter influences the spectrum and mean-square VCO phase noise and, undoubtedly, the PLL threshold. To this point in the development, the input filter has been neglected, mainly because there is no proven method of including input filter effects on PLL performance. However, some qualitative evaluation is possible, and the following comments apply.

The predetection filter (as discussed in Chapter 3) is designed to contain the incoming signal, and normally follows the Carson bandwidth design rule (see also Eq. 3-35)

$$B_{IF} = 2(f_b + \Delta f_p) = 2f_b(1 + m_p) \qquad (6\text{-}64)$$

where B_{IF} is predetection filter 3-dB bandwidth (Hz), f_b is top baseband frequency of transmission (Hz), Δf_p is peak signal frequency deviation (Hz), and m_p is peak system modulation index.

In most applications, the signal experiences some attenuation at the peak frequency deviation. At the same time, the PLL experiences the largest modulation phase-error component, and becomes most susceptible to the input noise and to the generation of a loss-of-lock impulse. The loss in signal power then, at the peak deviation point, represents a direct increase in the effective threshold CNR.

The VCO phase noise is found to be a function of the composite filtering characteristics of the input filter and the PLL closed-loop response. For the case of input filter bandwidth substantially greater than the PLL closed-loop bandwidth, it is essentially correct to assume white input noise to the loop, with the threshold determined by the loop bandwidth. As the loop bandwidth approaches, or becomes larger than the input filter bandwidth, the white-noise assumption becomes invalid as do the formulas dependent on the loop noise bandwidth B_n.

The loop design is by analytical necessity based on white input noise. In general, the presence of the predetection filter, when discounting signal attenuation effects, serves to improve performance over that predicted by the white-input-noise design. The results to be developed, therefore, may be considered conservative.

B. Design for Low Threshold

Two design approaches may be applied to realize the minimization of PLL threshold. They are: (1) minimization of the mean-square phase error at threshold, and (2) minimization of the signal carrier power (or the CNR) at threshold. Both approaches result in the same loop design. To illustrate both with design examples, the design for a noise-loaded signal (such as FDM and speech) is based on mean-square error minimization, while the design for a test tone is based on CNR minimization. This section is based exclusively on the second-order type-one loop with $b \ll K$ and $b \ll a$ (see Section 5.3). Some consideration and remarks on higher-order loop design are given later.

1. MEAN-SQUARE PHASE-ERROR MINIMIZATION TECHNIQUE

The critical factor determining PLL threshold is seen to be the loop mean-square phase error, which consists of both modulation and noise-induced components. The noise component of phase error, Eq. (5-83), is seen to increase linearly with increasing loop noise bandwidth, while from Eq. (6-61) the modulation component is seen to be a decreasing function of the open-loop gain developed across the baseband frequency range. This is similar to the behavior of the ThI and LLI as a function of loop bandwidth, as discussed in Section 5.4. In the second-order loop, the interrelationship between loop noise bandwidth and open-loop gain are the constants a, b, and K (from Table II, Chapter 5) or, alternatively, the loop damping factor ξ

6.3. Threshold-Limited Region

[refer to Eqs. (5-35) and (5-37)]. The first step in loop design is to minimize the loop noise bandwidth for a fixed open-loop gain constant (Kb product, for example, in the second-order type-one loop) which is equivalent to a fixed natural frequency ω_n. This will minimize the noise component of the phase error for a fixed modulation phase error. The second step is to minimize the total phase error by selecting the proper open-loop gain constant or, equivalently, the proper loop natural frequency and noise bandwidth.

The *first step* is carried forth by noting from Chapter 5, Table II, that

$$B_n = Kb\left(\frac{Kb}{a} + a\right) \Big/ 4a\left(\frac{Kb}{a} + b\right) \cong \left(\frac{Kb}{a} + a\right) \Big/ 4 \quad \text{Hz} \qquad a \ll K \quad (6\text{-}65)$$

Letting $\partial B_n/\partial a = 0$, B_n minimum is obtained for

$$a = (Kb)^{1/2} = \omega_n \tag{6-66}$$

which is equivalent to a loop damping factor $\xi \cong \frac{1}{2}$ ($b \ll K$). The relationship between loop gain constant $(Kb)^{1/2}$, natural frequency ω_n, and noise bandwidth B_n at the noise bandwidth minimum is

$$B_n = (Kb)^{1/2}/2 = \omega_n/2 \quad \text{(Hz)} \tag{6-67}$$

and the optimum closed loop transfer function becomes [see Eq. (5-34)]

$$H(s) = \frac{(s/\omega_n) + 1}{(s/\omega_n)^2 + (s/\omega_n) + 1} \tag{6-68}$$

The noise-bandwidth minimization is illustrated by the Bode diagram in Fig. 6-16. The influence of a on the closed-loop response is shown in Fig. 5-5. Qualitatively speaking, $a > \omega_n$ increases B_n due to excessive peaking while $a < \omega_n$ increases B_n through broadening of the closed-loop response.

The *second step* in the low-threshold design is the minimization of the total mean-square phase error $\overline{\phi_e^2(t)}$ given by [Eq. (6-58)]

$$\overline{\phi_e^2(t)} = \overline{\phi_{rn}^2(t)} + \overline{\phi_{es}^2(t)} \tag{6-69}$$

where the noise-induced phase-error component is

$$\overline{\phi_{rn}^2(t)} = (2\eta/A^2)(\omega_n/2) \tag{6-70}$$

The signal component of the phase error is formulated by first specifying the modulation spectral distribution across the baseband range, i.e., $W_{\phi_i}(f)$. Applying the modulation phase-error transfer function, we have

$$\overline{\phi_{es}^2(t)} = \int_0^{f_b} W_{\phi_i}(f) \left|\frac{\Phi_e}{\Phi_i}(j\omega)\right|^2 df \tag{6-71}$$

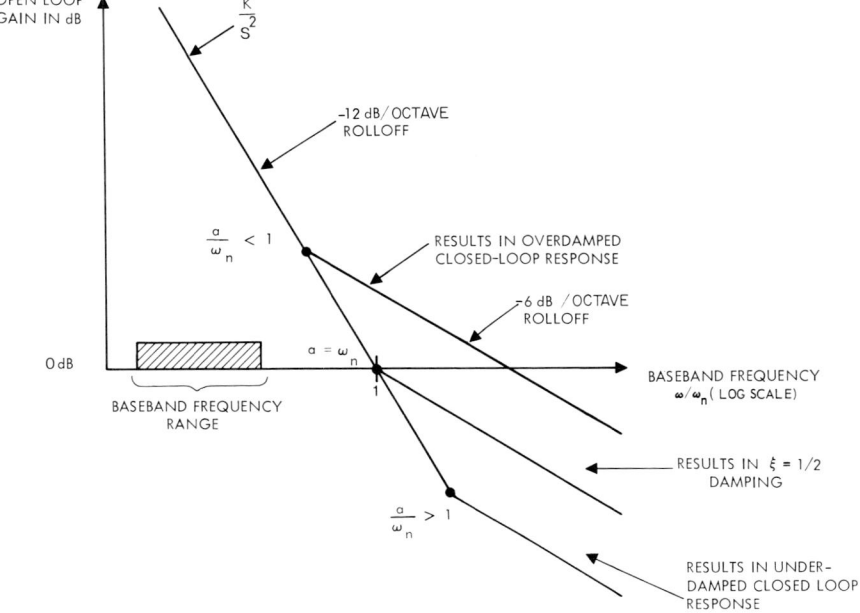

Fig. 6-16. Bode plot of second-order PLL. (Open-loop response.)

Using approximation (6-26) for the phase transfer function,

$$\left|\frac{\Phi_e}{\Phi_i}(j\omega)\right| \cong \left|\frac{s^2}{\omega_n^2}\right|_{s=j\omega} = \frac{\omega^2}{\omega_n^2} \tag{6-72}$$

we obtain

$$\overline{\phi_e^2(t)} = \frac{\eta \omega_n}{A^2} + \frac{1}{\omega_n^4}\int_0^{f_b} \omega^4 \, W_{\phi_i}(f) \, df \tag{6-73}$$

With

$$y = \int_0^{f_b} \omega^4 \, W_{\phi_i}(f) \, df \tag{6-74}$$

the total mean-square phase error may be minimized by setting

$$\frac{d}{d\omega_n}\overline{\phi_e^2(t)} = \frac{\eta}{A^2} - \frac{4y}{\omega_n^5} = 0 \tag{6-75}$$

which results in the condition

$$\eta\omega_n/A^2 = 4y/\omega_n^4 \tag{6-76}$$

Substituting Eq. (6-76) into (6-73) results in a minimum total mean-square phase error of

$$\overline{\phi_e^2(t)}_{\min} = (5/4)(\eta \omega_n / A^2) \qquad (6\text{-}77)$$

From Eqs. (6-73)–(6-77) it is seen that the optimum design, with noise-loaded baseband, requires that the total mean-square phase error at threshold be distributed as one-fifth toward the signal-induced components and four-fifths toward the noise-induced components. Equation (6-77) yields the threshold CNR by substituting the threshold index $\overline{\phi_e^2(t)} = \frac{1}{4}$ and the minimum noise bandwidth $B_n = \omega_n/2$

$$A^2/\eta \omega_n = A^2/(2\eta B_n) = \alpha = 5 \quad \text{or} \quad 7 \text{ dB} \qquad (6\text{-}78)$$

The noise bandwidth required to optimize the phase error at threshold may be derived by substituting Eqs. (6-77) [or (6-78)] and (6-76) into Eq. (6-67) resulting in

$$B_n = \frac{\omega_n}{2} = \left(\frac{5}{16\overline{\phi_e^2(t)}} y\right)^{1/4} = \left(\frac{5}{4} y\right)^{1/4} \qquad (6\text{-}79)$$

Note that the required loop natural frequency ω_n and the noise bandwidth depend upon the spectral distribution of the baseband signal.

As examples, several modulation cases are evaluated:

A. FLAT PHASE MODULATION (REPRESENTATIVE OF FDM-PM)

$$W_{\phi_i}(f) = \eta_m \quad \text{W/Hz}, \qquad 0 < f \le f_b \dagger \qquad (6\text{-}80)$$

B. FLAT FREQUENCY MODULATION (REPRESENTATIVE OF FDM-FM)

$$W_{\phi_i}(f) = \frac{\eta_m}{\omega^2} \quad \text{W/Hz}, \qquad 0 < f \le f_b \qquad (6\text{-}81)$$

C. SINGLE-CHANNEL SPEECH (FM)

$$W_{\phi_i}(f) = \frac{\eta_m}{\omega^4} \quad \text{W/Hz}, \qquad f_a \le f \le f_b \qquad (6\text{-}82)$$

where η_m is a modulation constant.

Evaluation of (6-74) results in

A.
$$y = \tfrac{1}{5}(2\pi)^4 \eta_m f_b^5 \qquad (6\text{-}83)$$

† For cases where the baseband does not begin at $f = 0$ and a finite lower limit is to be considered, the modification of the following is a straightforward matter. In most cases, however, this type of modification will have negligible effect on the following results.

B. $$y = \tfrac{1}{3}(2\pi)^2 \eta_m f_b^{\,3} \tag{6-84}$$

C. $$y = \eta_m[f_b - f_a] \tag{6-85}$$

It is convenient to reexpress the y parameter in terms of the mean-square frequency deviation $(\Delta\omega_{rms})^2$ of the signal. Note that†

$$(\Delta\omega_{rms})^2 = \int_0^{f_b} \omega^2 W_{\phi i}(f)\,df \quad \text{(rad/sec)}^2 \tag{6-86}$$

which results in

A. $$(\Delta\omega_{rms})^2 = \tfrac{1}{3}(2\pi)^2 \eta_m f_b^{\,3} \tag{6-87}$$

B. $$(\Delta\omega_{rms})^2 = \eta_m f_b \tag{6-88}$$

C. $$(\Delta\omega_{rms})^2 = \eta_m[(f_b - f_a)/f_a f_b]/(2\pi)^2 \tag{6-89}$$

Equations (6-83)–(6-85) are accordingly modified to read

A. $$y = \tfrac{3}{5}\omega_b^{\,2}(\Delta\omega_{rms})^2 \tag{6-90}$$

B. $$y = \tfrac{1}{3}\omega_b^{\,2}(\Delta\omega_{rms})^2 \tag{6-91}$$

C. $$y = \omega_a \omega_b (\Delta\omega_{rms})^2 \tag{6-92}$$

Substitution of Eqs. (6-90)–(6-92) into (6-79) yields the bandwidth design equations for the three modulation cases:

A. Flat Phase Modulation

$$\omega_n/\omega_b = (12)^{1/4}\sigma^{1/2} = 1.86\sigma^{1/2} \quad \text{or} \quad B_n/f_b = \pi(12)^{1/4}\sigma^{1/2} = 5.85\sigma^{1/2} \tag{6-93}$$

B. Flat Frequency Modulation

$$\omega_n/\omega_b = (20/3)^{1/4}\sigma^{1/2} = 1.61\sigma^{1/2} \quad \text{or} \quad B_n/f_b = \pi(20/3)^{1/4}\sigma^{1/2} = 5.05\sigma^{1/2} \tag{6-94}$$

C. Single-Channel Speech (FM)

$$\omega_n/\omega_b = (20)^{1/4}(\omega_a/\omega_b)^{1/4}\sigma^{1/2} = 2.12(\omega_a/\omega_b)^{1/4}\sigma^{1/2}$$
$$\text{or} \quad B_n/f_b = \pi(20)^{1/4}(\omega_a/\omega_b)^{1/4}\sigma^{1/2} = 6.65(\omega_a/\omega_b)^{1/4}\sigma^{1/2} \tag{6-95}$$

where σ is system rms modulation index $= \Delta\omega_{rms}/\omega_b$.

The design-normalized noise bandwidth (B_n/f_b) is summarized in Fig. 6-17 to facilitate design. The speech model assumes $\omega_b/\omega_a = 4$. The test-tone case

---

† Multiplication by ω^2 changes the spectral density of phase to that of frequency. Subsequent integration over the baseband gives the total power or the mean-square value.

6.3. Threshold-Limited Region

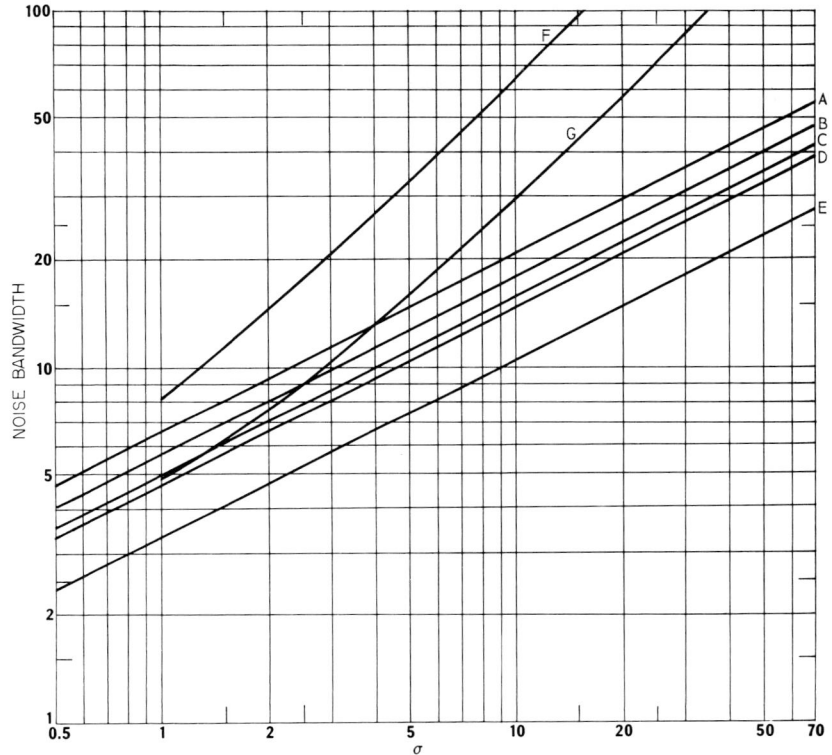

Fig. 6-17. Design PLL noise bandwidth (second-order loop), and predetection bandwidth as functions of modulation. [Curves A–E give normalized PLL noise bandwidth B_n/f_b, and curves F and G give normalized predetection bandwidth B_{IF}/f_b. A: test-tone (FM and PM, $\omega_T/\omega_b = 1$); B: FDM-PM; C: FDM-FM; D: speech model (FM, $\omega_b/\omega_a = 4$); E: test-tone (FM and PM, $\omega_T/\omega_b = \frac{1}{4}$); F: FDM and speech model; G: test-tone ($\omega_T/\omega_b = 1$).]

is derived presently. The predetection bandwidth curves apply to both FM and PM. They are a plot of Eq. (6-64), with the peak modulation index related to the rms modulation index (crest factor) by 10 dB for FDM and speech signals.

In terms of entire baseband or single-channel threshold, the normalized threshold CNR measured in $2f_b$, $(CNR)_{AM}$, is derived by substituting Eqs. (6-93)–(6-95) into (6-78), with the result:

A. Flat Phase Modulation

$$(CNR)_{AM} = (A^2/2)/2\eta f_b = 14.6\sigma^{1/2} \qquad (6-96)$$

B. Flat Frequency Modulation

$$(CNR)_{AM} = (A^2/2)/2\eta f_b = 12.6\sigma^{1/2} \qquad (6\text{-}97)$$

C. Single-Channel Speech (FM)

$$(CNR)_{AM} = (A^2/2)/2\eta f_b = 11.7\sigma^{1/2} \qquad (\text{Note: } f_a/f_b = \tfrac{1}{4}) \qquad (6\text{-}98)$$

The aforementioned results for threshold CNR can be applied in FDM to predict channel threshold. Usually the top-channel threshold is of interest, and we note from Fig. 6-15 that it is only slightly lower than the entire baseband threshold (the variation being on the order of 0.5 to 1.0 dB). Since the threshold index, taken as $\overline{\phi_e^2(t)} = \tfrac{1}{4}$, is considered slightly optimistic (refer to previous discussion), the application of the additional correction factor to translate from entire baseband threshold (predicted on the $\overline{\phi_e^2(t)} = \tfrac{1}{4}$ criterion) to top channel is not warranted. Therefore, in the FDM cases portrayed in Figs. 6-18 and 6-19, the entire baseband threshold, as predicted by Eqs. (6-96)–(6-98), is used interchangeably with top channel threshold.

To facilitate PLL and system design, the design equations are combined and plotted in Figs. 6-18 and 6-19. The bases for these figures are as follows:

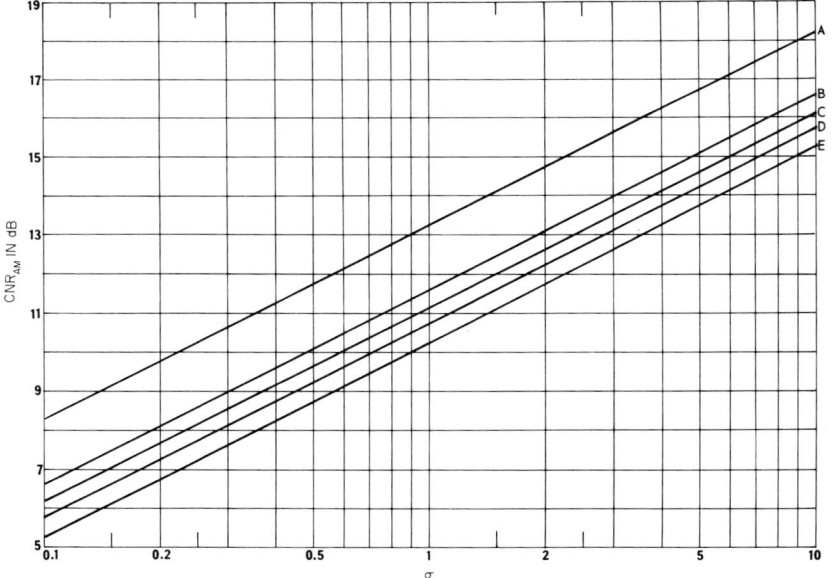

Fig. 6-18. Threshold CNR for the PLL for various modulations. [A: test-tone (FM and PM, $\omega_T/\omega_b = 1$); B: FDM-PM; C: FDM-FM; D: speech model-FM; E: test-tone (FM and PM, $\omega_T/\omega_b = \tfrac{1}{4}$).]

6.3. Threshold-Limited Region

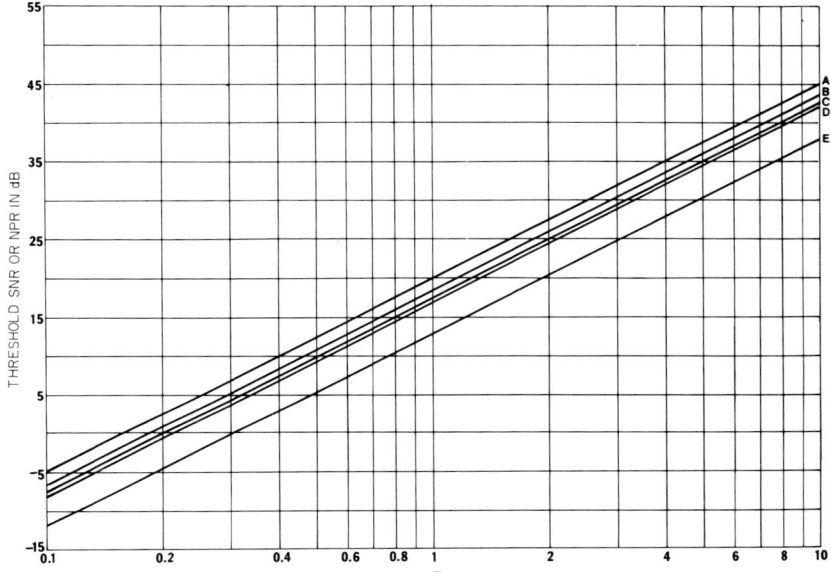

Fig. 6-19. Threshold SNR measured in base bandwidth, or top-channel NPR, for the PLL for various modulations. [A: test-tone (SNR) (FM, $\omega_T/\omega_b = 1$); B: FDM-PM (NPR); C: speech model (FM) (SNR); D: test-tone (SNR) (FM, $\omega_T/\omega_b = \frac{1}{4}$); E: FDM-FM (NPR).] Note: 1-dB threshold degradation in SNR (and NPR) is included.

A.

Figure 6-18—Threshold CNR for the PLL. This figure presents the normalized threshold CNR's based on Eqs. (6-96)–(6-98) and the results for TT modulation, Eq. (6-105).

B.

Figure 6-19—Threshold SNR for the PLL. This figure combines the SNR and NPR equations (6-1)–(6-7) with the CNR equations (6-96)–(6-98), and Eq. (6-105). In the case of FDM, top-channel performance in terms of NPR is considered, whereas in TT and single-channel speech cases, entire baseband performance in terms of SNR is considered. Output SNR and NPR are reduced by 1 dB over that predicted by the ideal above-threshold equations, to account for the performance degradation introduced by the threshold effect. The equations are:

$$(\text{SNR})_{\text{TT-FM}} = 20 + (10/2) \log_{10}(\omega_T/\omega_b) + 25 \log_{10} \sigma \quad \text{dB} \quad \text{(at threshold)} \tag{6-99}$$

$$(\text{NPR})_{\text{FDM-PM}} = 18.4 + 25 \log_{10} \sigma \quad \text{dB} \quad (\text{at threshold}, \omega_a/\omega_b \ll 1) \tag{6-100}$$

$(SNR)_{SP-FM} = 17.5 + 25 \log_{10} \sigma$ dB (at threshold, $\omega_a/\omega_b = \frac{1}{4}$) (6-101)

$(NPR)_{FDM-FM} = 13 + 25 \log_{10} \sigma$ dB (at threshold, $\omega_b/\omega_{CH} = 1$, $\omega_a/\omega_b \ll 1$) (6-102)

2. Threshold CNR Minimization Technique

Abstracting from Eq. (6-63), we have the threshold CNR (referred to B_n) for the TT case as

$$\alpha = 1 \bigg/ \left[\frac{1}{2} - \frac{\Delta\omega_p \,\omega_T}{\pi\omega_n^2}\right]^2 \qquad (6\text{-}63)$$

This may be translated to a normalized CNR referred to $2f_b$ (twice the baseband) by setting $B_n = \omega_n/2$ and

$$(CNR)_{AM} = \frac{A^2/2}{2\eta f_b} = \frac{B_n}{2f_b} \cdot \alpha = B_n \bigg/ \left\{ 2f_b \left[\frac{1}{2} - \frac{\Delta\omega_p \,\omega_T}{4\pi B_n^2}\right]^2 \right\} \qquad (6\text{-}103)$$

It is now minimized by setting $\partial(CNR)_{AM}/\partial B_n = 0$, resulting in

$$B_n/f_b = (10\pi)^{1/2}(\omega_T/\omega_b)^{1/2}(m_p)^{1/2} = 5.6(\omega_T/\omega_b)^{1/2}(m_p)^{1/2} \qquad (6\text{-}104)$$

where $m_p = \Delta\omega_p/\omega_b$ is system peak modulation index.

From Eqs. (6-104) and (6-103)

$$(CNR)_{AM} = 3.125 B_n/f_b = 17.5(\omega_T/\omega_b)^{1/2}(m_p)^{1/2} \qquad (6\text{-}105)$$

Note that both threshold and loop design depend upon the TT frequency. The equations are plotted in Figs. 6-17–6-19, with the rms modulation index $(m_p/2^{1/2})$ as the independent parameter.

3. Concluding Remarks—First-Order Correction Techniques

The system designer may wish to incorporate first-order corrections to the threshold results that have been established. This may be accomplished by referring to the threshold spread curves based on the threshold impulse model given earlier, or adaptation of other threshold criteria in terms of a threshold index. In terms of the threshold-spread curves presented here, the actual loop bandwidth is rather insensitive to the choice of threshold index, since the B_n/f_b ratio is related by a fourth-root factor to the threshold index Eq. (6-79). Therefore, the choice $\overline{\phi_e^2(t)} = \frac{1}{4}$ is adequate in determining the B_n/f_b ratio. The threshold point $(CNR)_{AM} = (B_n/2f_b)\overline{\phi_e^2(t)}$, however, is a direct function of the threshold index once the B_n/f_b ratio is determined. The first-order correction is achieved, therefore, by shifting the threshold point by the ratio of $\frac{1}{4}$ to the actual threshold index (established from the threshold-spread curve). The following example serves to illustrate this consideration:

Example: Given FDM-FM, $\sigma = 10$

(1) From Fig. 6-17 we have $B_n/f_b = 16$.

(2) From the threshold spread curve (Fig. 6-14, curve I), we have entire baseband threshold $\alpha = 7.2$ dB. Therefore, the first-order correction for entire baseband threshold is 7.2 dB − 6 dB = 1.2 dB.

(3) From the channel threshold-spread curve (Fig. 6-15) we have a relative top channel threshold of −0.5 dB, resulting in a net increase in top channel threshold, to that presented in Fig. 6-18, of 1.2 − 0.5 or 0.7 dB. This also illustrates the point that usually these two effects tend to cancel each other.

C. Effect of Delay on Threshold

Loop delay in the PLL demodulator causes the threshold and intermodulation distortion performance to deteriorate. Uncompensated delay reduces the phase margin in the open-loop response, resulting in greater closed-loop noise bandwidth and threshold CNR. It further leads to a larger error signal in the loop and, hence, greater distortion in the closed-loop system.

Compensation for the delay in the PLL is achieved by either or both of the following two steps: (1) reduction of the zero frequency a in the baseband filter, with a commensurate reduction of the gain constant, and (2) introduction of a zero-pole pair at frequencies above the 0-dB open-loop gain point in the open-loop response characteristic. These are illustrated in Fig. 6-20. The basis for the first technique was developed in Chapter 5, where minimum noise bandwidths, as functions of delay and zero constant a, are established (Fig. 5-16). These results are now incorporated with the PLL design theory to illustrate PLL performance and design in the presence of delay. The second technique results in a third-order loop, the optimization and design of which has not been sufficiently developed. For this reason, we will not consider this technique beyond indicating its applicability to delay compensation.

The evaluation of the effect of delay in PLL performance by utilizing the first compensation technique is carried forth as follows: It is convenient to refer the delay τ to the top baseband frequency in terms of excess phase shift ϕ_b; thus, we have

$$\phi_b = \tau \omega_b, \qquad \tau = \phi_b/\omega_b \qquad (6\text{-}106)$$

In Section 5.5, Fig. 5-16, τ is defined in terms of excess phase shift ϕ at ω_n

$$\tau = \phi/\omega_n \qquad (6\text{-}107)$$

From Eqs. (6-106) and (6-107)

$$\phi = \phi_b(\omega_n/\omega_b) \qquad (6\text{-}108)$$

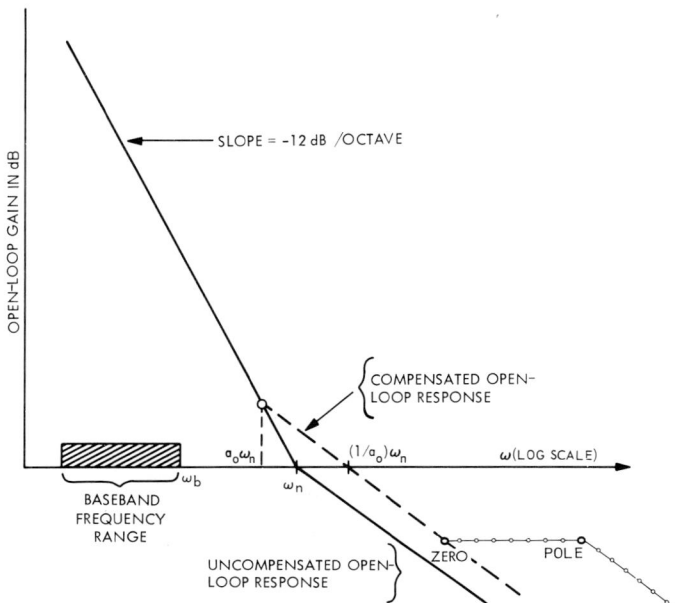

Fig. 6-20. Asymptotic open-loop response compensated for excess loop delay.

or by substitution of $\omega_n = 2B_n$ for the minimum bandwidth of the second-order loop,

$$\phi = (\phi_b/\pi)(B_n/f_b) \tag{6-109}$$

From Eqs. (6-93)–(6-95) and (6-104) (or Fig. 6-17) we have the optimum B_n/f_b ratio as a function of modulation index. Taking the case of FDM-FM as an example, from Eqs. (6-94) and (6-109) we obtain

$$\phi = \phi_b(20/3)^{1/4}\sigma^{1/2} = 1.61\sigma^{1/2}\phi_b \tag{6-110}$$

The system performance may be modified to include the effects of delay by: (1) evaluating Eq. (6-110) for fixed values of ϕ_b over the range of σ (rms modulation index); (2) noting the relative increase in noise bandwidth predicted by Fig. 5-16 and translating this directly to an increase in threshold CNR, i.e.,

$$\Delta\mathrm{CNR} = 10\log_{10} B_n(\tau)/B_n(0) \tag{6-111}$$

The foregoing fixes the threshold point for ϕ_b, in terms of σ and $\Delta\mathrm{CNR}$. Figure 6-21 presents the threshold boundaries for FDM-FM over a range of delay. The dashed lines are the above-threshold performance curves, Eq. (6-5) with $\omega_a/\omega_b \ll 1$. The solid lines are the loci of the threshold points. The intersection of a solid and dashed curve gives the threshold performance for a given modulation index.

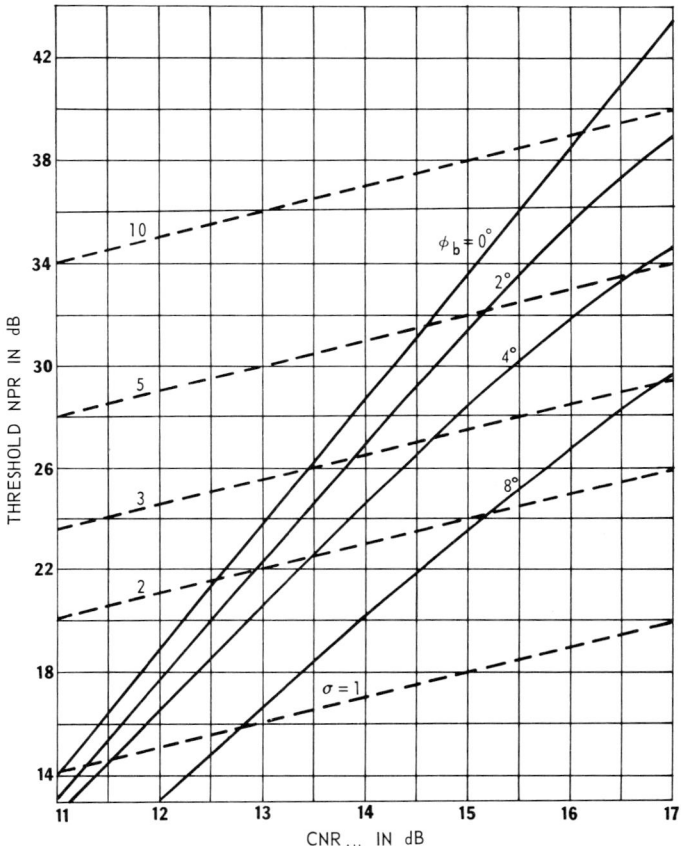

Fig. 6-21. Threshold NPR measured in top FDM-FM channel for the PLL with loop delay as a parameter (– – –: above threshold; —: loci of threshold points).

D. Effect of VCO Noise and Drift

The effect of VCO frequency instability, noise, or drift in the operation of a PLL manifests itself in several ways. The primary effect is to create within the PLL a phase-error component which results in both SNR and threshold CNR degradation. The sources of this frequency instability lie in the natural noise found in the VCO control element, the power supply or, equivalently, the baseband amplifying stages which precede the VCO. The noise may be represented by an equivalent input generator to a noiseless VCO, in a manner similar to distortion by an equivalent distortion generator, as indicated in Fig. 6-10. The characteristics of the equivalent generator are derived by measurement of the open-loop VCO characteristics. Formulization of the loop equations based on this equivalent input generator driving function indicates that the same effect may be achieved by introducing an equivalent modulation component in the input signal. For example, by specifying the input signal phase modulation to be

$$\phi_i(t) = \phi_{is}(t) + \mu \int^t e_n(\tau)\, d\tau \tag{6-112}$$

where $\phi_{is}(t)$ is received signal phase modulation, μ is VCO sensitivity in radians per volt-second, and $e_n(t)$ is equivalent noise input generator that accounts for VCO noise, and removal of the equivalent generator from the loop fully accounts for the effects of VCO drift. With this approach it is possible to apply the methods of analysis and design developed for modulation tracking.

Example: It is desired to extract a reference spectral component from additive noise utilizing a PLL which contains a noisy VCO. For this case, $\phi_{is}(t) = 0$ and the equivalent input which represents the VCO noise is

$$\phi_i(t) = \mu \int^t e_n(\tau)\, d\tau \tag{6-113}$$

By measurement,† the equivalent VCO noise generator is determined to be Gaussian noise, with a power spectral distribution of $E_n(f)$ of the form

$$E_n(f) = \eta_n/[(\omega/\omega_b)^2 + 1] \tag{6-114}$$

which results in an equivalent phase modulation power spectral density of

$$W_{\phi_i}(f) = \frac{\mu^2 E_n(f)}{\omega^2} = \frac{\mu^2 \eta_n}{\omega^2[(\omega/\omega_b)^2 + 1]} \tag{6-115}$$

The PLL mean-square phase error, as before, consists of "modulation" and additive noise induced components

$$\overline{\phi_e^2(t)} = \overline{\phi_{em}^2(t)} + \overline{\phi_{rn}^2(t)} \tag{6-116}$$

$$\overline{\phi_e^2(t)} = \int_0^\infty W_{\phi_i}(f)|1 - H(j\omega)|^2\, df + (2\eta/A^2)B_n \tag{6-117}$$

Evaluating the "modulation" induced component $\overline{\phi_{em}^2(t)}$ for the second-order loop with $\xi = \tfrac{1}{2}$, we have, using Eq. (6-25)

$$\overline{\phi_{em}^2(t)} = \frac{\mu^2 \eta_n}{(2\pi)\omega_n} \int_0^\infty \frac{(\omega/\omega_n)^2\, d(\omega/\omega_n)}{[(\omega/\omega_b)^2 + 1][(\omega/\omega_n)^4 - (\omega/\omega_n)^2 + 1]} \tag{6-118}$$

† See Chapter 10 for test procedures.

6.3. Threshold-Limited Region

Assuming $\omega_n \ll \omega_b$, which is usually the case in a "narrow-band" reference extraction tracking loop, results, using Eq. (6-117), in

$$\overline{\phi_e^2(t)} = \frac{\mu^2 \eta_n}{(2\pi \omega_n)} (\pi/2) + \frac{2\eta}{A^2} \frac{\omega_n}{2} \tag{6-119}$$

The mean-square error is minimized by setting $\partial \overline{\phi_e^2(t)}/\partial \omega_n = 0$ to yield an optimum loop bandwidth of

$$B_n = \frac{\omega_n}{2} = \frac{\mu}{2} \left(\frac{\eta_n A^2}{4\eta} \right)^{1/2} \tag{6-120}$$

The mean-square open-loop frequency jitter of the VCO is $(\Delta\omega_{\text{rms}})^2$ obtained by integrating $E_n(f)$ [Eq. (6-114)] for the total power and multiplying by the gain factor μ^2 to refer it to the VCO output,

$$(\Delta\omega_{\text{rms}})^2 = \frac{\mu^2 \eta_n \omega_b}{2\pi} \int_0^\infty \frac{d(\omega/\omega_b)}{(\omega/\omega_b)^2 + 1} = \frac{\mu^2 \eta_n \omega_b}{4} \tag{6-121}$$

Substituting Eq. (6-121) into (6-120) gives the optimum bandwidth in terms of $(\Delta\omega_{\text{rms}})$

$$B_n = (\Delta\omega_{\text{rms}}/2\pi^{1/2})(A^2/2\eta f_b)^{1/2} \tag{6-122}$$

or

$$B_n/f_b = (1/2\pi)^{1/2} \sigma [(\text{CNR})_{\text{AM}}]^{1/2} = 0.4\sigma [(\text{CNR})_{\text{AM}}]^{1/2} \tag{6-123}$$

where σ is equivalent modulation index of VCO jitter and $(\text{CNR})_{\text{AM}}$ is equivalent input CNR referred to $2f_b$.

Another manifestation of VCO frequency instability is the presence of a slowly varying frequency drift which gives rise to a slowly varying phase error in the type-one PLL. (Note: The type-two PLL integrates out slow drifts, rendering a zero phase error.) Whereas in linear demodulation the loop output SNR in analog systems is unaffected (the drift components are significantly below the baseband frequency range), the threshold and distortion performance can be significantly degraded. This is understood by noting that the drift-induced phase-error component reduces the phase-error margin to unlock, sensitizing the loop to the input additive noise, and resulting in an increase in the probability of LLI. Also, the operation is shifted to a less linear region and lower sensitivity of the phase detector. In digital systems requiring DC transmission, it produces additional performance degradation, as the DC output voltage from the loop tracks the drift, producing baseline uncertainties in the demodulated data waveform, thus biasing the data detector.

Threshold degradation is formulated by identifying a peak phase error that for a given input CNR is indicative of unlock. The degradation is related to the reduction of the available phase-error margin by the drift-induced phase error, as described in the example below. The distortion performance is analyzed as in Section 6.2, with the nonlinear phase-detector characteristic expanded about the new operating point.

Example

Consider a relatively small VCO drift of $\Delta\omega_p$ rad/sec. The drift-induced phase-error component is calculated from the DC response, based on the linear model. For the second-order type-one loop [see Eqs. (5-28) and (5-37)]

$$\Delta\phi_e \cong \Delta\omega_p/K = b\,\Delta\omega_p/\omega_n^2 \qquad (6\text{-}124)$$

Assuming $\phi_e = \pi/2$ is indicative of unlock, the resulting normalized "phase margin" is

$$\Lambda = \frac{(\pi/2) - \Delta\phi_e}{\pi/2} = 1 - \frac{\Delta\phi_e}{\pi/2} \qquad (6\text{-}125)$$

For small drift-induced phase errors, threshold degradation is given approximately by

$$\Delta \text{CNR}_{\text{TH}} = -20\log_{10}\Lambda = -20\log_{10}\left(1 - \frac{\Delta\phi_e}{\pi/2}\right) \cong 40\,\frac{\Delta\phi_e}{\pi} \quad \text{dB} \qquad (6\text{-}126)$$

Applying Eq. (6-124), the threshold degradation in terms of loop design and peak VCO drift is

$$\Delta \text{CNR}_{\text{TH}} = \frac{40}{\pi}\,\frac{b\,\Delta\omega_p}{\omega_n^2} \quad \text{dB} \qquad (6\text{-}127)$$

For $b = 0$, implying the type-two loop, the penalty is the expected 0 dB [see also the discussion relating to the derivation of Eq. (6-60)].

E. Design of Higher-Order Loops

Design and performance of PLL's of order greater than two are as yet largely unexplored. Two sources[6,7] examine the characteristics of the third-order loop and indicate that threshold improvement beyond that achieved by the second-order loop is available but only above a specific system modulation index. This section reviews some of these results and considers some characteristics of the nth-order loop.

The nth-order type-n loop is theoretically formed by incorporating the loop filter

$$H_2(s) = (a_1 s^{n-1} + a_2 s^{n-2} + \cdots + 1)/s^{n-1} \qquad (6\text{-}128)$$

6.3. Threshold-Limited Region

This filter is impractical however (due to the $n-1$-order integration required), but may be approximated by the filter having

$$H_2(s) = (a_1 s^{n-1} + a_2 s^{n-2} + \cdots + 1)/(b_1 s^{n-1} + b_2 s^{n-2} + \cdots + 1) \quad (6\text{-}129)$$

where the denominator represents a cluster (possibly Butterworth) of poles relatively close to the origin. In the region of open-loop unity gain, Eq. (6-128) is a suitable representation of the actual filter characteristic, with b_1 made part of the loop gain constant. The basic advantage of higher-order loops is that, for a given noise bandwidth, greater low-frequency gain may be developed. Modulation compression is thereby achieved more efficiently; or, alternatively, narrower loop bandwidth achieves the required modulation compression or tracking. This is illustrated in Fig. 6-22, where the error responses of second- and third-order loops are plotted as a function of frequency normalized with respect to noise bandwidth. The normalized

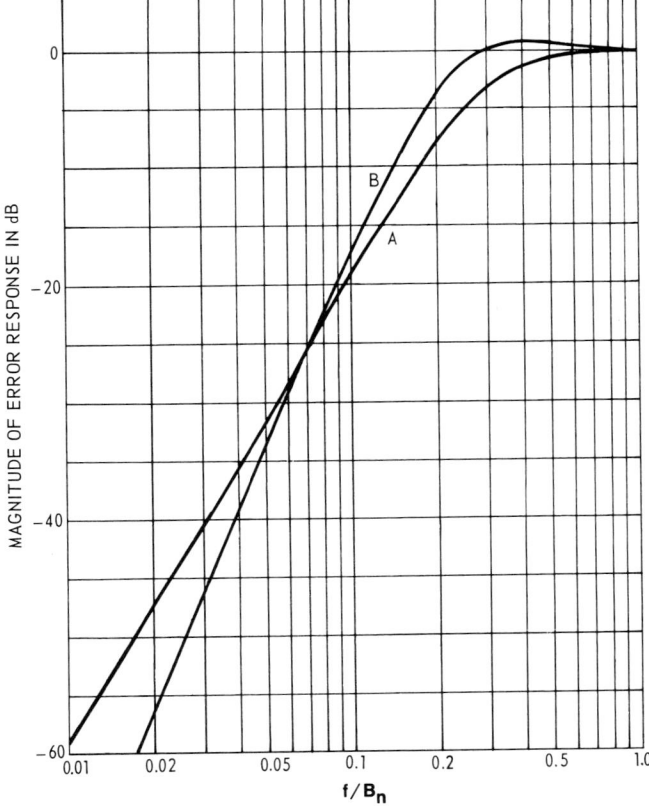

Fig. 6-22. Magnitude of error response as a function of normalized frequency for (A) second- and (B) third-order type-one loops.

crossover frequency is 0.07. For smaller ratios of signal frequency to PLL noise bandwidth, the third-order loop gives better compression. The loop error responses are of the Butterworth type.

F. Other Loop Designs

Several researchers provide further material from which low-threshold design procedures may be established. This section briefly reviews and comments on some of the available material. Develet,[8,9] by a technique developed by Booton, develops and analyzes a quasi-linear model of the PLL. The results are in terms of threshold boundary curves, where "threshold" is defined as the point where the model fails to provide a bounded solution for the mean square phase error. This definition of "threshold" is different from ours, with the result that Develet's "threshold" curves are at a substantially lower CNR (about 6 dB lower) than those presented here.

Examination of Develet's results[9] indicates that his quasi-linear model for the second-order loop predicts a 1-dB departure in mean-square phase error with respect to the linear model at $\alpha = 4$. Since this is the basic threshold criterion incorporated here, substitution of this criterion into Develet's performance equations results in the same performance characteristics as presented here. The analysis is limited to Gaussian signals and Gaussian noise.

The design and analysis presented by Heitzman[4] for sinusoidal modulation is very similar in method to that presented here for a Gaussian baseband. The mainstay of the Heitzman paper is the minimization of the total mean-square phase error, which consists of modulation and noise-induced components. A threshold criterion of $\alpha = 4$ is utilized; however, a loop damping factor $\xi = 0.707$ is specified arbitrarily.

Thomas[10] bases PLL analysis and design on a single-channel speech model resembling the model utilized herein. The threshold analysis is based on the probability that the total loop phase error exceeds $\pi/2$ rad a fixed percentage of the time. Although analysis and design are based on a different threshold criterion, the results are quite close to those given here for the single-channel speech case.

A departure in PLL theory is presented by Schilling;[11] his loop design procedure based primarily on the minimization of the loss of lock rate, does not lead to a minimum noise bandwidth PLL.

6.4. Step-by-Step Design Procedure and Examples

This section consolidates the design data developed in the chapter by first reviewing the design and performance charts and equations. Examples are then considered which illustrate the means to "walk-through" the charts

6.4. Step-by-Step Design Procedure

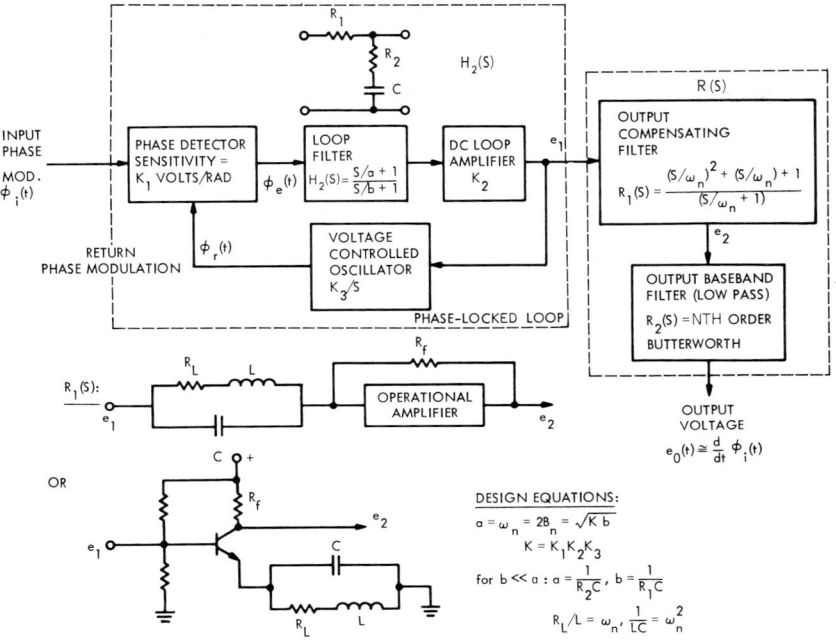

Fig. 6-23. Functional block diagram of phase-locked demodulator.

and establish loop design and system performance on a step-by-step basis, with all calculations carried through to completion. The section is organized so as to be maximally self-contained and to require least back-tracking to other sections. Table IV summarizes the equations necessary for system and loop design. The designs considered in this section are based exclusively on the second-order type-one loop with high gain ($b \ll K, a$), the functional block diagram of which is given in Fig. 6-23. Before considering design in detail, the following general design points are to be noted.

A. Introduction and Summary of Design

The first phase in PLL design is to establish the required noise bandwidth B_n of the loop (Fig. 6-17). This calculation is usually based on the modulation characteristics of the signal, which in turn are based on system considerations such as a required loop output SNR per input CNR (Figs. 6-19 and 6-18). Once the noise bandwidth is established, the gain constant K, loop filter parameters a and b, and the natural frequency ω_n, may be determined (Table IV). From these factors, and fixed interface considerations, the VCO center frequency, together with sensitivity and linearity, is determined. With regard to available input-signal power and input-noise characteristics, the phase detector sensitivity, along with required VCO drive level, may be determined. Finally, with VCO and phase detector sensitivities specified, the required DC amplifier gain is computed to realize the required gain constant K (note K is

TABLE IV SUMMARY OF

	Parameter	Single-channel Speech	Eq. no.	Fig. no.
Above threshold[b]	FM: SNR_{SP}	$6\sigma^2(CNR)_{AM}$	(6-1)	3-8
	PM: SNR_{SP}	$2(\omega_b/\omega_a)\sigma^2(CNR)_{AM}$	(6-2)	
	FM: SNR_{TT}	$3m_p^2(CNR)_{AM}$	(6-3)	3-4
	PM: SNR_{TT}	$m_p^2(\omega_b/\omega_T)^2(CNR)_{AM}$	(6-4)	
	Correction factor P	FM: $(3/f_b^3)\int_0^\infty f^2\|M[j(\omega-\omega_i)]\|^2 \cdot$ $\|H(j\omega)\|^2 \cdot \|R(j\omega)\|^2\,df$ PM: $(1/f_b)\int_0^\infty \|M[j(\omega-\omega_i)]\|^2 \cdot$ $\|H(j\omega)\|^2 \cdot \|R(j\omega)\|^2\,df$	(6-17′) (6-17a′)	6-3 6-4 6-5 6-6
Distortion limit	(Phase det.) $(S/D)_{IM}$	$(1024/m_p^2)(\omega_n/\omega_b)^6$	(6-32)	6-9
Threshold	Speech model FM: $(CNR_{AM})_{TH}$ Test-tone[c] FM or PM: $(CNR_{AM})_{TH}$	$11.7\sigma^{1/2}$ $8.75m_p^{1/2}$	(6-98) (6-105)	6-18 6-18
PLL design equations	Speech model FM: B_n/f_b Test-tone[c] FM or PM: B_n/f_b	$4.7\sigma^{1/2}(\omega_a/\omega_b = \frac{1}{4})$ $2.8m_p^{1/2}$	(6-95) (6-104)	6-17 6-17

[a] For second-order type-one loop (damping factor $= \frac{1}{2}$), $a = \omega_n = 2B_n = (Kb)^{1/2}$, Eq. (6-70).

[b] Threshold is defined as the point where NPR and SNR degrades by 1 dB below these values.

[c] Test-tone frequency $\omega_T = \omega_b/4$.

6.4. Step-by-Step Design Procedure

the product of VCO, phase detector sensitivities, and DC loop gain).

Not all specifications for the loop elements can be uniquely determined. Some factors are a function of engineering judgment and available technology. The following rules-of-thumb based upon currently common transistor circuit characteristics serve as guides, and are not to be considered inviolable.

PLL Performance and Design Equations[a]

	Parameter	Frequency Division multiplex	Eq. no.	Fig. no.
Above threshold[b]	FM: NPR	$\dfrac{2(\omega_b/\omega_{CH})^2 \sigma^2}{[1-(\omega_a/\omega_b)]}(CNR)_{AM}$	(6-5)	
	PM: NPR	$\dfrac{6\sigma^2}{[1-(\omega_a/\omega_b)^3]}(CNR)_{AM}$	(6-6)	
	FM & PM: SNR_{TT}	$(f_b/B)(\omega_b/\omega_{CH})^2 m_p^2 (CNR)_{AM}$	(6-7)	
Distortion limit	VCO: NPR	$10^5/[2\Delta^2 W_{dn}(f)]$	(6-43)	6-12
Threshold	FM: $(CNR_{AM})_{TH}$	$12.6\sigma^{1/2}$	(6-97)	6-18
	PM: $(CNR_{AM})_{TH}$	$14.6\sigma^{1/2}$	(6-96)	6-18
PLL design equations	FM: B_n/f_b	$5.05\sigma^{1/2}$	(6-94)	6-17
	PM: B_n/f_b	$5.85\sigma^{1/2}$	(6-93)	6-17

A. VCO Center Frequency

In order to restrict carrier frequency components and harmonics (due to phase detector unbalance) from propagating around the loop, and causing undesirable effects (amplifier saturation and spurious VCO modulation), the VCO frequency, as well as the input center frequency, should be 30–100 times the noise bandwidth of the loop. Furthermore, the VCO linearity and sensitivity will depend upon the center frequency ω_i, as described in Section 3.5, where typical available values are listed. The linearity is specified in terms of the peak-to-peak deviation, and more conservatively (especially for smaller modulation indices), in terms of Carson's rule bandwidth.

B. Phase Detector Sensitivity

Sufficient dynamic range to handle noise peaks is achieved by requiring the input drive level from the VCO to be at least 13 dB above the input signal-plus-noise rms level, the level being set at the threshold point. Assuming a PLL operating at an input threshold CNR of 5 dB, the signal-plus-noise rms level is 1.2 dB above the signal rms level, indicating a maximum input signal level at threshold of 13 + 1.2 = 14.2 dB below the input level from the VCO. Nominal inputs from the VCO of 5 V peak-to-peak restrict signal inputs to approximately 1 V peak-to-peak. Typically a phase detector has a signal loss of about 6 dB, resulting in a sensitivity of about 250 mV/rad.

C. Amplifier Gain, Dynamic Range, and Bandwidth

The amplifier must be DC coupled to achieve the locked condition in the PLL. Because of this factor, amplifier gain is normally restricted to a range of 10 to 100. The dynamic range is determined by the peak voltage available from the phase detector, which is nominally one-half the peak value of the loop input signal-plus-noise for a passive phase detector. By placing the loop filter before the DC amplifier, appreciable noise as well as rf may be eliminated from the amplifier input, thus requiring a dynamic range based primarily on the peak input signal level to the amplifier. For the example cited, a peak signal of 250 mV is indicated, and for an amplifier gain of 20, a requirement of ± 5 V output dynamic range is established. The amplifier 3-dB bandwidth f_{3dB} must be wide enough to pass the baseband with negligible phase shifting. The angle at the top of the baseband θ_b is given approximately by

$$\theta_b \cong (f_b/f_{3dB}) \times 57.4° \tag{6-130}$$

assuming $\tan \theta_b \cong \theta_b$, and a single dominant pole.

D. Filter Components

The final step in the design is to establish the filters required within and following the loop. The filter utilized to achieve the second-order type-one

6.4. Step-by-Step Design Procedure

loop is of the form shown in Fig. 6-23, with the parameters related to the loop design parameters by

$$a = 1/R_2 C \quad \text{and} \quad b = 1/(R_1 + R_2)C \cong 1/R_1 C$$

A unique solution for the filter does not exist, as there are three unknowns and two fixed parameters (a, b). Normally, the ratio R_1/R_2 is determined from

$$R_1/R_2 \cong a/b$$

Values of R_1 and R_2 are then chosen, together with consideration of the resulting value of C and interface impedances of the phase detector and DC amplifier.

The filter that follows the loop should usually contain a section which compensates for the closed-loop response of the PLL. This may be achieved by the filter shown in Fig. 6-23, with the parameters related to the PLL parameters by

$$R_L/L = \omega_n \quad \text{and} \quad 1/LC = \omega_n^2$$

Here again there is no unique solution. Choosing one parameter for convenience, fixes also the other two.

With these factors in mind, a brief review of the PLL design curves is in order before going on to detailed design examples. The key figures necessary to carry out a basic PLL design are Figs. 6-17–6-19. Refinements in the design or system calculations may be obtained from other figures appearing in the chapter but they are not repeated here. The key figures are summarized below.

Figure 6-17: The PLL closed-loop noise bandwidth may be determined for the various modulation cases covered in this chapter, in terms of the rms modulation index, which is defined as the rms frequency deviation/top baseband frequency. As an example, consider $\sigma = 10$ for FDM-FM; the design noise bandwidth is found to be $16 f_b$.

Figure 6-19: This curve presents the output SNR or worst-channel NPR (for FDM), with the 1-dB threshold degradation factor included, as a function of system rms modulation index for the various modulation cases considered. The curve may be used in two ways: First, given the modulation index, the output SNR (or NPR) may be determined. As an example, for $\sigma = 1$, the NPR for FDM-PM in the top channel at threshold is 18 dB. As an alternate use of this figure, the required system modulation index may be determined given the desired threshold SNR (or NPR).

Figure 6-18: This curve is similar to Fig. 6-19, except that its purpose is to provide threshold CNR. As an example, for speech modulation with $\sigma = 5$, the threshold CNR (referred to twice baseband bandwidth) is 14.2 dB.

B. Design Examples: Single-Channel Signals

This section consolidates the design data developed in the chapter by examining design problems on a step-by-step basis. The first set of design problems deals with PLL design for single-channel speech applications.

Design Problem I

For a second-order type-one PLL FM (speech) demodulator: (1) evaluate threshold performance, both SNR and CNR, for a 1-kHz test tone; (2) specify the design parameters; and (3) calculate the distortion characteristics.

A. GIVEN

Baseband	300–4000 Hz
Peak signal deviation	10 kHz
Receiver IF noise bandwidth	35 kHz

B. SOLUTION

Step 1: System modulation index.

$$m_p = 10\text{kHz}/4\text{kHz} = 2.5$$

$$\sigma_{TT} = 2.5/2^{1/2} = 1.77$$

$$\sigma_{sp} \text{ (speech model)} = m_p/10^{1/2} = 0.79$$

Step 2: Threshold CNR. From Fig. 6-18 directly (using σ_{TT}), or Eq. (6-105), we have as the TT CNR threshold

$$(\text{CNR}_{AM})_{TH} = 8.75 m_p^{1/2} = 8.75(2.5)^{1/2} = 13.8 \quad \text{or} \quad 11.4 \quad \text{dB}$$

Referring this threshold CNR to the 35-kHz IF bandwidth, we have

$$(\text{CNR}_{IF})_{TH} = (\text{CNR}_{AM})_{TH} \cdot (2f_b/B_{IF}) = 13.8(8\text{kHz}/35\text{kHz})$$
$$= 3.16 \quad \text{or} \quad 5 \quad \text{dB}$$

This represents a *5-dB improvement* with respect to the normally encountered 10-dB threshold for the LD.

Step 3: Threshold SNR. From Fig. 6-19 directly, or Eq. (6-3) (after deducting 1 dB) we have as the TT SNR at threshold

$$(\text{SNR}_{TT})_{TH} = (3m_p^2/1.26)(\text{CNR}_{AM})_{TH} = 205 \quad \text{or} \quad 23.1 \quad \text{dB}†$$

† This calculation assumes ideal postdemodulation filtering, which provides for perfect compensation for the peaking in the PLL response, and results in rectangular filtering. See Section 6.1 for the effect of nonideal postdemodulation filtering. For example, if $R(j\omega)$ is simply a sixth-order Butterworth type, the penalty in SNR is about 2 dB [see Eq. (6-18)].

6.4. Step-by-Step Design Procedure

Step 4: Loop design and calculation of loop parameters. From Fig. 6-17, or Eq. (6-104), the PLL design noise-bandwidth based on TT modulation is

$$B_n = 2.8 f_b\, m_p^{1/2} = 2.8(4\text{kHz})(2.5)^{1/2} = 17.7 \quad \text{kHz}$$

The gain constant and filter characteristic for minimum noise bandwidth are now obtained as

$$K = K_1 K_2 K_3, \qquad H_2(s) = (s/a + 1)/(s/b + 1)$$

with

$$a = \omega_n = 2B_n = (Kb)^{1/2} \qquad (6\text{-}131)$$

Therefore,

$$a = 2(17.7 \times 10^3) = 35.4 \quad \text{krad/sec}$$

The bottom of the baseband is 300 Hz and, therefore, maximum gain across the baseband is guaranteed by setting the pole at 300 Hz or less, thus

$$b = 1.88 \quad \text{krad/sec}$$

Finally, the gain constant for the loop is established from Eq. (6-131)

$$K = 4B_n^2/b = (4)(17.7)^2 \times 10^6/(1.88)(10^3) = 6.62 \times 10^5 \quad \text{sec}^{-1} \qquad (6\text{-}132)$$

Note that K represents the product of the phase detector sensitivity, amplifier gain, and VCO sensitivity. The loop design parameters are reduced to the circuit design specifications given below by referring to the general comments made earlier.

i. VCO

Type	Varactor controlled
Center frequency	1 MHz
1% linear range	±10 kHz
Sensitivity	25 kHz/V
Output level	5 V peak-to-peak

ii. INPUT

Center frequency	1 MHz
Level	1 V peak-to-peak (signal only)
Phase detector sensitivity	250 mV/rad

iii. DC AMPLIFIER

Gain	$\dfrac{K}{K_1 K_3} = \dfrac{6.62 \times 10^5}{(0.25)(2\pi)(25)10^3} = 16.9$ or 24.6 dB
Output dynamic range	13.5 V peak-to-peak
3-dB bandwidth	DC to 500 kHz

iv. FILTERS

Loop filter $H_2(s)$:
For convenience, choose $C = 0.03 \ \mu F$.

Then $R_1 = \dfrac{1}{bC} = \dfrac{1}{1.88 \times 10^3 \times 3 \times 10^{-8}} = 17.7 \ \text{k}\Omega$

and $R_2 = \dfrac{1}{aC} = \dfrac{1}{35.4 \times 10^3 \times 3 \times 10^{-8}} = 0.94 \ \text{k}\Omega$

Compensating filter $R_1(s)$:
For convenience, choose $C = 0.1 \ \mu F$.

Then $L = \dfrac{1}{\omega_n^2 C} = \dfrac{1}{(35.4)^2 \times 10^6 \times 10^{-7}} = 8 \ \text{mH}$

and $R_L = \omega_n L = 35.4 \times 10^3 \times 8 \times 10^{-3} = 280 \ \Omega$

Step 5: Maximum SNR. The maximum SNR is determined by the distortion characteristics of the phase detector and the VCO. The maximum SNR allowed by the phase detector is derived by referring to Fig. 6-9, or Eq. (6-32),

$$\left(\dfrac{S}{D}\right)^\dagger_{\text{IM}} = \dfrac{1024}{m_p^2}\left(\dfrac{\omega_n}{\omega_b}\right)^6 = \dfrac{1024}{(2.5)^2}\left(\dfrac{35.4 \times 10^3}{2\pi \times 4 \times 10^3}\right)^6 = 1.3 \times 10^3 \ \text{or} \ 62 \ \text{dB}$$

This indicates that the VCO will be the limiting element at high SNR, since its linearity was specified as 1% only.

For comparison, we repeat the calculations based now on the speech model.

Step 2: Threshold CNR. From Fig. 6-18 directly, or Eq. (6-98), we have as the speech model CNR threshold

$$(\text{CNR}_{\text{AM}})_{\text{TH}} = 11.7\sigma^{1/2} = 10.4 \ \text{or} \ 10.2 \ \text{dB}$$

which is 1.2 dB lower than the TT case.

Step 3: Threshold SNR. From Fig. 6-19 directly, or Eq. (6-1) after deducting 1 dB, we have

$$(\text{SNR}_{\text{SP}})_{\text{TH}} = \dfrac{6\sigma^2}{1.26}(\text{CNR}_{\text{AM}})_{\text{TH}} = \dfrac{(6)(0.79)^2(10.4)}{1.26} = 30.8 \ \text{or} \ 14.9 \ \text{dB}$$

which is *8.2 dB lower* than for TT. (Note that the speech model, for the same peak deviation, has 7 dB less average power than the test-tone.)

Step 4: Loop design. From Fig. 6-17 directly, or Eq. (6-95), the PLL design noise bandwidth, based on speech model modulation is

$$B_n = 4.7 f_b \sigma^{1/2} = 4.7(4\text{kHz})(0.79)^{1/2} = 16.7 \ \text{kHz}$$

† The term $(S/D)_{\text{IM}}$ is given as a voltage ratio.

6.4. Step-by-Step Design Procedure

Note the closeness of the design noise bandwidths based on TT and speech model (17.7 kHz vs 16.7 kHz, respectively). This result indicates one aspect in the compatibility of a 1-kHz TT for single-channel systems which use PLL demodulators.

The remainder of the design proceeds similarly to that for the TT case.

Design Problem II

For a second-order type-one PLL FM demodulator, establish loop and system design parameters to achieve a 13-dB output TT SNR at threshold.

A. GIVEN

Baseband   300–4000 Hz
Test-tone frequency 1 kHz

B. SOLUTION

Step 1: System modulation index. Anticipating nonideal filter characteristics following the PLL, we apply a 0.5-dB penalty and, therefore, require 13.5-dB threshold SNR. From Fig. 6-19 directly, or Eqs. (6-3) and (6-105), with inclusion of the 1-dB SNR degradation factor, the system modulation index is derived as follows:

$$(SNR_{TT})_{TH} = 3m_p^2 (CNR_{AM})_{TH}/1.26 \qquad \text{at threshold}$$

$$(CNR_{AM})_{TH} = 8.75 m_p^{1/2} \qquad \text{at threshold}$$

Combining the equations,

$$m_p = \left[\frac{(SNR_{TT})_{TH}(1.26)}{(3)(8.75)}\right]^{0.4} = \left[\frac{(22.4)(1.26)}{(3)(8.75)}\right]^{0.4} = 1.03$$

To use Fig. 6-19 set $m_p = 2^{1/2}\sigma$.

Step 2: Threshold CNR. From Fig. 6-18 directly, or Eq. (6-105), we have

$$(CNR_{AM})_{TH} = 8.75 m_p^{1/2} = 8.9 \quad \text{or} \quad 9.5 \quad dB$$

By Carson's rule, Eq. (3-25), the recommended IF bandwidth preceding the PLL is obtained from the following equation (or Fig. 6-17)

$$B_{IF} = 2f_b(1 + m_p) = 8(2.03) = 16.3 \quad kHz$$

The threshold CNR referred to this bandwidth is

$$(CNR_{IF})_{TH} = (CNR_{AM})_{TH} \cdot \frac{2f_b}{B_{IF}} = (8.9)(8/16.3) = 4.38 \quad \text{or} \quad 6.4 \quad dB$$

Step 3: Loop design. From Fig. 6-17 directly, or Eq. (6-104), the PLL design noise bandwidth is derived as

$$B_n = 2.8 f_b m_p^{1/2} = (2.8)(4\text{kHz})(1.03)^{1/2} = 11.4 \quad \text{kHz}$$

The second-order type-1 PLL requires the following gain constant and filter characteristics for minimum noise bandwidth,

$$K = K_1 K_2 K_3, \qquad H_2(s) = (s/a + 1)/(s/b + 1)$$

with

$$a = \omega_n = 2B_n = (Kb)^{1/2} = 22.8 \times 10^3 \quad \text{rad/sec}$$

The bottom of the baseband is 300 Hz and, therefore, maximum gain across the baseband is derived by setting the pole at 300 Hz or less; thus, with

$$b = 1.88 \times 10^3 \quad \text{rad/sec,}$$

the gain constant may be established as

$$K = 4B_n^2/b = (4)(11.4)^2 10^6/(1.88)10^3 = 2.75 \times 10^5 \quad \text{sec}^{-1}$$

Similarly to design problem I, the following circuit design specifications may be set down.

i. VCO

Type	Varactor controlled
Center frequency	500 kHz
1% linear range	± 4.15 kHz
Sensitivity	12 kHz/V
Output level	5 V peak-to-peak

ii. INPUT

Center frequency	500 kHz
Level	1 V peak-to-peak (signal only)
Phase detector sensitivity	250 mV/rad

iii. DC AMPLIFIER

Gain	$\dfrac{K}{K_1 K_3} = \dfrac{2.75 \times 10^5}{(0.25)(2\pi)(12)10^3} = 14.7$ or 23.4 dB
Output dynamic range	8 V peak-to-peak
3-dB bandwidth	DC to 250 kHz

iv. FILTER

Loop filter $H_2(s)$
$R_1 = 10.5$ kΩ, $R_2 = 870$ Ω
$C = 0.05$ μF

Compensating filter $R_1(s)$
$R_2 = 160$ Ω, $L = 7$ mH
$C = 0.275$ μF

C. Design Examples: FDM Signals

This section illustrates the PLL design curves and equations as applied to frequency (and phase) modulated signals with FDM basebands. The following example considers FDM-FM primarily; however, FDM-PM is amenable to the same design approach and analysis, with somewhat different performance results.

The problem is to specify signal parameters and PLL design parameters, and to determine (1) worst-channel threshold NPR, TT SNR and CNR; (2) VCO linearity to achieve worst-channel high-level NPR = 45 dB; and (3) evaluate performance degradation in the presence of loop delay.

A. Given

Modulation	600-channel FDM-FM (4 kHz bandwidth/channel)
Baseband	60 kHz to 2.54 MHz, 10-dB peak-to-rms amplitude ratio
System IF bandwidth	50 MHz

B. Solution

Step 1: System modulation index. The given IF bandwidth, in conjunction with the Carson's rule bandwidth design [Eq. (3-25)], predicts the system modulation index

$$B_{IF} = 2f_b(10^{1/2}\sigma + 1) \quad \text{or} \quad \sigma = \frac{1}{10^{1/2}}\left(\frac{B_{IF}}{2f_b} - 1\right) = \frac{8.85}{10^{1/2}} = 2.8$$

The peak and rms frequency deviations of the carrier are

$$\Delta\omega_{rms}/2\pi = f_b \sigma = (2.54)(2.8) = 7.1 \quad \text{MHz}$$

$$\Delta\omega_p/2\pi = 10^{1/2}(\Delta\omega_{rms}/2\pi) = 22.5 \quad \text{MHz}$$

Step 2: Threshold CNR. From Fig. 6-18 directly, or Eq. (6-97), the threshold CNR for the top (in this case, the worst FDM) channel is

$$(CNR_{AM})_{TH} = 12.6\sigma^{1/2} = 12.6(2.8)^{1/2} = 21.2 \quad \text{or} \quad 13.2 \quad \text{dB} \qquad (6\text{-}133)$$

which, when referred to the IF bandwidth, results in an IF CNR of

$$(CNR_{IF})_{TH} = (2f_b/B_{IF})(CNR_{AM})_{TH} = (21.2)(5.08/50) = 2.15 \quad \text{or} \quad 3.3 \text{ dB}$$
$$(6\text{-}134)$$

Step 3: Threshold NPR and TT SNR. From Fig. 6-19 directly, or Eq. (6-5), the NPR at threshold, after applying the 1-dB threshold degradation factor for the top channel is

$$(NPR)_{TH} = 2(\omega_b/\omega_{CH})^2 \sigma^2 (CNR_{AM})_{TH}/[(1.26)(1 - \omega_a/\omega_b)]$$
$$\cong [2(2.8)^2/1.26](21.2) = 264 \quad \text{or} \quad 24.2 \quad \text{dB}$$

6. Design of PLL for FM Demodulation

We note the relation between psophometrically weighted TT SNR and NPR,[12]

$$\text{SNR}_{TT} = \text{NPR} + \text{BWR} - L + 2.5 \quad \text{dB} \qquad (6\text{-}135)$$

where SNR_{TT} is channel TT SNR, NPR is noise power ratio, BWR is base bandwidth to channel bandwidth ratio, and L is multiplex noise loading ratio.†

Referring to the CCIR recommended loading formula for FDM,

$$L(\text{dB}) = -15 + 10 \log_{10} N$$

where N is the number of FDM channels, we have

$$L = -15 + 10 \log_{10} 600 = 12.8 \quad \text{dB}$$

$$\text{BWR} = 10 \log_{10} \frac{f_b - f_a}{B} = 10 \log_{10} \frac{2.54 - 0.06}{0.004} = 27.92 \quad \text{dB}$$

where B is the channel bandwidth in Hz, resulting in

$$\text{SNR}_{TT} = 24.2 + 27.92 - 12.8 + 2.5 = 41.8 \quad \text{dB} \qquad \text{(at threshold)}$$

Step 4: PLL design. From Fig. 6.17 directly, or Eq. (6-94), the PLL design noise bandwidth for FDM-FM is

$$B_n = 5.05 f_b \sigma^{1/2} = (5.05)(2.54)(2.8)^{1/2} = 21.5 \quad \text{MHz}$$

The second-order type-one PLL has the following gain constant and filter characteristic

$$K = K_1 K_2 K_3, \qquad H_2(s) = (s/a + 1)/(s/b + 1)$$

For a minimum-bandwidth PLL

$$a = \omega_n = 2B_n = (Kb)^{1/2} = 43 \quad \text{Mrad/sec}$$

To realize maximum gain across the baseband frequency components, b should not be greater than 376 krad/sec. With $b = 376$ krad/sec the gain constant K is established as

$$K = 4B_n^2/b = (4)(21.5)^2 \times 10^{12}/(3.76 \times 10^5) = 4.9 \times 10^9 \quad \text{sec}^{-1}$$

Step 5: VCO linearity. The required worst-channel NPR at high input CNR is given as 45 dB. The minimum VCO linearity may be derived by application of Eq. (6-43) and reference to Fig. 6-12. We have

$$\text{NPR} = 10^5/[2\Delta^2 W_{dn}(f)] \qquad \text{or} \qquad \Delta = [10^5/2\text{NPR}\, W_{dn}(f)]^{1/2} \%$$

Note from Fig. 6-12 that the bottom channels experience the largest values of

---

† This is the dB increase of baseband noise loading power over a 1-mW TT at the 0-dBm0 point in the system.[12]

6.4. Step-by-Step Design Procedure

$W_{dn}(f)$ and, thus, the lowest NPR for a given Δ. For the bottom channels, we have $W_{dn}(f) \cong 1$, and

$$\Delta = (10^5/6.32 \times 10^4)^{1/2} = 1.26\%$$

indicating that the VCO must be within $\pm 1.26\%$ of linear deviation over the required peak-to-peak deviation of 45 MHz. An order of magnitude check in Fig. 6-19 indicates that phase detector distortion will not be the limiting factor.

Step 6: Circuit design. The loop design parameters may be reduced to the following circuit design specifications by referring to the general comments made earlier.

i. VCO

Type	Varactor controlled
Center frequency	4 GHz
1.26% linear range	± 22.5 MHz
Sensitivity	100 MHz/V
Output level	2 V peak-to-peak (10 dBm across 50 Ω)

ii. INPUT

Center frequency	4 GHz
Level	0.4 V peak-to-peak (signal only)
Phase detector sensitivity	0.1 V/rad

iii. DC AMPLIFIER

Gain $\quad \dfrac{K}{K_1 K_3} = \dfrac{4.9 \times 10^9}{(2\pi)100 \times 10^6 \times 0.1} = 78$

Output dynamic range 16 V peak-to-peak
3-dB bandwidth 72.5 MHz (for 2° excess phase shift at 2.54 MHz, assuming a single dominant pole).

Before proceeding it is worthwhile to note that the DC amplifier specification is unattainable because of the large gain and bandwidth required (i.e., 5.7 GHz gain-bandwidth product). The solution is to lower the required gain constant of the loop by increasing the pole frequency b of the loop filter. With the availability of 2-GHz transistors, an increase in b by a factor of 5 to 1.88×10^6 rad/sec will do, and is permissible, as the filter pole is still more than a decade below the filter zero a. The closed-loop response and, hence, noise bandwidth as well as the modulation phase-error component, are only slightly affected. The readjusted specifications are

loop gain constant $K = 9.85 \times 10^8$
loop filter pole frequency = 300 kHz
loop filter zero frequency = 6.88 MHz

Thus, the DC amplifier specifications are now

Gain $\quad\dfrac{K}{K_1K_3} = \dfrac{9.85 \times 10^8}{(2\pi)(10^8)(0.1)} = 15.6$

Output dynamic range $\quad$ 3.1 V peak-to-peak

The circuit design parameter calculations are concluded with the filter constants.

iv. FILTER

Loop filter $H_2(s)$ $\quad\quad\quad\quad$ *Compensating filter $R_1(s)$*
$R_1 = 1.05\ \text{k}\Omega,\ R_2 = 50\ \Omega$ $\quad\quad R_L = 430\ \Omega,\ L = 10\ \mu\text{H}$
$C = 465\ p\text{F}$ $\quad\quad\quad\quad\quad\quad\quad C = 54\ \text{pF}$

Step 7: Delay calculations. Both threshold CNR and the loop design are functions of the parasitic delay found in the PLL implementation. Recall that in realizing the DC amplifier, an inherent $2°/f_b$ excess phase shift was accepted. If we assume that the additional delay contributed by the other loop elements and the electrical path of the loop results in an additional $2°/f_b$ excess phase shift, then we have an overall delay of $4°/f_b$, and a threshold degradation (from Fig. 6-21) of 1.2 dB. The loop may be optimized by readjusting the loop filter zero a by a factor of 0.975 (from Fig. 5-16, set delay phase $\phi = 4° \times \omega_n/\omega_b = 4 \times 6.88/2.54 = 10.8°$) resulting in $a = 42$ Mrad/sec. Slight readjustment of loop filter circuit parameters is indicated, specifically $R_2 = 51\ \Omega$. In the presence of delay, the PLL performance is modified to be

$(\text{CNR}_{\text{AM}})_{\text{TH}} = 14.55 \quad \text{dB} \quad$ [Fig. 6-21]

$(\text{CNR}_{\text{IF}})_{\text{TH}} = 4.6 \quad \text{dB} \quad$ [Eq. (6-134)]

$(\text{NPR}_{\text{TH}}) = 25.15 \quad \text{dB} \quad$ [Fig. 6-21 less 1 dB]

$(\text{SNR}_{\text{TT}})_{\text{TH}} = 43.1 \quad \text{dB} \quad$ [Eq. (6-135)]

References

1. H. Chestnut and R. W. Mayer, "Servomechanism and Regulating System Design," Section 11.3. Wiley, New York, 1951.
2. Transmission Systems for Communication, p. 534. Bell Telephone Labs., Murray Hill, N.J., 1964.
3. W. B. Davenport and W. L. Root, "Random Signals and Noise," Section 12-2. McGraw-Hill, New York, 1958.
4. R. E. Heitzman, A study of the threshold power requirements of FMFB receivers. *IRE Trans.* **SET-8**, 249–256 (1962).
5. A. J. Viterbi, Phase-locked loop dynamics in the presence of noise by Fokker-Planck techniques. *Proc. IRE* **51**, No. 12 1737–1753 (1963).

6. S. C. Gupta and R. J. Solem, Optimum filters for second- and third-order phase-locked loops by an error-function criterion. *IEEE Trans.* **SET 11**, 54–62 (1965).
7. E. J. Baghdady *et al.*, *Advanced threshold reduction techniques study*. 1st Quart. Rep., Adcom, Inc., NASA Rep. N65-30842, 1964.
8. J. A. Develet, A threshold criterion for phase-locked demodulation. *Proc. IEEE* **51**, 349–356 (1963).
9. J. A. Develet, An analytic approximation of phase-lock receiver threshold, *IEEE Trans.* **SET 9**, 9–12 (1963).
10. C. M. Thomas, Optimization of phase-lock demodulator for single channel voice. *Proc. IEEE Int. Comm. Conf.*, June 1966.
11. B. S. Abrams, J. F. Oberst, M. Berkoff, and D. Schilling, *Phase-locked loop threshold investigations*. No. 73 PIBMRI-1274-65. Polytech. Inst. of Brooklyn, Brooklyn, New York, June 1965.
12. Int. Radio Consultative Comm. (CCIR), Recommendation #393. *Docu. Plenary Assembly, 10th, 1963*, **4**. Int. Telecommun. Union, Geneva, 1963.

CHAPTER

7

Design of Frequency-Feedback Loops for FM Demodulation

Paralleling the format of Chapter 6, we present the design concepts supplemented by the necessary design curves to carry forth low-threshold FMFB loop design. The design procedure is based on the "two-thresholds" concept developed by Enloe,[1a,b] with some modifications based on recent experimental results. A discussion is included of factors not adequately covered by theory, with a qualitative evaluation of their influence on loop performance and design. The chapter concludes with several inclusive examples.

By referring to the reference receiving system discussed in Chapter 6, and depicted in Fig. 7-1, we can evaluate the performance characteristics of the FMFB in terms of system performance and loop design. Similar to the conventional FM detector (Chapter 3) and the PLL (Chapter 6), the system will display the distinctive performance curve shown in Fig. 7-2, which illustrates the three basic regions of operation.

7.1. FM Improvement Region

The central portion of the performance curve given in Fig. 7-2 is called the "linear SNR-CNR" or "FM improvement" region, which is derived from the above-threshold detection characteristic of the FMFB. This is seen in Chapter 4 to be equivalent to that of a frequency detector (operating above its threshold) followed by a filter that represents the FMFB closed-loop response, $H(j\omega)$. The same result was found for the PLL, indicating that the

7.1. FM Improvement Region

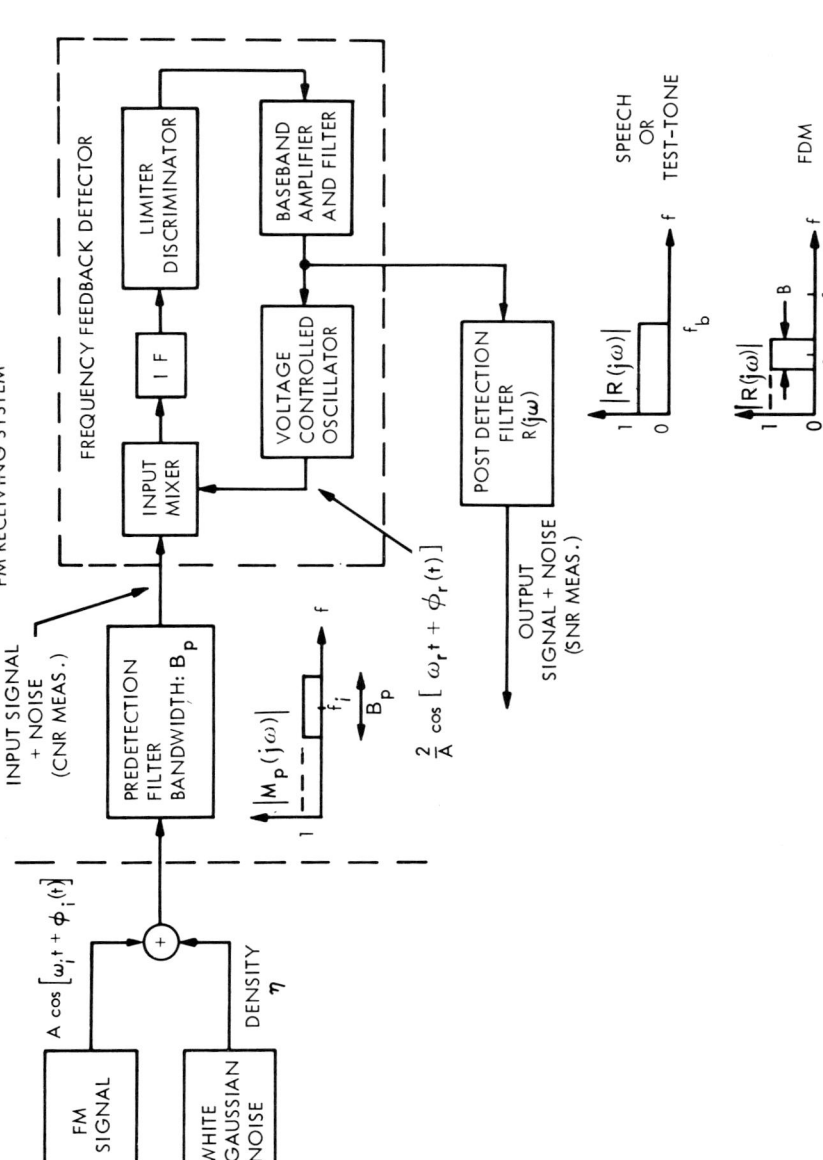

Fig. 7-1. FM receiving system employing frequency feedback demodulator.

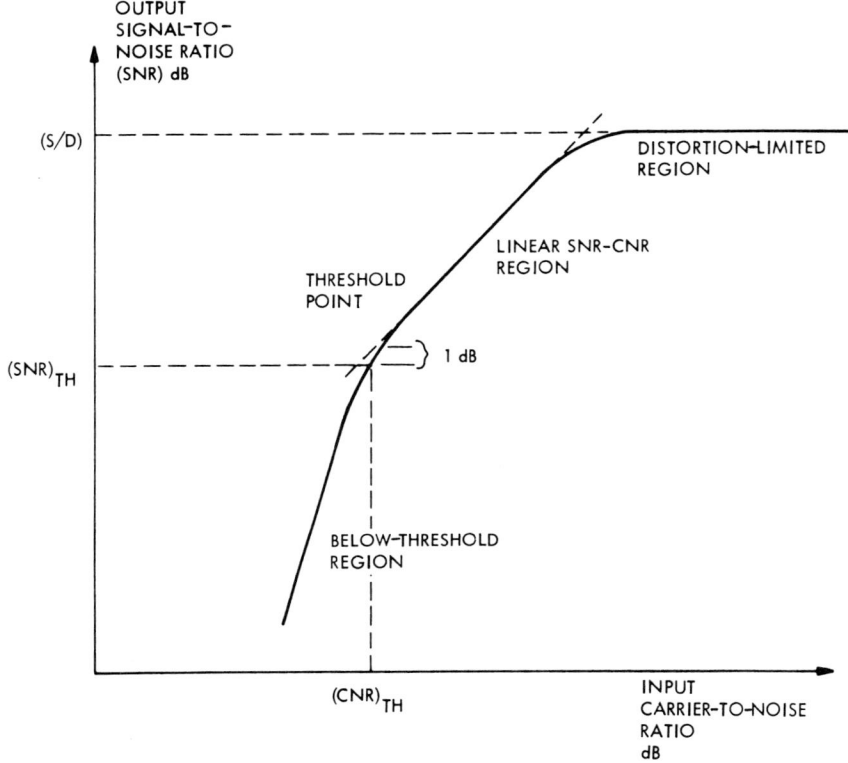

Fig. 7-2. Typical performance characteristic of FM receiving system.

comment and discussion given there apply also to this case, with the performance characteristics summarized by:

A. SINGLE-CHANNEL SPEECH (FM AND PM)

$$\text{SNR}_{\text{SP}} = 6\sigma^2(\text{CNR})_{\text{AM}} \qquad \text{(FM)} \qquad (7\text{-}1)$$
$$\text{SNR}_{\text{SP}} = 2(\omega_b/\omega_a)\sigma^2(\text{CNR})_{\text{AM}} \qquad \text{(PM)} \qquad (7\text{-}2)$$
$$\text{SNR}_{\text{TT}} = 3m_p^2(\text{CNR})_{\text{AM}} \qquad \text{(FM)} \qquad (7\text{-}3)$$
$$\text{SNR}_{\text{TT}} = m_p^2(\omega_b/\omega_T)^2(\text{CNR})_{\text{AM}} \qquad \text{(PM)} \qquad (7\text{-}4)$$

B. FREQUENCY DIVISION MULTIPLEX† (FM AND PM)

$$\text{NPR} = \frac{2(\omega_b/\omega_{\text{CH}})^2\sigma^2}{[1 - \omega_a/\omega_b]}(\text{CNR})_{\text{AM}} \qquad \text{(FM)} \qquad (7\text{-}5)$$

† High-Q channels are assumed: $2\pi B \ll \omega_{\text{CH}}$.

7.1. FM Improvement Region

$$\text{NPR} = \frac{6\sigma^2}{[1 - (\omega_a/\omega_b)^3]} (\text{CNR})_{AM} \quad (\text{PM}) \qquad (7\text{-}6)$$

$$\text{SNR}_{TT} = \left(\frac{f_b}{B}\right)\left(\frac{\omega_b}{\omega_{CH}}\right)^2 m_p{}^2(\text{CNR})_{AM} \quad \binom{\text{FM}}{\text{PM}} \qquad (7\text{-}7)$$

The results for output SNR correction factors that account for nonideal filtering characteristics are not identical for the PLL and FMFB, since their closed-loop responses $H(s)$ differ:

$$H(s) = \frac{[(s/\omega_n) + 1]}{[(s/\omega_n)^2 + (s/\omega_n) + 1]} \quad (\text{PLL}) \qquad (6\text{-}68) \quad (7\text{-}8)$$

$$H(s) = \frac{K_0[(s/a_0\omega_n) + 1]}{[(s/\omega_n)^2 + 2\xi_0(s/\omega_n) + 1]}, \quad (\text{FMFB}) \qquad (4\text{-}52) \quad (7\text{-}9)$$

the difference being contained in the constants a_0, K_0, and ξ_0 given in Eqs. (7-11), (7-12), and (7-13) (derived in Appendix F). It is observed that in the limit of very high gain, the three constants of the FMFB approach the values of their counterparts in the PLL ($a_0 \to 1$, $K_0 \to 1$, $\xi_0 \to \frac{1}{2}$). Of more immediate interest, however, is the fact that even at moderate gains the value $\xi_0 = \frac{1}{2}$ is a good approximation.

The following example calculates the appropriate SNR correction factors in terms of FMFB characteristics:

Example: Output noise correction factor (FMFB)

Consider the following FM single-channel system characteristics

$$H(s) = \frac{K_0[(1/a_0)(s/\omega_n) + 1]}{(s/\omega_n)^2 + 2\xi_0(s/\omega_n) + 1} = \text{second-order FMFB closed-loop response (Enloe design)} \qquad (7\text{-}10)$$

$$a_0 = 2^{1/2}\frac{F^{1/2}}{F-1}\left\{1 + \left[1 + \frac{(F-1)^2}{2F}\right]^{1/2}\right\} = \frac{a}{F^{1/2}} \quad \text{(see Appendix F)} \qquad (\text{F-9}) \quad (7\text{-}11)$$

$$K_0 = K/(1 + K), \qquad F = 1 + K \qquad (7\text{-}12)$$

$$\xi_0 = \frac{1}{2}\left[\frac{2^{1/2} + (F-1)/(a_0 F^{1/2})}{F^{1/2}}\right] \qquad (\text{F-10}) \quad (7\text{-}13)$$

$$|R(j\omega)|^2 = \frac{1}{1 + (\omega/\omega_b)^{2n}} = \text{post-FMFB filter, } n\text{th order Butterworth with cutoff frequency } \omega_b \qquad (7\text{-}14)$$

and

$$|M_{\text{pL}}\,j(\omega)|^2 = \begin{cases} 1, & 0 \leq f \leq B_p/2 \\ 0, & f > B_p/2 \end{cases} = \text{predetection filter} \quad (7\text{-}15)$$
(crystal) lowpass
equivalent

From Eq. (6-9) and letting $\xi_0 = \frac{1}{2}$, the output noise is derived similarly to Eq. (6-14) as

$$N_{\text{po}} = \frac{4\pi(2\eta)f_b{}^3 K_0}{A^2\mu^2} \int_0^{b'} \frac{[(x/a_0 a')^2 + 1]}{[1 - (x/a')^2]^2 + (x/a')^2} \cdot \frac{x^2\,dx}{1 + (x)^{2n}} \quad (7\text{-}16)$$

where

$$x = \omega/\omega_b$$

$$a' = \omega_n/\omega_b = F^{1/2} = \text{ratio of loop natural frequency}$$
to base bandwidth (4-53) (7-17)

$$b' = B_p/2f_b \quad = \text{ratio of predetection semi-bandwidth} \quad (7\text{-}18)$$
to base bandwidth

Normalization of this result to the quantity†

$$\frac{4\pi^2(2\eta)f_b{}^3}{3A^2\mu^2} \quad (7\text{-}19)$$

provides a direct measure of the noise penalty (or increase) due to nonideal filtering, and gives the factor P

$$P = 3\int_0^{b'} \frac{[(x/a_0 a')^2 + 1]}{[1 - (x/a')^2]^2 + (x/a')^2} \cdot \frac{x^2\,dx}{1 + (x)^{2n}} \quad (7\text{-}20)$$

P is plotted over a range of n and a' in Fig. 7-3‡. Essentially the same comments and observations apply as for the PLL.

For PM transmission, an integrator is utilized after the FMFB to render the detected voltage proportional to phase modulation. This deemphasis of the high-frequency noise makes nonideal filtering a less critical factor in system performance. Analogous to the results presented for FM, a set of correction factor curves is derived for PM and presented in Fig. 7-4. The equation for the PM output noise is

$$N_{\text{po}} = 2\eta f_b P K_0/A^2\mu^2 \quad (7\text{-}21)$$

† This is the above-threshold output noise in a system with perfect filtering [see Eq. (3-18)], where we have set $\mu = 1/\lambda K_0$ (λ is ideal frequency detector sensitivity in volt-seconds per radian).

‡ In Fig. 7-3, b' is assumed infinite. Observe in Figs. 6-3–6-6 plotted for the PLL, that the effect of b' is generally negligible for a practical range of parameters ($b' > 4.5$, $n > 3$).

7.1. FM Improvement Region

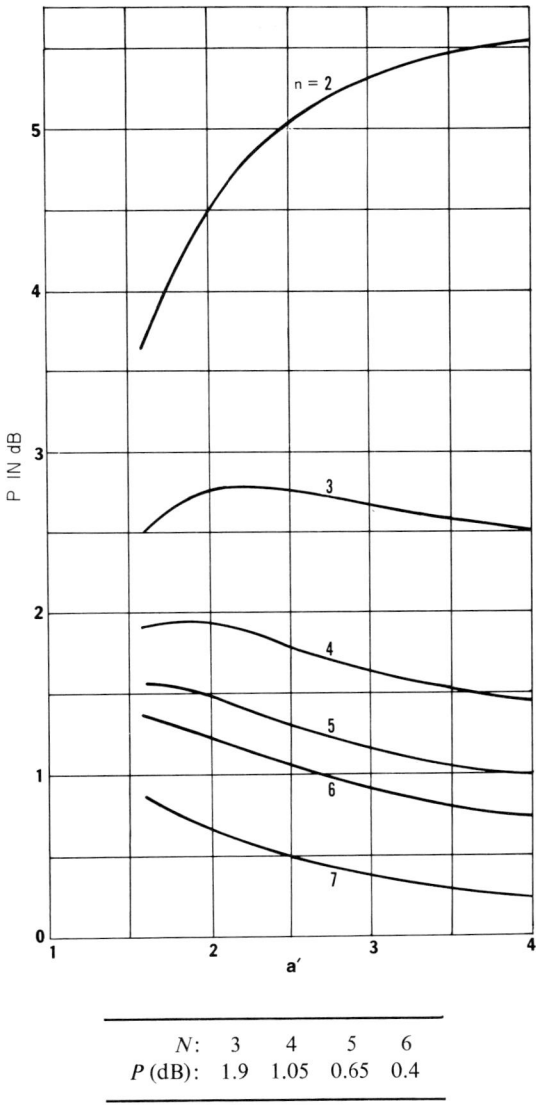

N:	3	4	5	6
P (dB):	1.9	1.05	0.65	0.4

Fig. 7-3. Correction factor for FMFB FM output noise. (No compensation for FMFB closed-loop response.) Asymptotic values ($a' = \infty$) are shown in the accompanying tabulation.

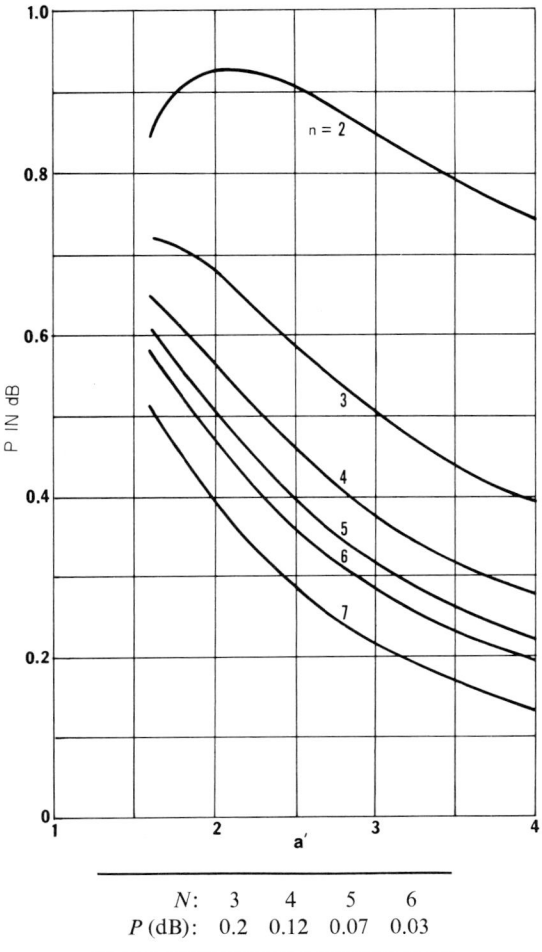

Fig. 7-4. Correction factor for FMFB PM output noise. (No compensation for FMFB closed-loop response.) Asymptotic values ($a' = \infty$) are shown in the accompanying tabulation.

and the plotted penalty factor is

$$P = \int_0^\infty \frac{[x/(a_0 a')^2 + 1]}{[1 - (x/a')^2]^2 + (x/a')^2} \cdot \frac{dx}{1 + x^{2n}} \tag{7-22}$$

To conclude, we apply these results to calculate the corrections necessary to find the actual TT output SNR in a single-channel FMFB-FM detection system. A typical design, discussed further in Section 7.4 has $m_p = 2.5$, $a' = 2.36$, and $b' = 4.5$; with $n = 6$, $P = 1.1$ dB from Fig. 7-3. If suitable compensation is employed for $H(j\omega)$, by including a suitable complex zero and real pole in the design of $R(j\omega)$, the correction factor P is reduced to 0.45 dB.† Without compensation, the relative increase of TT power is $|H(j\omega_T)|^2/|H(j0)|^2$, which for a 1-kHz TT in a 4-kHz baseband is 0.05 dB [Eqs. (7-10)–(7-13), and (7-17) with $a' = 2.36$ and $s = j\omega_b/4$]. With compensation, there is no correction to the relative TT level (i.e., $|H(j\omega_T)|^2/|H(j0)|^2 = 1$). Finally, we have as the output TT SNR a modified form of Eq. (7-3)

$$\text{SNR}_{TT} = \frac{3m_p^2 |H(j\omega_T)|^2}{P \cdot |H(j0)|^2} (\text{CNR})_{AM} \qquad (7\text{-}23)$$

which, for the compensated and uncompensated cases under evaluation, reduces to

$$\text{SNR}_{TT} = 10 \log_{10}[18.75(\text{CNR})_{AM}] - 0.45 \quad \text{dB} \quad \text{(compensated)} \qquad (7\text{-}24)$$

$$\text{SNR}_{TT} = 10 \log_{10}[18.75(\text{CNR})_{AM}] - 1.05 \quad \text{dB} \quad \text{(uncompensated)} \qquad (7\text{-}25)$$

7.2. Distortion-Limited Region

With a reduction of the additive input noise to zero (or conversely, with the input signal becoming arbitrarily large), Eqs. (7-1)–(7-7) predict that the FMFB output SNR grows without bound. Practical limitations in the receiving system, however, produce a residual noise that results in a maximum attainable output SNR as indicated in the performance curve of Fig. 7-2. This residual noise stems from two major sources: (1) The idle noise inherent in the circuit elements, and (2) intermodulation (and harmonic) distortion of the signal, characteristic of dynamic nonlinearities in the system elements. Idle noise is primarily a function of oscillator frequency jitter, which is mainly contributed by the VCO's employed in the transmitter (FM modulator) and receiver. Discussion of VCO frequency jitter is reserved for Section 7-3. The discussion here will be concerned primarily with harmonic and intermodulation distortion noise.

For conditions of low distortion, i.e., less than 1% of the primary signal component, the distortion introduced by an individual loop element may be included in the loop analysis and design by substituting for the nonlinear element a linear equivalent element that contains a dependent generator, as indicated in Fig. 7-5. The resulting output distortion is then calculated from the linear transfer characteristic of the FMFB.

Consider the effect of introducing the $\phi_d'(t)$ generator (which represents a distortion source) into the FMFB linear model, as indicated in Fig. 7-6. The loop output will consist of two components, signal $e_s(t)$ and distortion $e_d(t)$,

† The effect of compensation is obtained by letting $a' \to \infty$.

7. Design of FMFB for FM Demodulation

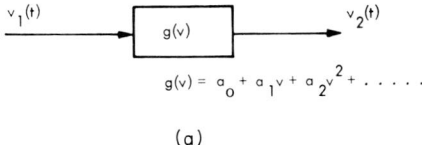

(a)

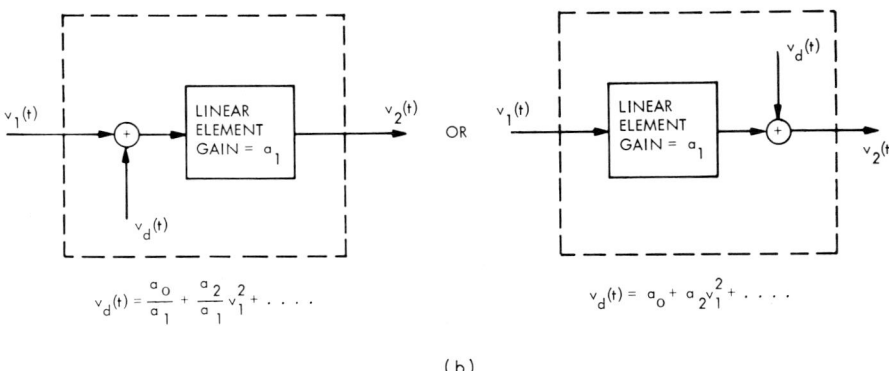

(b)

Fig. 7-5. Linear element representation for nonlinear loop element (v is total voltage across nonlinear element). (a) Nonlinear element. (b) Linear element with dependent generator (approximation to nonlinear element).

which arise from the excitations, $\phi_i'(t)$ (signal), and $\phi_d'(t)$ (distortion). The output components are calculated as

$$E_s(s) = K_1 K_2 s H_2(s) \hat{\Phi}_e'(s) \tag{7-26}$$

with

$$\frac{\hat{\Phi}_e'}{\Phi_i'}(s) = \frac{H_{b1}(s)}{1 + K_1 K_2 K_3 H_{b1}(s) H_2(s)} \tag{7-27}$$

$$E_d(s) = \frac{K_1 K_2 s H_2(s)}{1 + K_1 K_2 K_3 H_{b1}(s) H_2(s)} \Phi_d'(s) \tag{7-28}$$

For FDM, both $\hat{\Phi}_e'(s)$ and $\Phi_d'(s)$ are taken as random signals having continuous spectral distributions. The loop performance measure is NPR, for which we have at the loop output†

$$\text{NPR} = \left|\frac{E_s}{E_d}(j\omega_{\text{CH}})\right|^2 = |1 + K_1 K_2 K_3 H_{b1}(j\omega_{\text{CH}}) H_2(j\omega_{\text{CH}})|^2 \cdot \left|\frac{\hat{\Phi}_e'}{\Phi_d'}(j\omega_{\text{CH}})\right|^2 \tag{7-29}$$

where ω_{CH} is baseband channel center frequency.

† High-Q narrow-band FDM channels are assumed; i.e., the ratio of power spectral densities adequately determines channel NPR.

7.2. Distortion-Limited Region

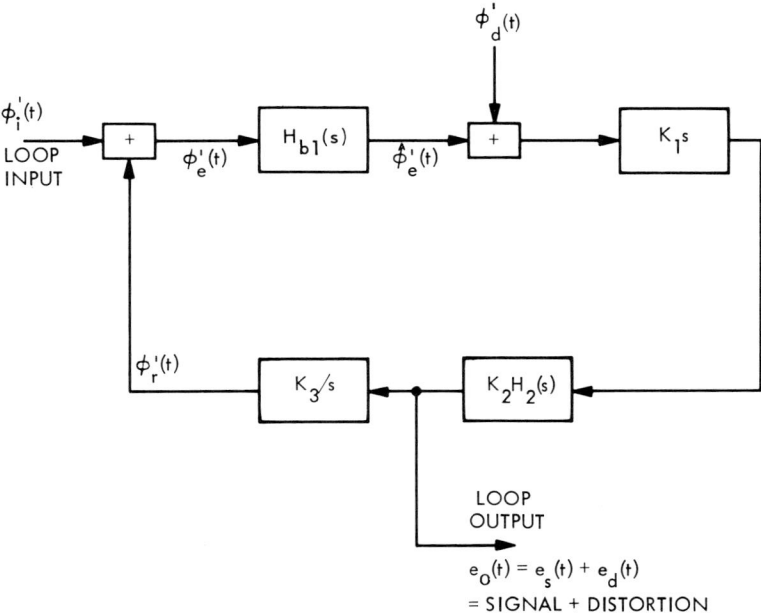

Fig. 7-6. FMFB linear model with equivalent distortion generator.

For TT modulation (i.e., $\phi_i'(t) = (\Delta\omega_p/\omega_T)\sin\omega_T t$), the output signal to harmonic distortion ratio is desired, and from Eqs. (7-26) and (7-28)

$$\left(\frac{S}{D}\right)_N = \left|\frac{E_s(j\omega_T)}{E_d(jN\omega_T)}\right|^2 = \frac{1}{N^2}\left|\frac{H_2(j\omega_T)}{H_2(jN\omega_T)}\right|^2$$

$$\cdot\,|1 + K_1 K_2 K_3 H_{b1}(jN\omega_T)H_2(jN\omega_T)|\cdot\left|\frac{\hat{\Phi}_e'(j\omega_T)}{\Phi_d'(jN\omega_T)}\right|^2 \quad (7\text{-}30)$$

where N is harmonic number of the distortion component, $|\phi_d'(jN\omega_T)|$ is the amplitude of the Nth order harmonic in the equivalent distortion generator $\phi_d'(t)$, and $(S/D)_N$ is signal to Nth order harmonic distortion power ratio.

Determination of the equivalent distortion generator for either internal IF or LD induced distortion requires computation of the internal IF filter input signal. This is accomplished by recourse to the linear models and transfer functions developed in Chapter 4. From Figs. 4-3 or 7-6 and Eq. (2-21) we have as the IF input signal

$$\Phi_e'(s) = \frac{\Phi_i'(s)}{1 + K_1 K_2 K_3 H_{b1}(s)H_2(s)} \quad (7\text{-}31)$$

For the second-order FMFB, whose design is covered more fully in Section 7.3 [Eqs. (7-94) and (7-96)], we have

$$1 + K_1 K_2 K_3 H_{b1}(s)H_2(s) = 1 + G(s)$$
$$= (1 + K) \frac{s^2/[(1 + K)^{1/2}\omega_b]^2 + 2\xi_0 s/[(1 + K)^{1/2}\omega_b] + 1}{[(s/\omega_b)^2 + 2^{1/2}(s/\omega_b) + 1]} \quad (7\text{-}32)$$

The internal IF input signal "power spectral density" in terms of FM, i.e., the derivative of $\phi_e'(t)$, is of the form†

$$W_{\dot\phi_e'}(f) = \omega^2 W_{\phi_e'}(f) = \frac{\omega^2}{(1+K)^2} |T(j\omega)|^2 W_{\phi_i'}(f) \quad (7\text{-}33)$$

with

$$|T(j\omega)|^2 = \frac{(1+K)^2}{|1 + K_1 K_2 K_3 H_{b1}(j\omega)H_2(j\omega)|^2}$$
$$= \frac{(1-x^2)^2 + 2x^2}{[1 - x^2/(1+K)]^2 + x^2/(1+K)} \ ; \quad x = \frac{\omega}{\omega_b} \quad (7\text{-}34)$$

for the second-order Enloe design loop with ξ_0 approximated by $\frac{1}{2}$.

The factor $(1 + K)$ (or loop feedback factor F) is seen from Eqs. (7-33) and (7-34) to directly control the internal IF input signal spectrum amplitude and shape. Note that the term $\omega^2 W_{\phi_i'}(f)$ is the loop input modulation density directly in terms of FM; hence, $T(j\omega)/F^2$ represents the FM transfer function from loop input to IF input. The magnitude of the normalized transfer function $|T(j\omega)|$, is given in Fig. 7-7 over a range of F. Note that $|T(j\omega)|$ is relatively flat over the baseband range ($x = 0$ to 1), is relatively insensitive to F, and has a characteristic rise of approximately 3–4 dB at the top baseband frequency.

The rise in the IF input spectrum distribution is often neglected in the FMFB literature; $W_{\dot\phi_e'}(f)$ is usually taken as the loop input modulation spectral distribution reduced by F^2. For FDM-FM, or TT at one quarter the top baseband frequency, such an approximation has little effect on the analytical accuracy of the IF distortion calculation. However, for cases where the signal modulation energy is concentrated at the top of the baseband, failure to include the spectral rise due to $|T(j\omega)|$ in the calculation results in a substantial error.

† There are two FM carriers in this discussion—the wide-band FM carrier at the loop input, and the compressed FM carrier of the internal IF input. To differentiate between the quantities relating to the two carriers, the subscripts i and e are used to signify loop input and internal IF input, respectively.

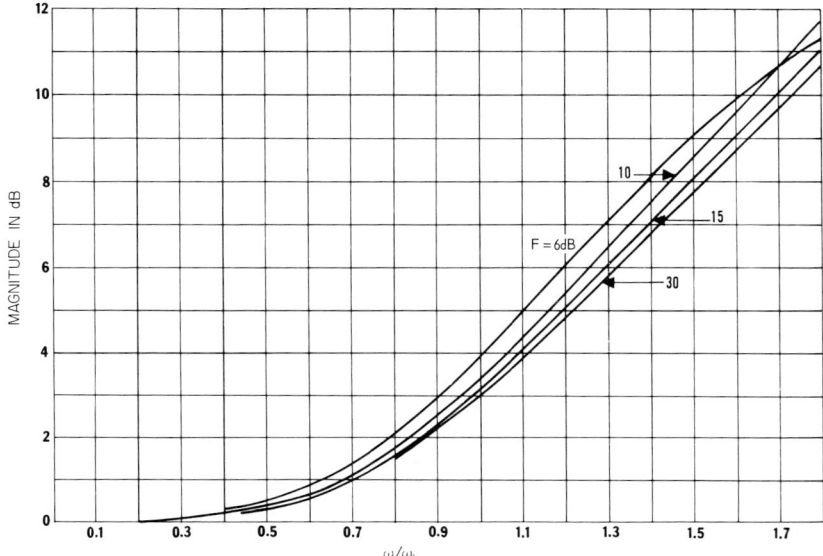

Fig. 7-7. Plot of $20 \log_{10} |T(j\omega)|$ as a function of normalized frequency with F as a parameter.

A. IF Network Distortion

The fundamental source of signal distortion in the FMFB is the internal narrowband IF filter. For both random and TT modulation, the general analysis of filtering and distortion is very complex and difficult to apply, especially to the intermediate range of modulation indices found inside the FMFB loop (refer to Section 3.4). For specific cases, however, such as the high-Q single-pole filter (which is often used as the internal FMFB IF filter), useful results are available from Izatt,[2] and Bedrosian and Rice,[3] with Izatt's data applicable to TT distortion calculation and Bedrosian and Rice's data applicable to noise-loading distortion calculations. The calculation proceeds by computing the characteristics of the equivalent distortion generator $\phi_d'(t)$ which is equivalent to computing the distortion existing at the IF output on an open-loop basis. The data are available from Bedrosian and Rice (see also Figs. 3-9 and 3-10 for Izatt's), where normalized modulation cross-talk curves are presented for the high-Q single-pole filter excited by a *flat* noise loaded FM signal. The curves are reproduced as Fig. 7-8; note from Eq. (7-33) that their direct application requires $\omega^2 W_{\phi_i'}(f) \cdot |T(j\omega)|^2$ to be flat over the baseband, which may be satisfied by requiring $W_{\phi_i'}(f)$ to contain the factor $1/|T(j\omega)|^2$.

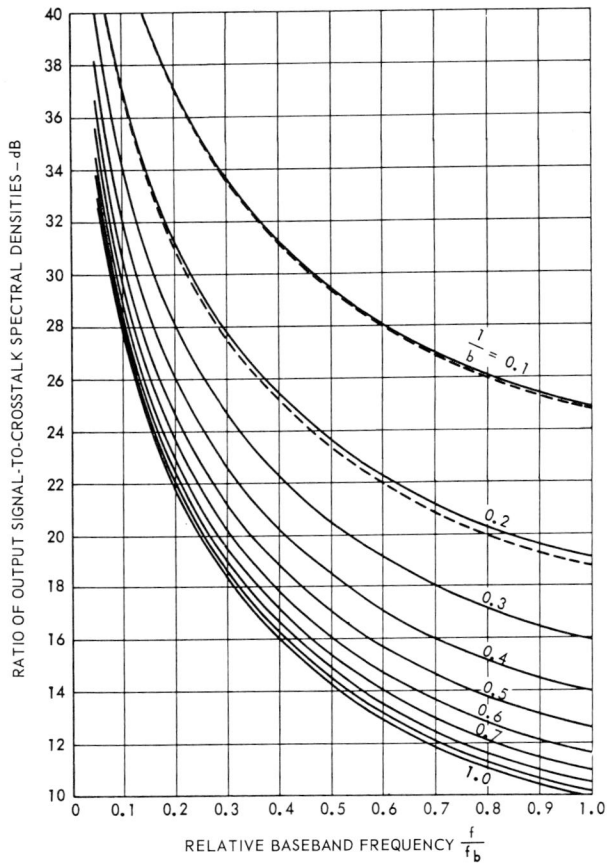

Fig. 7-8. Single-pole IF output modulation cross-talk spectral densities (from Bedrosian and Rice[3]). It is a plot of $[W_{\dot{\theta}}^s(f)/W_{\dot{\theta}}^c(f)][D_e/(B_{IF}/2)]^4$ in decibels versus ω/ω_b. Input: FM with rms frequency deviation D_e using a Gaussian baseband uniform in $(0, f_b)$; Filter: single pole, bandpass with 3-dB half-bandwidth $B_{IF}/2$ (– – –: approximation for $1/b$ small).

To demonstrate the analytical method, a first-order calculation is carried out, with $\omega^2 W_{\phi_i'}(f)$ suitably distorted so as to render $W_{\dot{\phi}_e'}(f)$ flat over the baseband. Correction factors are then derived to extend the results to an FDM-FM input (i.e., $\omega^2 W_{\phi_i'}(f)$ flat over the baseband).

The characteristics of the equivalent distortion generator $\phi_d'(t)$ are computed by representing the loop input signal as noise modulation having a power spectral density $W_{\dot{\phi}_i'}(f)$

$$W_{\dot{\phi}_i'}(f) = \omega^2 W_{\phi_i'}(f) = \eta_m/|T(j\omega)|^2, \quad 0 \le \omega \le \omega_b \quad (7\text{-}35)$$
$$= 0, \quad \omega > \omega_b$$

7.2. Distortion-Limited Region

where

$$(\Delta\omega_{rms})_i^2 = \int_0^{f_b} W_{\dot{\phi}_i'}(f)\, df = \eta_m \cdot f_b \cdot I_1 \quad (7\text{-}36)$$

with η_m a constant and

$$I_1 = \int_0^1 [1/|T(j\omega)|^2]\, dx, \qquad x = \omega/\omega_b \quad (7\text{-}37)$$

The integral I_1 is given (in decibels) in Table V over a range of loop feedback factors. By application of Eqs. (7-33), (7-35), and (7-36), the IF input signal power spectral density is

$$W_{\dot{\phi}_e'}(f) = \eta_m/F^2 = (\Delta\omega_{rms})_i^2/(f_b \cdot I_1 \cdot F^2) = \omega^2 W_{\phi_e'}(f) \quad 0 \le \omega \le \omega_b$$
$$= 0, \qquad \omega > \omega_b \quad (7\text{-}38)$$

with the IF output signal power spectral density

$$\omega^2 W_{\hat{\phi}_e'}(f) = |H_{b1}(j\omega)|^2 W_{\dot{\phi}_e'}(f)$$
$$= [(\Delta\omega_{rms})_i^2/(f_b \cdot I_1 \cdot F^2)]|H_{b1}(j\omega)|^2, \quad 0 \le \omega \le \omega_b \quad (7\text{-}39)$$
$$= 0, \qquad \omega > \omega_b$$

We now compute the spectral characteristics of the equivalent distortion generator $\phi_d'(t)$ from the IF open-loop crosstalk curves in Fig. 7-8 in terms of the quantity $[W_\theta^s(f)/W_\theta^c(f)][D_e/(B_{IF}/2)]^4$. The parameters found in these curves are defined as, and related to the parameters used herein, by

$$D_e = \Delta\omega_{rms}/2\pi = \text{rms frequency deviation (Hz)} \quad (7\text{-}40)$$

$$W_\theta^s(f) = \omega^2 W_{\hat{\phi}_e'}(f) = \text{modulation power spectral density at IF output} \quad (7\text{-}41)$$

$$W_\theta^c(f) = \omega^2 W_{\phi_d'}(f) = \text{crosstalk power spectral density at IF output} \quad (7\text{-}42)$$

The term $[D_e/(B_{IF}/2)]^2$ is obtained as follows:

$$D_e^2 = (1/2\pi)^2 \int_0^{f_b} W_{\dot{\phi}_e'}(f)\, df = W_{\dot{\phi}_e'}(f) f_b/(2\pi)^2 \quad (7\text{-}43)$$

since we assumed $W_{\dot{\phi}_e'}$ to be flat over the baseband to permit the application of the results of Bedrosian and Rice. In terms of the loop input quantities, we have from Eq. (7-38)

$$D_e^2 = (\Delta\omega_{rms})_i^2/[I_1 F^2 (2\pi)^2] \quad (7\text{-}44)$$

On the other hand, the IF semibandwidth and the base bandwidth are related by

$$(B_{IF}/2)^2 = b^2 f_b^2 \quad (4\text{-}61)$$

TABLE V

TABULATION OF INTEGRALS RELATED TO THE FMFB MODULATION ERROR TRANSFER FUNCTION $T(jx)$

Feedback factor F (dB)	$10\log_{10}(I_1)$		$10\log_{10}(I_2)$
10	−0.952		1.276
15	−0.820	$\|T(jx)\|^2 = \dfrac{(1-x^2)^2 + 2x^2}{(1-x^2/F)^2 + x^2/F}$; $\quad I_1 = \int_0^1 \dfrac{dx}{\|T(jx)\|^2}$; $\quad I_2 = \int_0^1 \|T(jx)\|^2\, dx$	1.083
20	−0.737		0.960
25	−0.687		0.887
30	−0.658		0.846

7.2. Distortion-Limited Region

Therefore,

$$\left(\frac{D_e}{B_{IF}/2}\right)^2 = \frac{(\Delta\omega_{rms})_i^2}{b^2 I_1 F^2 \omega_b^2} = \frac{\sigma_i^2}{b^2 I_1 F^2} \quad (7\text{-}45)$$

The FMFB output NPR is calculated from Eq. (7-29)

$$\text{NPR} = |1 + K_1 K_2 K_3 H_{b1}(j\omega)H_2(j\omega)|^2 \left.\left|\frac{\hat{\Phi}_e'}{\Phi_d'}(j\omega)\right|^2\right|_{\omega=\omega_{CH}} \quad (7\text{-}29)\ (7\text{-}46)$$

The first term, by application of Eq. (7-31), reduces to

$$|1 + K_1 K_2 K_3 H_{b1}(j\omega)H_2(j\omega)|^2 = F^2/|T(j\omega)|^2 \quad (7\text{-}47)$$

and the second term, by application of Eqs. (7-41) and (7-42) reduces to

$$\left|\frac{\hat{\Phi}_e'}{\Phi_d'}(j\omega)\right|^2 = \frac{W_{\dot{\theta}}^s(f)}{W_{\dot{\theta}}^c(f)} \quad (7\text{-}48)$$

Combining Eqs. (7-43)–(7-48) we have, as the FMFB output NPR,

$$\text{NPR} = 10 \log_{10} \left[\frac{F^6 I_1^2 b^4}{|T(j\omega)|^2 \sigma_i^4} \frac{W_{\dot{\theta}}^s(f)}{W_{\dot{\theta}}^c(f)} \left(\frac{D_e}{B_{IF}/2}\right)^4\right] \quad (7\text{-}49)$$

Example

Consider the following set of parameters:

$$\sigma_i = 3.14, \quad b = 2.05, \quad F = 4.8 \quad (\text{or} \quad 13.6 \quad \text{dB})$$

Find the top-channel NPR. From Fig. 7-8 for $1/b = 0.512$, we have for $(f/f_b) = 1$

$$\frac{W_{\dot{\theta}}^s(f)}{W_{\dot{\theta}}^c(f)} \left(\frac{D_e}{B_{IF}/2}\right)^4 = 12.5 \quad \text{dB} \quad (\text{or} \quad 17.8)$$

and from Fig. 7-7, we obtain at the top channel $|T(j\omega)|^2 = 3.3$ dB. Therefore,

$$\begin{aligned}\text{NPR (top channel)} &= 40 \log_{10}(b/\sigma_i) + 60 \log_{10} F - 10 \log_{10}|T(j\omega)|^2 \\ &\quad + 20 \log_{10} I_1 + 12.5 \quad \text{dB} \\ &= 40 \log_{10}(2.05/3.14) + 40.8 - 3.3 - 1.7 \\ &\quad + 12.5 \quad \text{dB} \\ &= 40.7 \quad \text{dB at FMFB output} \quad (7\text{-}50)\end{aligned}$$

Finally, consider the case where the input modulation $\dot{\phi}_i'$ has a flat spectral distribution (i.e., multiply $W_{\dot{\phi}_i'}(f)$ in Eq. (7-35) by the factor $|T(j\omega)|^2$).

Assuming the same density level η_m [see Eq. (7-35)], the input mean-square frequency deviation will increase by the factor $1/I_1$ [see Eq. (7-36)] which, for this numerical example, is equivalent to +0.74 dB (see Table V). The signal power spectral density at the top channel (see Fig. 7-7) will increase by 3.1 dB, which is reflected in the FMFB output as a direct 3.1-dB top-channel NPR improvement. The increase in the input deviation, or the boost in the high frequencies of $\dot{\phi}_e'(t)$ result in an increase in the single-pole IF input signal frequency deviation [see Eq. (7-44)] by a factor of $I_2 = \int_0^1 |T(j\omega)|^2 \, dx$, $x = \omega/\omega_b$, which is 0.96 dB for $F = 20$ dB (Table V). Because of the increased IF input frequency deviation, there will be an increase in the IF output distortion, which may be evaluated by the following: (1) The in-band distortion spectral shape for mildly differing signal spectral shapes remains essentially fixed as the distortion spectrum arises from a triple convolution of the signal spectrum upon itself; (2) the spectrum level will increase by the fourth power (see Fig. 7-8) for increases in the single-pole IF input rms frequency deviation. These two factors indicate an increase in FMFB top channel output distortion on the order of $4(0.96 \text{ dB}) = 3.84$ dB, which essentially cancels the 3.1-dB factor discussed above. We conclude, therefore, that the NPR indicated by Eq. (7-50) is representative of FDM-FM signals without further modification.

B. Distortion Due to Limiter-Discriminator

The LD is another possible source of distortion in an FMFB. We refer to a study of design and dynamic nonlinearity of an optimized frequency discriminator (Fig. 7-10) by Fancourt and Skwirzynski.[4] Their result, in terms of open-loop distortion for TT modulation, is now applied to calculate FMFB output distortion arising from discriminator nonlinearity. The procedure is to first relate the discriminator open-loop distortion to the equivalent distortion generator $\phi_d'(t)$ and then, by Eq. (7-30) to relate $\phi_d'(t)$ to the FMFB output.

The applicable open-loop data, following Fancourt and Skwirzynski, are summarized in Fig. 7-9, and are interpreted by reference to the following equations. For an FM TT signal,

$$c(t) = \cos[\omega_c t + \hat{\phi}_e'(t)] \qquad (7\text{-}51)$$

with

$$\hat{\phi}_e'(t) = (\Delta\omega_e/\omega_T) \sin \omega_T t \qquad (7\text{-}52)$$

at the discriminator input, the discriminator output wave takes the form

$$d(t) = K_1 \Delta\omega_e [\cos \omega_T t + \delta_2 \cos 2\omega_T t + \delta_3 \cos 3\omega_T t + \cdots] \qquad (7\text{-}53)$$

7.2. Distortion-Limited Region

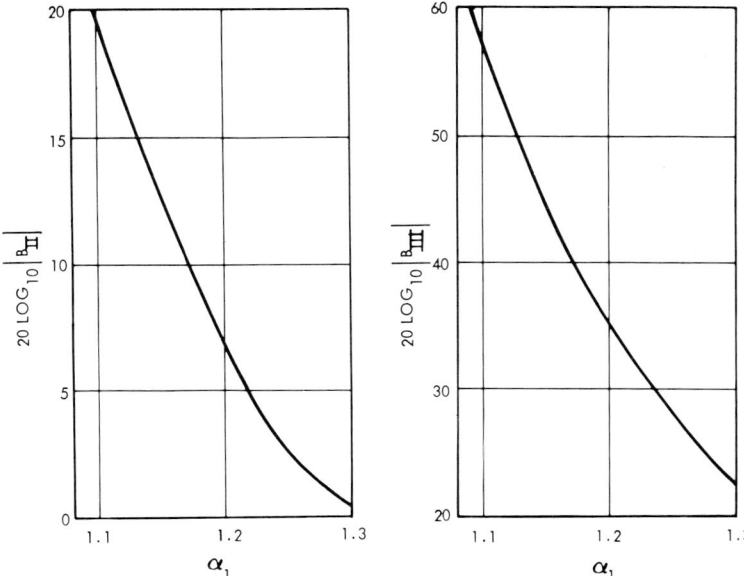

Fig. 7-9. Discriminator distortion factors (from Fancourt and Skwirzynski[4]).

An equivalent output wave may be obtained from a distortionless discriminator having a gain of K_1 V-sec/rad, by the addition (Fig. 7-6b) of the equivalent distortion generator $\phi_d'(t)$ having the form

$$\phi_d'(t) = (\Delta\omega_e/2\omega_T)(\delta_2) \sin 2\omega_T t + (\Delta\omega_e/3\omega_T)(\delta_3) \sin 3\omega_T t + \cdots \quad (7\text{-}54)$$

Following Fancourt and Skwirzynski, the (δ_2) and (δ_3) factors are given by

$$\delta_2 = B_{II}(\Delta\omega_e/\omega_e)^3 \quad (7\text{-}55)$$

and

$$\delta_3 = B_{III}(\Delta\omega_e/\omega_e)^4 \quad (7\text{-}56)$$

with B_{II} and B_{III} plotted in Fig. 7-9. These factors are seen to depend on the discriminator design factor α_1 which is the fractional discriminator upper-peak-frequency-to-center-frequency ratio, i.e., the upper peak in the discriminator "S" curve is at frequency $\alpha_1 \omega_e$ rad/sec. The α_2 factor (the discriminator lower-peak-frequency-to-center-frequency ratio) is specified in terms of α_1, resulting in a peak-to-peak discriminator bandwidth of $(\alpha_1 - \alpha_2) \cdot \omega_e$ rad/sec. The design elements for the discriminator are given in Fig. 7-10.

Continuing with the FMFB output signal-to-distortion ratio calculation, we note that the desired answer is contained in Eq. (7-30), the last term of which may be reduced by application of Eqs. (7-52), (7-54)–(7-56) to read

$$\left|\frac{\hat{\Phi}_e{}'(j\omega_T)}{\Phi_d{}'(jN\omega_T)}\right|^2_{N=2} = \frac{(\Delta\omega_e/\omega_T)^2}{(\Delta\omega_e/2\omega_T)^2\delta_2{}^2} = \frac{4}{\delta_2{}^2} = \frac{4\omega_e{}^6}{B_{\text{II}}^2(\Delta\omega_e)^6} \qquad (7\text{-}57)$$

$$\left|\frac{\hat{\Phi}_e{}'(j\omega_T)}{\Phi_d{}'(jN\omega_T)}\right|^2_{N=3} = \frac{(\Delta\omega_e/\omega_T)^2}{(\Delta\omega_e/3\omega_T)^2\delta_3{}^2} = \frac{9}{\delta_3{}^2} = \frac{9\omega_e{}^8}{B_{\text{III}}^2(\Delta\omega_e)^8} \qquad (7\text{-}58)$$

By application of Eq. (7-34) we may reduce the second term of Eq. (7-30) to yield

$$|1 + K_1 K_2 K_3 H_{b1}(jN\omega_T) H_2(jN\omega_T)|^2 = F^2/|T(jN\omega_T)|^2 \qquad (7\text{-}59)$$

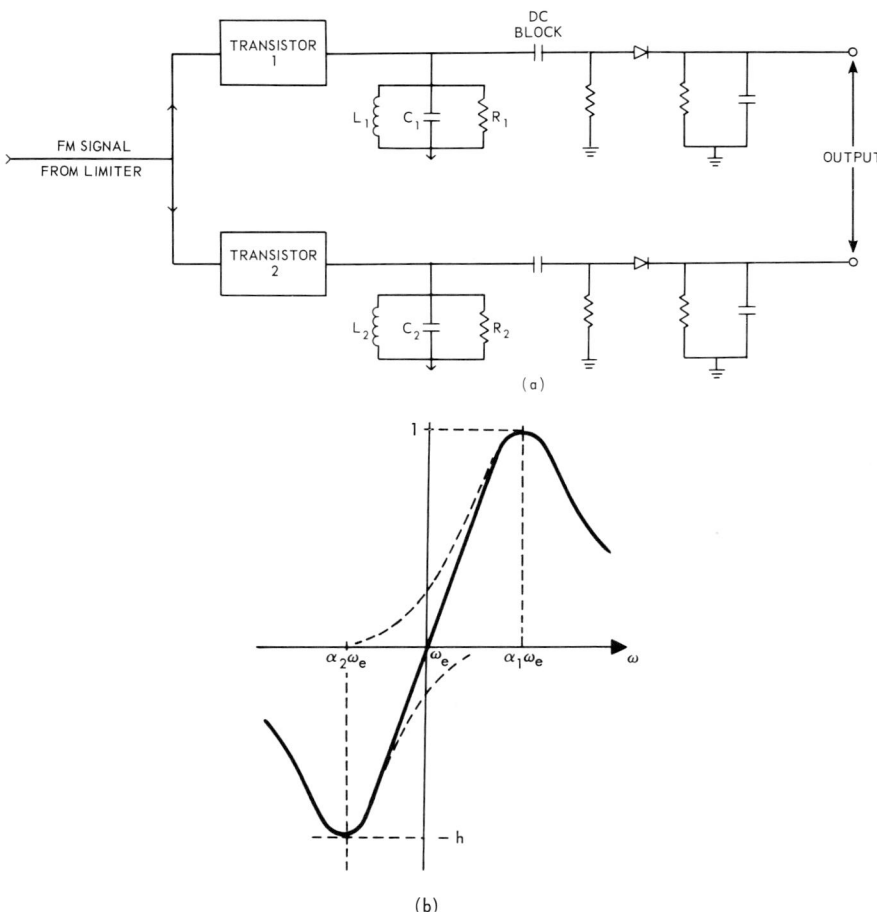

Fig. 7-10. Optimized discriminator (from Fancourt and Skwirzynski[4]).

Using Eqs. (7-57)–(7-59), Eq. (7-30) reduces to

$$\left(\frac{S}{D}\right)_2 = \frac{\omega_e^6 F^2}{B_{II}^2 (\Delta\omega_e)^6} \cdot \frac{|H_2(j\omega_T)|^2}{|T(j2\omega_T)|^2 \cdot |H_2(j2\omega_T)|^2} \tag{7-60}$$

$$\left(\frac{S}{D}\right)_3 = \frac{\omega_e^8 F^2}{B_{III}^2 (\Delta\omega_e)^8} \cdot \frac{|H_2(j\omega_T)|^2}{|T(j3\omega_T)|^2 \cdot |H_2(j3\omega_T)|^2} \tag{7-61}$$

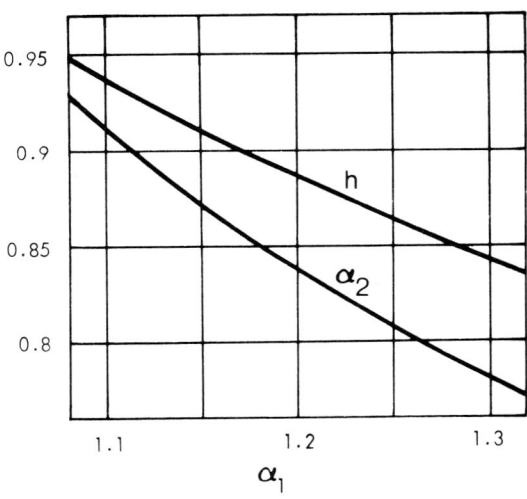

(c)

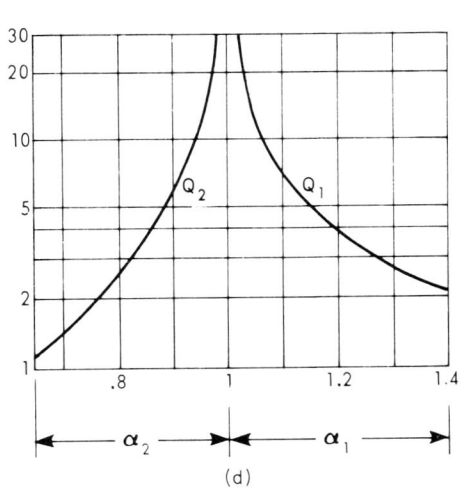

(d)

(a) Network. (b) Response. (c) Gain and detuning ratios. (d) Q factors.

7. Design of FMFB for FM Demodulation

Note from Eqs. (7-34) and (2-20) that

$$\left(\frac{\Delta\omega_e}{\Delta\omega_i}\right)^2 = \left|\frac{\hat{\Phi}_e{'}}{\Phi_i{'}}(j\omega_T)\right|^2 = \frac{|T(j\omega_T)|^2 \cdot |H_{b1}(j\omega_T)|^2}{F^2} \quad (7\text{-}62)$$

When applied to Eqs. (7-60) and (7-61), this results in signal-to-distortion ratios as a function of the input deviation $\Delta\omega_i$

$$\left(\frac{S}{D}\right)_2 = \left[\frac{\omega_e^6 F^8}{B_{\text{II}}^2(\Delta\omega_i)^6}\right]$$

$$\cdot \left[\frac{|H_2(j\omega_T)|^2}{|T(2j\omega_T)|^2 \cdot |H_2(2j\omega_T)|^2 \cdot |T(j\omega_T)|^6 \cdot |H_{b1}(j\omega_T)|^6}\right] \quad (7\text{-}63)$$

$$\left(\frac{S}{D}\right)_3 = \left[\frac{\omega_e^8 F^{10}}{B_{\text{III}}^2(\Delta\omega_i)^8}\right]$$

$$\cdot \left[\frac{|H_2(j\omega_T)|^2}{|T(3j\omega_T)|^2 \cdot |H_2(3j\omega_T)|^2 \cdot |T(j\omega_T)|^8 \cdot |H_{b1}(j\omega_T)|^8}\right] \quad (7\text{-}64)$$

Under representative conditions, especially when compensation is employed after the loop, the second factor in Eqs. (7-63) and (7-64) may be neglected, with the first factor essentially determining the FMFB output signal to distortion ratio.

Example

Consider the following set of parameters:

A. SIGNAL PARAMETERS

$TT = \begin{cases} \Delta\omega_i = 2\pi(10K) \text{ rad/sec} \\ \omega_T = 2\pi(1K) \text{ rad/sec} \end{cases}$
Baseband: $0\text{-}2\pi(4K)$ rad/sec
System modulation index $= m_p = 2.5$

B. LOOP PARAMETERS

Feedback factor $F = 15$ dB $= 5.6$ (numeric)
Zero constant $a = 4.6$
IF constant $b = 0.7$
Baseband filter damping factor $\xi = 2^{1/2}/2$

C. DISCRIMINATOR DESIGN

$\omega_e = 2\pi(200K)$ rad/sec
$\alpha_1 = 1.29$, $\alpha_2 = 0.79$; bandwidth $= (\alpha_1 - \alpha_2)\omega_e = 2\pi(100K)$ rad/sec

From Fig. 7-9, $20 \log_{10} B_{\text{II}} = 0.7$ dB; $20 \log_{10} B_{\text{III}} = 23$ dB. When these parameters are applied to Eqs. (7-63), (7-64), (4-61), and (4-62) and recalling

7.2. Distortion-Limited Region

that in this case $H_{b1}(j\omega)$ is the lowpass equivalent $H_{L1}(j\omega)$ given by Eq. (4-61), the result is

$$10 \log_{10} \left[\frac{\omega_e^6 F^8}{B_{II}^2 (\Delta\omega_i)^6} \right]$$

$$= 60 \log_{10} \left(\frac{\omega_e}{\Delta\omega_i} \right) + 80 \log_{10} F - 20 \log_{10} B_{II} \quad (7\text{-}65)$$

$$= 137.3 \quad \text{dB}$$

$$10 \log_{10} \left[\frac{|H_2(j\omega_T)|^2}{|T(2j\omega_T)|^2 |H_2(2j\omega_T)|^2 |T(j\omega_T)|^6 |H_{b1}(j\omega_T)|^6} \right]$$

$$= -0.1 \quad \text{dB} \quad (7\text{-}66)$$

$$10 \log_{10} \left[\frac{\omega_e^8 F^{10}}{B_{III}^2 (\Delta\omega_i)^8} \right]$$

$$= 156 \quad \text{dB} \quad (7\text{-}67)$$

$$10 \log_{10} \left[\frac{|H_2(j\omega_T)|^2}{|T(3j\omega_T)|^2 |H_2(3j\omega_T)|^2 |T(j\omega_T)|^8 |H_{b1}(j\omega_T)|^8} \right]$$

$$= -1.3 \quad \text{dB} \quad (7\text{-}68)$$

with the final result

$$(S/D)_2 = 137.2 \quad \text{dB} \quad (7\text{-}69)$$

and

$$(S/D)_3 = 154.7 \quad \text{dB} \quad (7\text{-}70)$$

The analysis of Fancourt and Skwirzynski is of the "quasi-stationary" type, intended for large modulation index FM waves. The distortion levels may be higher for the small modulation index FM waves usually found in FMFB loops, and a linearity safety factor is appropriate, depending upon the parameters of the signal and the loop.

C. VCO Distortion

As with the PLL, a possible major source of distortion in the FMFB is the nonlinear modulation characteristic of the VCO. Basically, we follow the same development as in Chapter 6, except that the example here concerns itself with the distortion arising from a test tone. It should be noted that this example and that worked out for the PLL are interchangeable. By this we mean that, since the closed-loop responses for the PLL and the FMFB are similar, the results from each example essentially apply to both systems.†

In general, the dynamic modulation equation of the VCO is of the form

$$\Delta\omega(t) = a_1 v(t) + a_2 [v(t)]^2 + a_3 [v(t)]^3 + \cdots \quad (7\text{-}71)$$

where $\Delta\omega(t)$ is VCO frequency deviation (radians per second) about quiescent center frequency, $v(t)$ is VCO input control voltage, and a_1, a_2, etc., are coefficients determined by the oscillator circuit characteristics. The distortion coefficients are related to the parameters of a varactor-controlled oscillator in Appendix D.

For low-distortion conditions, a VCO can usually be characterized by

$$\Delta\omega(t) = a_1 \{v(t) + (a_2/a_1)[v(t)]^2\} \quad (7\text{-}72)$$

from which the dependent distortion generator (as indicated in Fig. 7-11) can be derived as

$$v_d(t) = (a_2/a_1)[v(t)]^2 \quad (7\text{-}73)$$

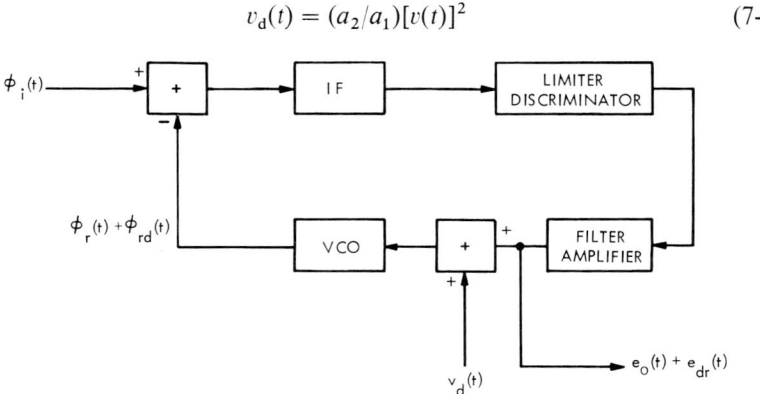

Fig. 7-11. FMFB linear model with equivalent VCO distortion input.

For TT, the modulating voltage is represented by

$$v(t) = V_p \cos \omega_T t \quad (7\text{-}74)$$

which, when applied to Eq. (7-73) results in

$$\begin{aligned} v_d(t) &= V_p^2 (a_2/a_1) \cos^2 \omega_T t \\ &= \tfrac{1}{2}(V_p^2)(a_2/a_1) + \tfrac{1}{2}(V_p)^2 (a_2/a_1) \cos 2\omega_T t \end{aligned} \quad (7\text{-}75)$$

The second term in the above equation is taken as the equivalent distortion

---

† A difference occurs if the loop output is taken after the VCO; this is discussed in the next section which deals with techniques to reduce distortion.

7.2. Distortion-Limited Region

generator, and the first term, representing a "DC shift," is neglected. The distortion power P_{do} appearing at the loop output is

$$P_{do} = \tfrac{1}{8}(V_p)^4(a_2/a_1)^2|H(j2\omega_T)|^2 \qquad (7\text{-}76)$$

where $H(j\omega)$ is the FMFB closed-loop response and, therefore, represents the closed-loop response between the VCO output and the VCO input.

For a design condition where $\omega_T/\omega_b = \tfrac{1}{4}$, the distortion component at frequency $2\omega_T$ essentially falls on the flat portion of the FMFB closed-loop response (see Fig. 4-4), resulting in

$$|H(j2\omega_T)|^2 = [(F-1)/F]^2 \qquad (7\text{-}77)$$

reducing Eq. (7-76) to

$$P_{do} = \tfrac{1}{8}(V_p)^4(a_2/a_1)^2[(F-1)/F]^2 \qquad (7\text{-}78)$$

Note that the FMFB output signal power P_{so} at frequency ω_T is by Eq. (7-74)

$$P_{so} = \tfrac{1}{2}(V_p)^2 \qquad (7\text{-}79)$$

Therefore, the output signal-to-distortion ratio $(S/D)_2$ becomes

$$\left(\frac{S}{D}\right)_2 = \frac{P_{so}}{P_{do}} = \frac{\tfrac{1}{2}(V_p)^2}{\tfrac{1}{8}(V_p)^4(a_2/a_1)^2}\left(\frac{F}{F-1}\right)^2 \qquad (7\text{-}80)$$

$$= 4\frac{[F/(F-1)]^2}{(V_p)^2(a_2/a_1)^2} \qquad (7\text{-}81)$$

When the oscillator is peak deviated by V_p volts, the peak error in deviation frequency (including the DC shift) to that predicted by the linear term in Eq. (7-72) is $a_2 V_p^2$, and the percentage error is

$$\Delta = \text{percentage error at peak deviation} = 100\,(a_2/a_1)V_p \qquad (7\text{-}82)$$

or

$$\Delta^2 = 10^4(a_2/a_1)^2 V_p^2 \qquad (7\text{-}83)$$

Substitution into Eq. (7-81) results in

$$\left(\frac{S}{D}\right)_2 = \frac{[F/(F-1)]^2}{\Delta^2} 10^4 \qquad (7\text{-}84)$$

which is plotted in Fig. 7-12.

D. Techniques to Reduce Distortion

The methods that may be employed to reduce signal distortion introduced by the FMFB are much the same as those discussed for the PLL in Chapter 6. Of first consideration is the design of suitably linear loop components. The VCO may represent the element that exhibits the greatest nonlinearity, unless

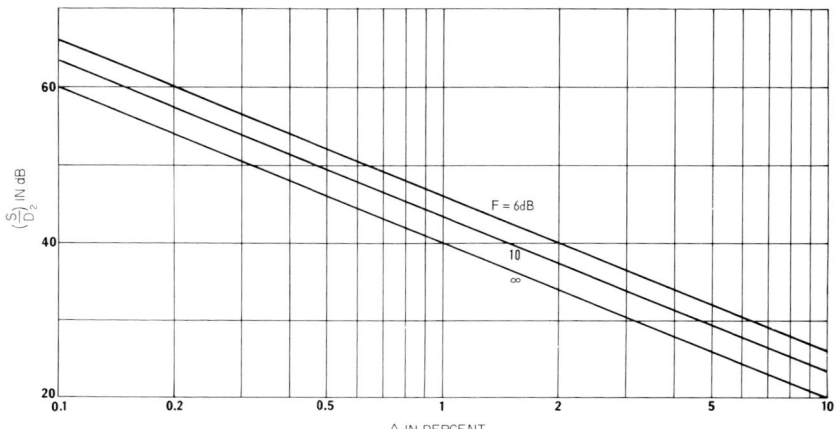

Fig. 7-12. FMFB output signal-to-distortion ratio as a function of VCO deviation linearity, with F as a parameter.

it is of suitable design. FMFB performance can, therefore, benefit greatly by attention to VCO linearity factors. For example, the varactor-controlled VCO exhibits primarily square-law distortion, which may be eliminated by the push–pull design discussed in Chapter 3. The result, of course, is a VCO primarily exhibiting cubic distortion at a much lower level. Other types of VCO's, such as voltage-controlled multivibrators, afford superior linearity but usually at a sacrifice of stability and internal noise.

The LD represents a secondary source of distortion, as evidenced in the example above. The discriminator is designed wideband to minimize loop delay contributions; wideband design generally provides adequate demodulation linearity.

The internal IF filter (usually a single pole), on the other hand, contains a fundamental source of distortion that arises from the filtering of FM signals, as discussed in Chapter 3. Although the distortion is basically related to the complete filter transfer function, components of the distortion may be identified, in an approximate manner, as arising from phase or amplitude nonlinearity. Thus, by applying "all-pass" phase correction networks after the internal IF filter, a significant portion of the modulation distortion may be suppressed. The main disadvantage in this technique is that excess delay is inadvertently introduced into the loop, resulting in threshold performance deterioration. The trade-off between distortion reduction and threshold degradation for this technique is not presently known. Another reported approach to reduce IF network distortion or, alternatively, to permit a narrowing of the IF network noise bandwidth (and thus reduce open-loop threshold), is the utilization of a double-pole IF filter.[5]

Finally, output distortion is seen to be controlled by the choice of the FMFB output point within the loop. As shown in Section 6.2, the loop gain effectively reduces the distortion with respect to the signal component when taking the signal output at the output, rather than the input, of the distorting element. Note that with FDM-FM operation and a predominance of baseband or VCO distortion, the loop output distortion peaks at low frequencies (see Fig. 6-12), resulting in poorest output NPR at the bottom of the baseband. This is true for both output points (before and after† the VCO), since the FMFB open-loop gain characteristic is relatively flat across the baseband. However, the output NPR (when dominated by VCO distortion) is improved by the loop feedback factor when the output is taken after the VCO. In the case of the PLL, the loop gain increases (12 dB/octave for the second-order loop) for decreasing baseband frequency, causing the poorest NPR (when dominated by VCO distortion) to exist in the top baseband channel when the loop output is taken after the VCO. In general, for competitive loop FMFB and PLL designs, VCO distortion (which manifests itself equally‡ in both circuits when taking the loop output before the VCO) will be more effectively suppressed (resulting in higher NPR's) in the PLL configuration with the output taken after the VCO. This is a result of the generally larger design loop gains found in the PLL.

7.3. Threshold-Limited Region

A. Factors Affecting Threshold

Before considering the design approach applicable to low-threshold FM demodulation, a discussion of the factors that influence threshold is in order. Threshold itself is a function of the additional output noise introduced by the nonlinear detection characteristics of the FMFB, as indicated in Chapter 4. We begin by considering some of the characteristics of this additional, or threshold-producing, noise.

The excess noise is generated from two sources, which appear to be independently controlled by the loop parameters. The foremost source, or threshold mechanism, is the threshold of the detector employed within the loop,

---

† The loop output is taken after the VCO by applying the VCO output signal to a linear LD. Overall threshold performance remains the same, being determined by the loop. The input to the auxiliary LD is an FM signal, with all noise components inscribed in phase and frequency.

‡ Equal VCO linearities are assumed; however, because the FMFB VCO deviates $(F-1)/F$ times the loop input deviation, somewhat less VCO-induced distortion is experienced for small values of F in the FMFB than in the PLL.

in this case an LD. In Chapter 3 we presented a physical picture and a model of the LD threshold mechanism which was applied successfully by Rice[6] to LD threshold noise calculations. The model, represented in Fig. 7-13, gives the total LD output noise as the sum of "linearly" demodulated and impulse noise. The impulses, called "threshold impulses," have 2π intensity, and are due to the 2π additional angular rotations of the resultant vector formed from input signal plus noise.

Although this model gives a physical picture of the LD threshold phenomenon, it fails when attempting to predict the threshold of an LD preceded by a single-pole filter; the single-pole is the most often used FMFB internal IF filter. Note that threshold is defined here as that operating point where the total output noise increases by 1 dB over that predicted by the "linearly" demodulated noise alone. Therefore, in terms of the impulse noise model, threshold implies a specific threshold impulse rate, which, by Eq. (3-28) is seen to depend directly on the radius of gyration (about the center frequency) of the LD input filter. As the single pole's radius of gyration is unbounded, the application of Eq. (3-28) is prohibited, thus voiding the model in this instance. Approximate analytical results may be obtained by substitution of other filter shapes for the single pole; this is done by Enloe, where threshold characteristics for the LD preceded by a Gaussian shaped filter are utilized in FMFB analysis and synthesis. The Gaussian filter threshold characteristics from Enloe (Fig. 7-14) are based on calculations contained in unpublished Bell Telephone Laboratories Memorandums, which are, in turn, based on early published material by Rice.[6a] The threshold criterion of Enloe is the point where the output SNR begins to drop off more rapidly than predicted

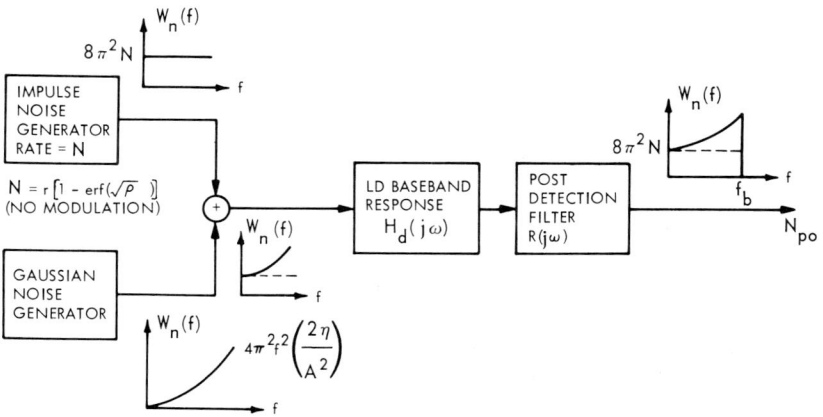

Fig. 7-13. Diagrammatical representation of threshold impulse noise model for LD. ($W_n(f)$ is noise power density in watts per hertz, r is IF filter radius of gyration, and ρ is input carrier-to-noise ratio.)

7.3. Threshold-Limited Region

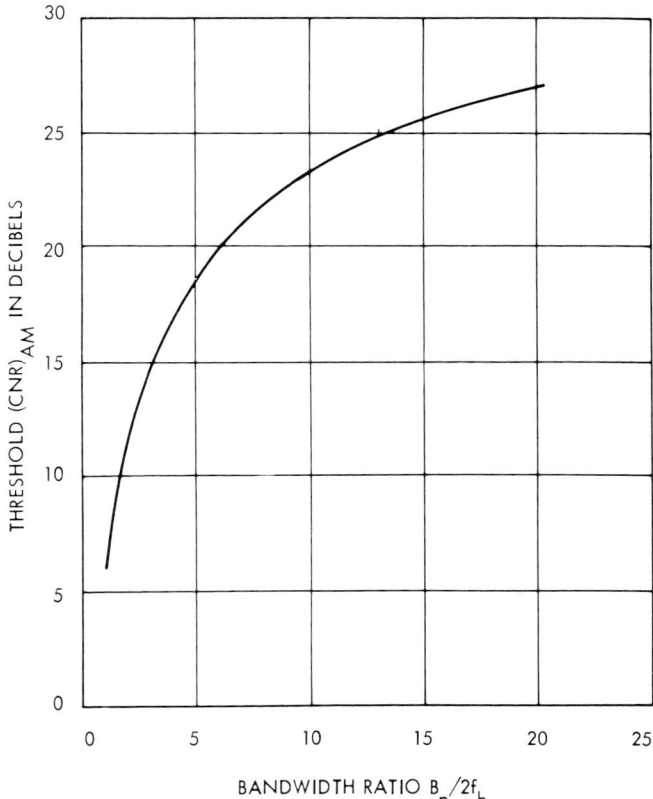

Fig. 7-14. Threshold $(CNR)_{AM}$ as a function of IF-to-base-bandwidth ratios (from Enloe[1b]).

by the FM improvement equations. The adequacy of this data to represent LD single-pole characteristics is evidenced by the comparison, in Fig. 7-15, to available FDM-FM experimental threshold data.[7] Note that the Enloe curve, lying between top and bottom channel experimental characteristics, appears to represent a "mean" baseband threshold.

The second source of nonlinear noise in the FMFB is the feedback threshold noise. Originally identified by Enloe as an intermodulation between "in-phase" input noise and the loop feedback noise, it gives rise to a feedback threshold mechanism which, from empirical experimental data, appears to be uniquely related to the FMFB closed-loop noise bandwidth.

For the case of an unmodulated signal and white Gaussian noise input to the FMFB, Enloe has determined that the feedback threshold is a function of the noise component of the mean square VCO phase jitter $\overline{\phi_{rn}^2(t)}$ (Section 4.4).

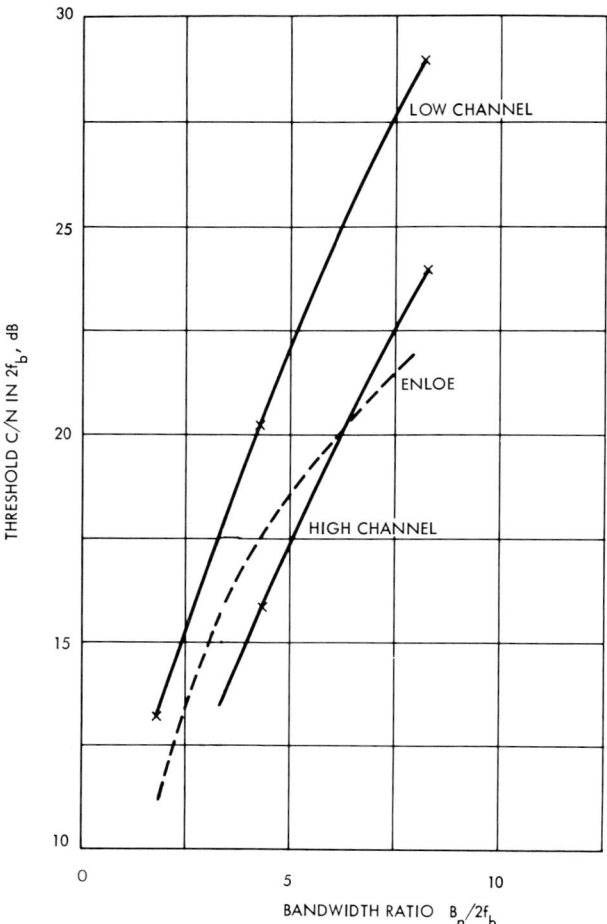

Fig. 7-15. Open-loop FDM-FM threshold CNR vs IF-to-base-bandwidth ratio (from Lefrak[7]).

The mean-square VCO jitter above threshold is related to the normalized input CNR by way of Eq. (4-74)

$$\overline{\phi_{rn}^2}(t) = \frac{2\eta B_n}{A^2}|H(0)|^2 = \frac{4\eta f_b}{A^2}|H(0)|^2 \frac{B_n}{2f_b} = \frac{1}{2(CNR)_{AM}}|H(0)|^2 \frac{B_n}{f_b}† \quad (7\text{-}85)$$

where

$$B_n \equiv \frac{\int_0^\infty |H(j\omega)|^2\, df}{|H(0)|^2} = \text{FMFB closed-loop noise bandwidth} \quad (7\text{-}86)$$

† Note that for the Enloe design, $|H(0)|^2 = [(F-1)/F]^2$, where $F = 1 + K$, the loop feedback factor [see Eqs. (7-9) and (7-12)].

7.3. Threshold-Limited Region

Applying the Enloe feedback threshold criterion $\overline{\phi_{rn}^2}(t) = 1/9.6$, we obtain

$$(\text{CNR})_{\text{AM}} = 4.8(B_n/f_b)|H(0)|^2 \qquad (7\text{-}87)$$

Equation (7-87) points out the important fact that to minimize the feedback threshold for a fixed closed-loop gain $|H(0)|^2$, we must minimize the closed-loop noise bandwidth B_n.

Recall that the above results and conclusions are based on an unmodulated input carrier and white Gaussian noise. Both modulation and pre-FMFB filtering are yet to be included in the feedback threshold characteristic. The precise way in which these two factors enter the picture is unknown; however, certain qualitative results are known, and are discussed next.

We saw in Section 4.4 that a quasi-static signal frequency deviation alters the closed-loop response and bandwidth of the FMFB resulting, for the particular case examined, in an increased mean square VCO phase jitter for increasing signal frequency deviation (given a fixed input $(\text{CNR})_{\text{AM}}$ white noise environment). It is thus reasoned that signal modulation tends to deteriorate the feedback threshold; the extent of deterioration found experimentally, however, is not fully accounted for by the noise bandwidth increase discussed above. Note that modulation also increases, or deteriorates, the LD (or open-loop) threshold. As modulation deviates the signal from center frequency, the LD input filter (commonly a single pole) increasingly attenuates the signal amplitude, increasing the probability or rate of threshold impulses. Even for an ideal rectangular IF filter, where no signal attenuation is experienced on a quasi-static frequency deviation basis, signal modulation increases the rate of threshold impulses (Chapter 3).

Pre-FMFB filtering primarily affects threshold in two ways. First, some loss in signal power is experienced when filtering an FM signal. This, however, may be held to an absolute minimum by suitable choice of bandwidth and filter response shape; usually, a Butterworth† shape with the Carson rule design bandwidth

$$B_{\text{CR}} = 2(\Delta f_p + f_b) \qquad (6\text{-}64) \quad (7\text{-}88)$$

is employed. Second, the noise in the loop, calculated from the linear model, is shaped by the pre-FMFB filter. Specifically, the phase noise spectrum at the VCO output for the case of an unmodulated input signal, is the product of the lowpass equivalent power response of the pre-FMFB filter and the FMFB closed-loop response at high loop input CNR (Section 4.2). Hence, the mean square VCO phase jitter (unmodulated signal input) is

$$\overline{\phi_{rn}^2}(t) = (2\eta/A^2)\int_0^\infty |M_{\text{pL}}(j\omega)|^2 |H(j\omega)|^2\, df \qquad (7\text{-}89)$$

† Phase correction networks following the filter may be utilized to reduce signal distortion; in this instance, threshold is unaffected, since the networks are applied ahead and not within the loop. The Bessel filter, which in itself exhibits excellent phase linearity, is an alternate as the pre-FMFB filter.

where $|M_{pL}(j\omega)|^2$ is the lowpass-equivalent power response of the pre-FMFB filter, and $|H(j\omega)|^2$ is FMFB closed-loop response (magnitude squared). We may deduce that the pre-FMFB filter, by constraining the integrand of Eq. (7-89), tends to reduce or inhibit the feedback threshold; the extent to which this occurs is related to the relative bandwidth and shapes of the FMFB closed-loop and pre-FMFB filter responses. We conclude that in the absence of modulation, the test for the effectiveness of the pre-FMFB filter to reduce the closed-loop threshold is given by evaluating Γ, the ratio of effective noise bandwidth to the loop noise bandwidth,

$$\Gamma = \frac{\int_0^\infty |H(j\omega)|^2 |M_{pL}(j\omega)|^2 \, df}{|H(j0)|^2 |M_{pL}(j0)|^2} \bigg/ \frac{\int_0^\infty |H(j\omega)|^2 \, df}{|H(j0)|^2} \quad (7\text{-}90)$$

Usually, Γ is close to unity, the numerator is determined primarily by $|H(j\omega)|^2$, and the presence or absence of a pre-FMFB filter is inconsequential as far as the feedback threshold is concerned (the pre-FMFB filter is still needed to prevent overloading of the mixer and to reject other interfering signals). However, for smaller modulation indices, and depending upon the type of modulation, Γ may be substantially less than unity. It would appear that the threshold $(CNR)_{AM}$ would in such cases be reduced by Γ. However, this has not yet been fully verified experimentally.

In the presence of modulation, the influence of the pre-FMFB filter is more difficult to assess. In this case, the fed-back noise is nonstationary, being a function of the instantaneous frequency deviation of the signal. Consider an extreme case, where the input signal has deviated to one side of the pre-FMFB filter response. The above-threshold noise, calculated on the basis of the FMFB linear model, is shaped such that it has a power spectral density η and bandwidth B_p, resulting in a mean square VCO phase jitter of

$$\overline{\phi_{rn}^2}(t) = (\eta/A^2) \int_0^\infty |M_p\{j[\omega + \omega_i - (B_p/2)]\}|^2 \cdot |H(j\omega)|^2 \, df\dagger \quad (7\text{-}91)$$

The significant test ratio is of the same form as Eq. (7-90), except that $|M_{pL}(j\omega)|^2$ is replaced by $|M_p\{j[\omega + \omega_i - (B_p/2)]\}|^2$. Note that

$$\int_0^\infty |M_p\{[\omega + \omega_i - (B_p/2)]\}|^2 \, df = 2\int_0^\infty |M_p[j(\omega - \omega_i)]|^2 \, df \quad (7\text{-}92)$$

† For simplicity, a rectangular shaped pre-FMFB filter with bandwidth B_p is assumed to prevent spectral foldover problems.

Thus with modulation, the predetection filter does not restrict the noise spectrum as much as without modulation. On the other hand, with the signal at the edge of its passband, the filter prevents noise foldover and this reduces the effective noise density within the loop by as much as 3 dB.

The exact relationship and interplay between the controlling factors have yet to be developed, but in a qualitative way we have indicated the major factors in FMFB threshold theory.

We conclude by commenting on an unexploited investigation made by Davis,[8] where he demonstrated, for the case of quasi-static modulation, that a dynamic instability exists in the FMFB. This instability is akin to a loss of synchronism (such as the loss-of-lock in a PLL) with respect to the signal, and is found to be a function of the signal modulation and fed-back noise, thus resembling a "feedback threshold." The concept that the feedback threshold is actually "a loss of synchronism phenomenon" was advanced earlier by Baghdady.[9] However, reduction of the phenomenon to a model that adequately includes the influence of modulation or pre-FMFB filtering is lacking. We note, in this respect, that the establishment of a loss-of-lock model in the case of the PLL facilitated considerably its analysis over a wide range of operating conditions.

B. Design for Low Threshold

The work of Enloe in 1962 represented a major breakthrough in FMFB low-threshold design analysis and procedure. Soon afterward, various analysis and design theories were proposed, falling basically into two schools of thought: (1) narrow-band design[1,10] and (2) wide-band design.[11,12]

The narrow-band design featured here and originated by Enloe develops from the two-threshold mechanism theory; the first threshold being that of the internal detector, and the second, a closed-loop (or feedback) threshold, being a function of the fed-back noise in the loop. The Enloe design follows by selecting the loop gain and filter parameters to independently control the threshold mechanisms. The two thresholds are made to occur at the same input CNR, resulting in a minimum-threshold design. The shortcomings in the available design method are that (1) modulation, in general, is inadequately considered, and (2) a white noise input is assumed, thereby forcing the design to neglect the influence of pre-FMFB filtering and raising the possibility of a false design optimization.

The wide-band design originates with Develet[11] followed by Schilling[12] with similar conclusions based, however, on a different analytical approach. This design essentially requires as large a loop gain as possible, so that the input FM signal is compressed to a very small frequency deviation. The signal may now be filtered by a very narrow-band IF filter, resulting in the elimination of the detection threshold. The final result is that the threshold determining factor, following the Schilling analysis, is a feedback type threshold controlled by the pre-FMFB filter.

Of the methods available, the Enloe procedure is considered of widest applicability. It begins with the objective of realizing a maximally-flat open-loop transfer function $G(s)$ across the baseband frequency range. The resulting closed-loop response is restricted to second-order, with a design for minimum closed-loop noise bandwidth. These objectives are attained by the use of the loop baseband filter $H_2(s)$,

$$H_2(s) = \frac{[s/(b\omega_b) + 1][s/(a\omega_b) + 1]}{[(s/\omega_b)^2 + 2^{1/2}(s/\omega_b) + 1]} \tag{7-93}$$

The first zero (at frequency $b\omega_b$) in $H_2(s)$ effectively cancels the pole created by the narrow-band IF single-pole filter used in the forward part of the loop.† The second zero (at frequency $a\omega_b$) is utilized to minimize the closed-loop noise bandwidth B_n for a fixed open-loop gain K. The resulting open- and closed-loop responses, $G(s)$ and $H(s)$ are, respectively,

$$G(s) = \frac{K\left[\dfrac{s}{a\omega_b} + 1\right]}{\left(\dfrac{s}{\omega_b}\right)^2 + 2^{1/2}\left(\dfrac{s}{\omega_b}\right) + 1} \tag{7-94}$$

$$H(s) = \frac{\dfrac{K}{1+K}\left(\dfrac{s}{a\omega_b} + 1\right)}{\left[\dfrac{s}{(1+K)^{1/2}\omega_b}\right]^2 + \left[\dfrac{2^{1/2} + (K/a)}{(1+K)^{1/2}}\right]\left[\dfrac{s}{(1+K)^{1/2}\omega_b}\right] + 1}$$

$$= \frac{\dfrac{K}{1+K}\left[\dfrac{s}{a_0\omega_n} + 1\right]}{\left(\dfrac{s}{\omega_n}\right)^2 + 2\xi_0\left(\dfrac{s}{\omega_n}\right) + 1} \tag{7-95}$$

with

$$\omega_n = (1+K)^{1/2}\omega_b, \quad a_0 = \frac{a}{(1+K)^{1/2}}, \quad \xi_0 = \frac{1}{2}\left[\frac{2^{1/2} + K/a}{(1+K)^{1/2}}\right], \quad b = \frac{B_{IF}}{2f_b}$$

(7-96)

† Refer to Section 4.3 for details on the equivalent response of a single-pole IF filter. Note that the Enloe design is predicated on "no modulation."

7.3. Threshold-Limited Region

For very wide-band basebands, it may be necessary also to cancel the baseband effect of the LD, as approximated by Eq. (3-34). A check on the relative location of the poles given by Eq. (3-34) will reveal whether such cancellation is needed.

The closed-loop noise bandwidth, from Eq. (7-86), is

$$B_n = \left(\frac{1+K}{K}\right)^2 \int_0^\infty |H(j\omega)|^2 \, df \qquad (7\text{-}97)$$

which is minimized, for each value of K, by adjusting a. The minimization is performed in Appendix F, giving the optimum value for

$$a = 2^{1/2} \frac{F}{F-1}\left[1 + \left(1 + \frac{(F-1)^2}{2F}\right)^{1/2}\right] \qquad (7\text{-}98)$$

and the minimum noise bandwidth

$$\frac{B_n}{f_b} = \frac{\pi F}{2}\left[\frac{F + a^2}{a(2^{1/2}a + F - 1)}\right] \qquad (7\text{-}99)$$

Equations (7-98) and (7-99) are plotted in Fig. 7-16.

By applying the minimum noise bandwidth to Eq. 7-87, the feedback threshold characteristic curve [(CNR)$_{AM}$ vs. F] is obtained as

$$(\text{CNR})_{AM} = \frac{2.4\pi(F-1)^2}{F}\left[\frac{F+a^2}{a(2^{1/2}a + F - 1)}\right] \qquad (7\text{-}100)$$

Equation (7-100) is plotted in Fig. 7-17.

The final step in the Enloe design is to establish the intersection of the open-loop and feedback thresholds, so that the optimum feedback factor may be determined. Note that the FMFB compresses the signal deviation by the factor F, resulting in a compressed modulation index of σ/F for FDM and speech signals, or m_p/F for TT signals within the loop IF. The optimum IF bandwidth and the resulting open-loop threshold based on the compressed signal are to be determined; unfortunately, these factors are not adequately known. Experience[7,13] indicates that the following IF design rules

$$B_{IF} = 2f_b(10)^{1/2}(\sigma/F) \qquad \text{FDM} \quad (0 \text{ to } \omega_b)\dagger \qquad (7\text{-}101)$$

$$B_{IF} = 2f_T[1 + \Delta f_p/(f_T F)] \qquad \text{TT} \quad \text{at } \omega_T \qquad (7\text{-}102)$$

where Δf_p is FMFB input peak frequency deviation, σ is FMFB input rms modulation index (rms frequency deviation/top baseband frequency),

---

† The $\sqrt{10}$ factor indicates the application of a 10 dB peak factor to convert rms frequency deviation to an equivalent peak frequency deviation.

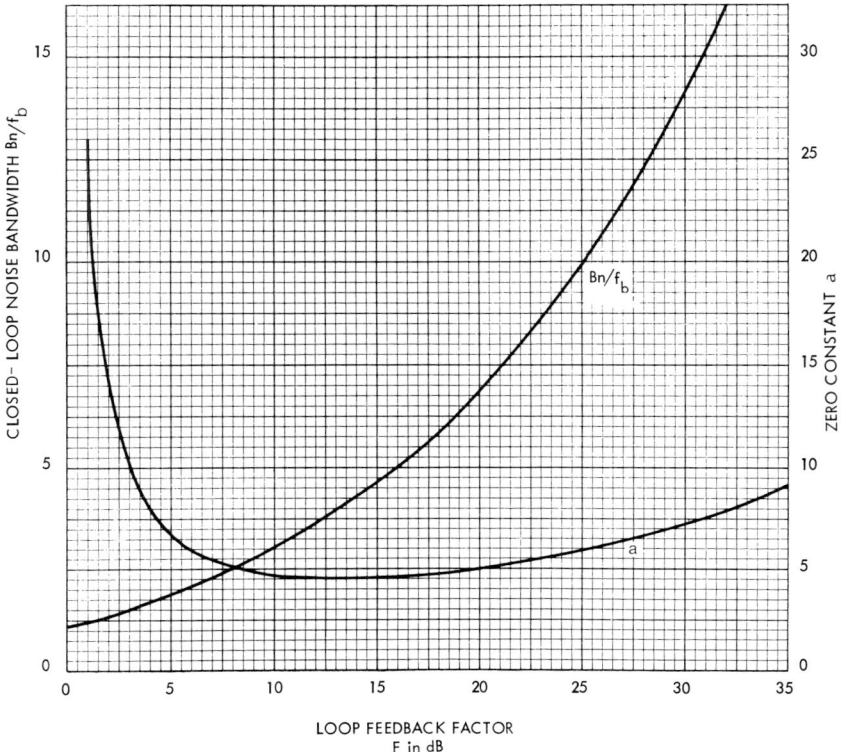

Fig. 7-16. FMFB minimum noise bandwidth and optimum zero constant as a function of loop feedback factor.

coupled with the threshold curve of Fig. 7-14, adequately predict the open-loop threshold characteristic curves. These are plotted in Figs. 7-18 and 7-19 with the peak modulation index as a parameter.†

Figures 7-18 and 7-19 also have a replot of the feedback threshold curve (Fig. 7-17) for the no-modulation case, to obtain the optimum feedback factor F and minimum threshold CNR, at its intersection with the open-loop threshold characteristics. The other loop parameters follow from Eqs. (7-101) or (7-102), and Fig. 7-16, accordingly. Experimental evidence[7,13] indicates that while threshold in the absence of modulation is accurately predicted by the foregoing method, threshold in the presence of modulation is considerably different. Experimental optimization by variation of loop parameters in the presence of modulation, as described in Chapter 10, indicates

---

† These curves are basically from Enloe, with the modulation index modified as per Eq. (7-101). Note that the single-pole IF noise bandwidth to 3 dB bandwidth relationship is $B_n(\mathrm{IF}) = (\pi/2)B_{\mathrm{IF}}$.

7.3. Threshold-Limited Region

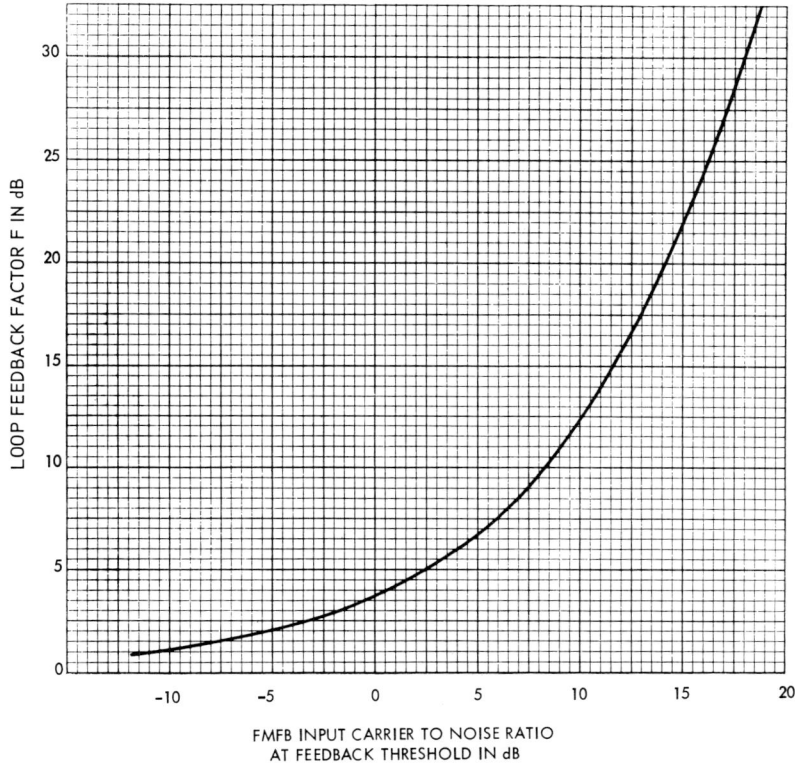

Fig. 7-17. FMFB feedback threshold characteristic (referred to twice base bandwidth).

that minimum threshold is achieved by decreasing the feedback factor and increasing the internal IF bandwidth beyond that indicated by the "no modulation" design. This signifies that the initial loop design is "feedback threshold dominated" in the presence of modulation; in other words, the modulation effectively shifts the feedback threshold curve (Figs. 7-18 and 7-19) to the right, which results in a higher feedback threshold for a given feedback factor F. The values of minimum threshold and optimum feedback factor F that correspond to the experimental results can be generally achieved by shifting the feedback threshold curve about 5 dB to the right (see Figs. 7-18 and 7-19) for the case of FDM and TT (at top baseband frequency) modulation. The reasons why these types of modulation deteriorate the feedback threshold by such a large factor is not known. Experience indicates that the effective shift in the feedback threshold curve depends directly on the relative concentration of modulating energy at the top of the baseband. For TT at $\frac{1}{4}$ baseband frequency (the typical test condition for a 4-kHz speech channel) the feedback threshold shift does not appear to be significant.

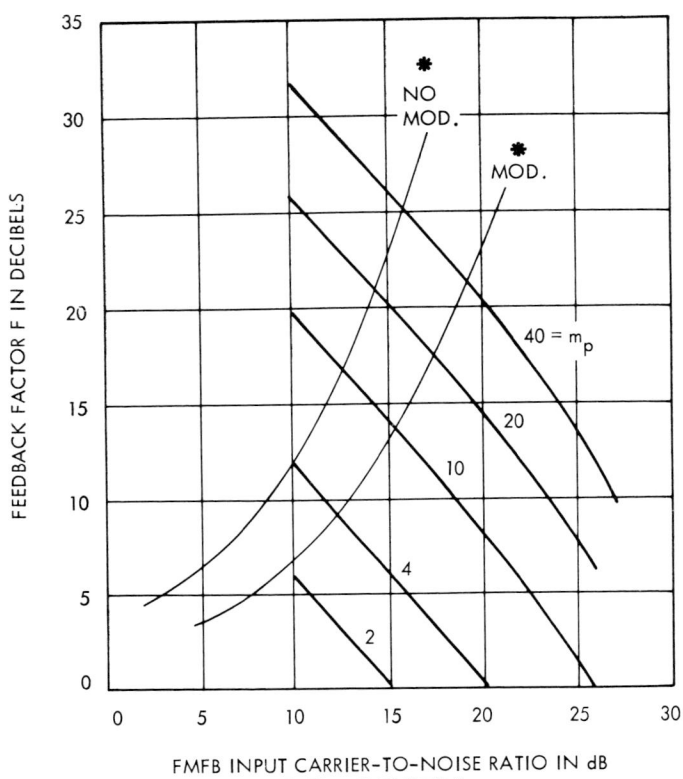

Fig. 7-18. Open-loop and feedback thresholds vs feedback factor F for FDM signals $[m_p = 10^{1/2}\sigma$; No mod. (no modulation) and open-loop threshold curves are from Enloe[1b]; * indicates feedback threshold].

For an experimental optimization under various modulation conditions, or where even small improvements are of significance, the reader is referred to procedures listed in Chapter 10.

For speech modulation, a design based on a 1-kHz TT having a peak deviation equal to the speech appears to be valid for all parameters except the cutoff frequency of the baseband filter which remains 4 kHz.[14] The internal IF bandwidth is determined by the TT parameters and application of Eq. (7-102), which results in

$$B_{IF} = 2f_T[1 + (\Delta f_p/Ff_T)] \quad \text{and} \quad b = B_{IF}/2f_T \quad (7\text{-}103)$$

The threshold CNR and the optimum feedback factor F are determined from Fig. 7-19 from the intersections of the "no modulation" feedback threshold curve and the family of open-loop threshold curves. The parameter

7.3. Threshold-Limited Region

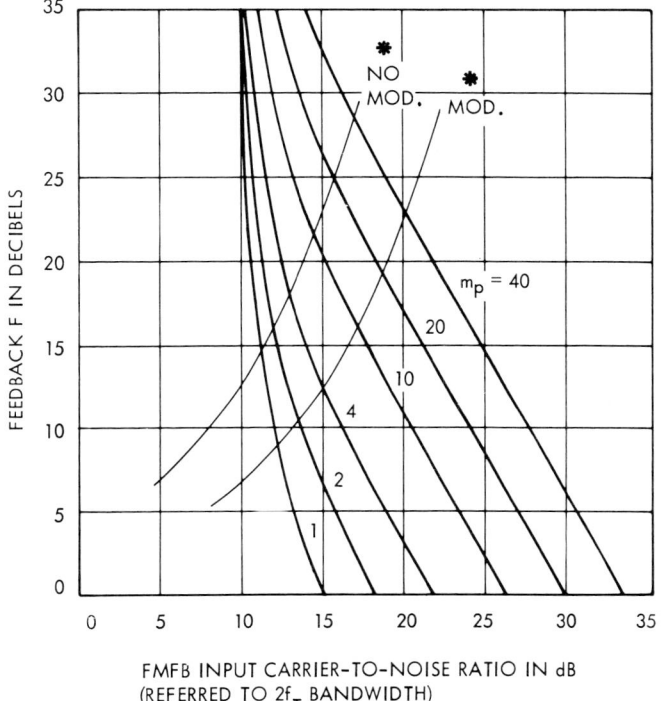

Fig. 7-19. Open-loop and feedback threshold characteristics for test-tone at top baseband frequency [No mod. (no modulation) and open-loop threshold curves are from Enloe[1b]; * indicates feedback threshold].

a, which is solely a function of the feedback factor F, is found by reference to Fig. 7-16.

The FMFB will actually be used to demodulate a speech signal which has a baseband of 4 kHz; it is therefore desirable to reference the threshold CNR to twice the base bandwidth $(2f_b)$, and the modulation index to the system peak index (m_p), resulting in

$$\text{CNR}_{2f_b} = (\text{CNR})_{\text{AM}} = \text{CNR}_{2f_T} - 6 \quad \text{dB} \tag{7-104}$$

and

$$m_p = \Delta f_p / 4 f_T \tag{7-105}$$

Although the same loop design is taken for both speech and TT ($f_T = f_b/4$) modulation, the threshold performance is not necessarily the same in the presence of either modulation. As seen for the PLL, the same loop design resulted (Section 6.4) in a threshold lower by 1.2 dB for speech than for TT of the same peak deviation. This 1.2-dB factor is also applied to the FMFB to project speech modulation threshold performance, as indicated in Fig. 7-20. Note that for equal peak deviations, the rms modulation indexes are related by

$$\sigma = m_{\text{rms}}/5^{1/2} \tag{7-106}$$

where σ is rms modulation index for speech and m_{rms} is rms modulation index for TT, because of the difference in peak-to-rms ratio.

The results for optimum feedback factor and minimum threshold CNR as a function of FMFB input signal modulation index are read from Figs. 7-18 and 7-19, and summarized in Figs. 7-20 and 7-21. By applying the IF bandwidth design Eqs. (7-101)–(7-103) to the results of Fig. 7-21, the loop filter zero constant b may be determined as a function of FMFB input modulation index. Similarly, by reference to Figs. 7-21 and 7-16, the loop filter zero constant a is determined. Both a and b are given in Fig. 7-22 as functions of FMFB input modulation parameters. As a further aid to loop and system

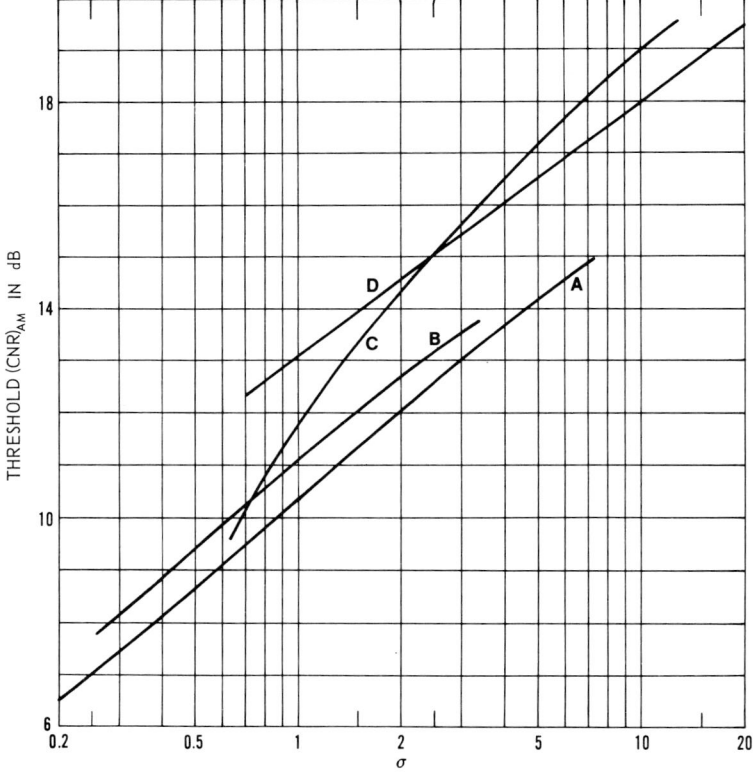

Fig. 7-20. FMFB threshold CNR as a function of input rms modulation index. A: test-tone ($f_T = 0.25 f_b$); B: speech (projected data); C: FDM-FM; D: test-tone ($f_T = f_b$).

7.3. Threshold-Limited Region

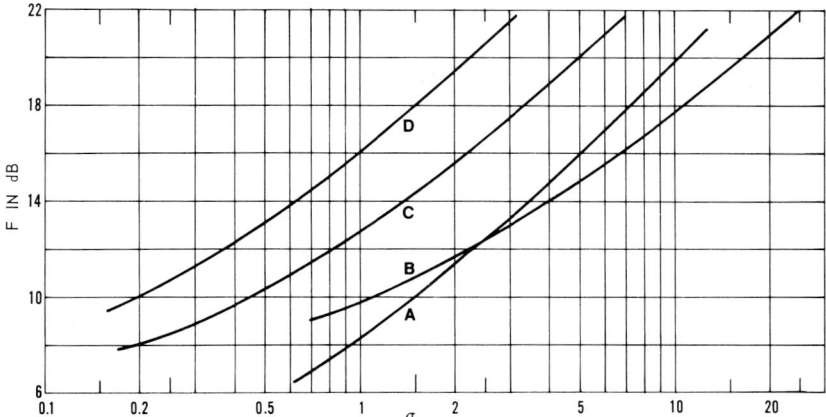

Fig. 7-21. FMFB design feedback factor as a function of input rms modulation index. A: FDM-FM; B: test-tone ($f_T = f_b$); C: test-tone ($f_T = 0.25 f_b$); D: speech.

design, Fig. 7-23 summarizes loop output SNR (or NPR) at threshold (assuming appropriate closed-loop response compensation by the output filter) as a function of loop input modulation parameters. The figure was obtained from Eqs. (7-1)–(7-5) by taking the CNR from Fig. 7-20 and introducing the 1-dB SNR degradation that defines the threshold point.

The reader has been made aware, throughout the development, of the various engineering approximations that were made. These, generally, have only secondary effects on the FMFB design parameters. However, the predicted CNR at threshold is more strongly influenced by the approximations made, and should, therefore, be used mainly in estimating the relative threshold improvement available with the FMFB technique. For the relative performance of the various demodulators, see Fig. 3-8.

C. Effect of Delay on Threshold

The Enloe analysis and synthesis procedure permits optimum loop design in the presence of excess delay. The open-loop transfer characteristic for this case is written as [compare with Eq. (7-94)]

$$G(s) = \frac{K\left(\dfrac{s}{a\omega_b} + 1\right) \exp\left(\dfrac{-s\phi_b}{\omega_b}\right)}{\dfrac{s^2}{\omega_b^2} + 2^{1/2}\dfrac{s}{\omega_b} + 1} \qquad (7\text{-}107)$$

where ϕ_b is loop delay phase shift at baseband frequency ω_b, with the result

210   7. Design of FMFB for FM Demodulation

that the closed-loop noise bandwidth B_n is a function of F (loop feedback factor is $1 + K$), a (compensating zero constant), and ϕ_b (delay constant). For specific values of ϕ_b and over a range of F, B_n is minimized by adjustment of a. The result is given by Enloe and reproduced here as Fig. 7-24. Taking the feedback threshold equation (7-87), and applying the results of Fig. 7-24, a family of feedback threshold characteristics may be derived, as shown in Fig. 7-25. In a manner similar to the design procedure discussed earlier in this section, the open-loop threshold characteristic (Fig. 7-18, for example) may be overlayed to obtain the optimum feedback factor and minimum threshold. As a by-product of the noise bandwidth minimization, we determine the compensating zero constant a as a function of F. The resulting optimum design, taken as the intersection of the open-loop and feedback threshold characteristics, determines the internal IF bandwidth and, hence, the zero constant b; these factors are given in Fig. 7-25 as a function of feedback factor

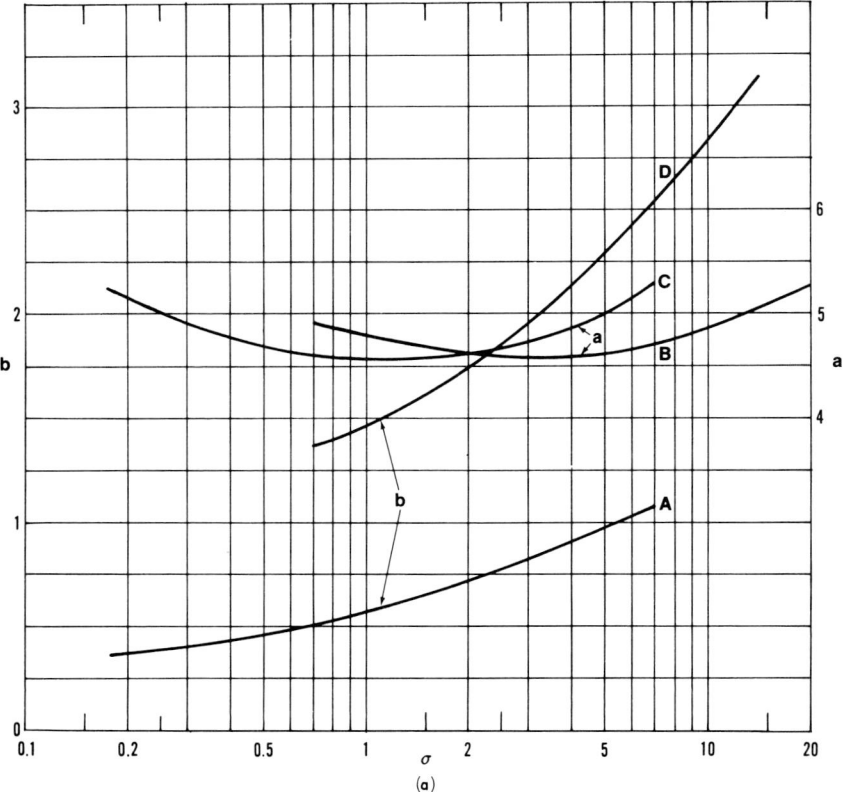

Fig. 7-22. FMFB filter parameters as functions of rms modulation index. (a) TT modulation (curves A and C are for $f_T = 0.25f_b$. Curves B and D are for $f_T = f_b$). (b) Speech

7.3. Threshold-Limited Region

F. As discussed previously, the Enloe feedback threshold characteristic, does not include the influence of modulation which, as indicated earlier can take on significant proportions and far outweigh design modifications (from the results of Fig. 7-24) dictated by delay alone. An accurate loop design in the presence of delay is not possible without a more concrete feedback threshold theory which encompasses modulation and delay simultaneously. With the lack of this information, the best alternative is to consider the results of Figs. 7-25 and 7-26 as a guide to the direction that loop parameters must be adjusted, from the "nominal" delayless design, to compensate for delay.

Consider a signal with a peak modulation index $m_p = 10$. For a loop design based on the premise of no delay (and no modulation) a feedback factor $F = 17$ dB, with a resulting threshold $(CNR)_{AM} = 12.5$ dB, is indicated by Fig. 7-25. In fact, if the loop should contain excess phase shift, say $\phi_b = 8°$, the optimum design (with no modulation) calls for a feedback factor $F = 15.5$

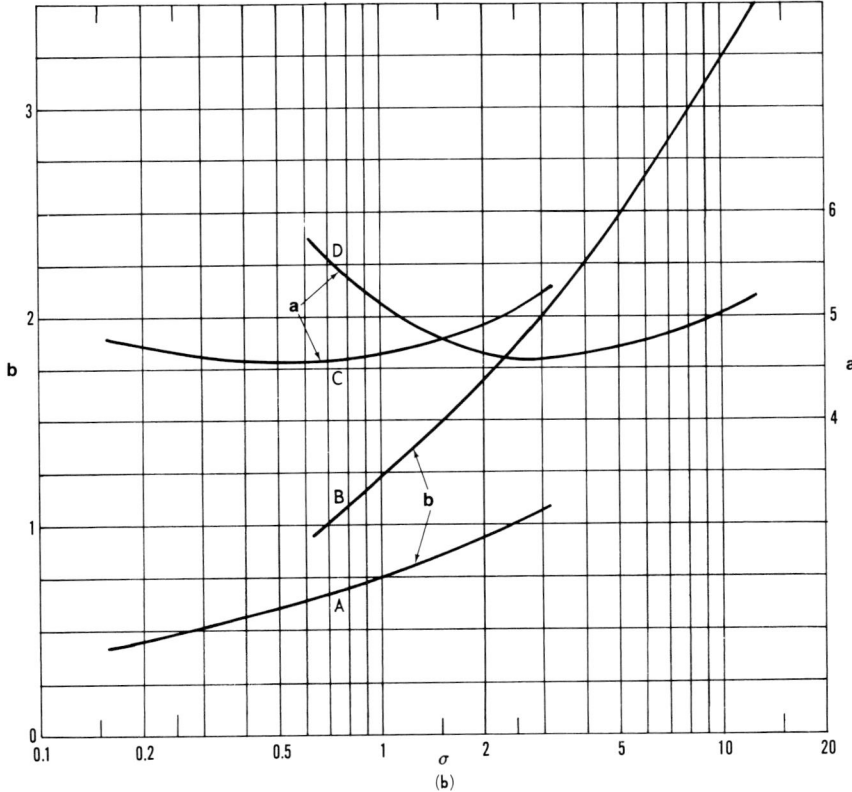

modulation (curves A and C), and FDM-FM modulation (curves B and D).

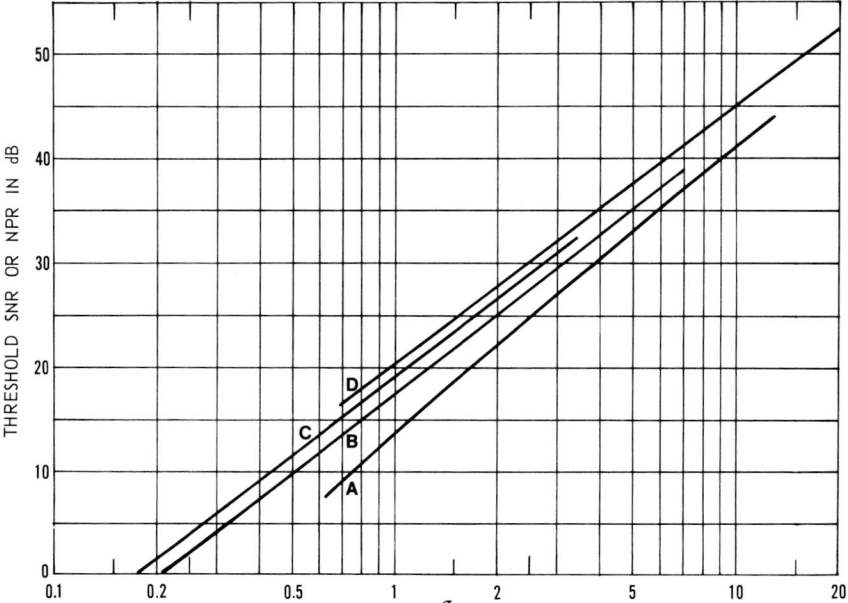

Fig. 7-23. FMFB threshold SNR measured in base-bandwidth, or top channel NPR, as a function of rms modulation index. A: FDM-FM; B: test-tone ($f_T = 0.25 f_b$); C: speech; D: test-tone ($f_T = f_b$).

dB, with a resulting $(CNR)_{AM} = 13.5$ dB. The other parameters in the loop design follow accordingly from Fig. 7-26, with the presence of delay requiring a lower compensating zero constant a, wider internal IF bandwidth, and larger zero constant b.

D. Effect of VCO Noise and Drift

The effect of VCO frequency instability, noise, or drift in the operation of an FMFB demodulator manifests itself in several ways. The primary effect is to create within the FMFB a frequency error component which results in both SNR and threshold CNR degradation. The sources of this frequency instability lie in the natural noise found in the VCO control element, the power supply or, equivalently, the amplifying stages which precede the VCO. The noise may be represented by an equivalent input generator to a noiseless VCO, in a manner similar to distortion represented by an equivalent distortion generator, as was indicated in Fig. 7-11. The characteristics of the equivalent generator are derived by measurement of the open-loop VCO characteristics. Examining the resulting loop equations indicates that the same loop

7.3 Threshold-Limited Region

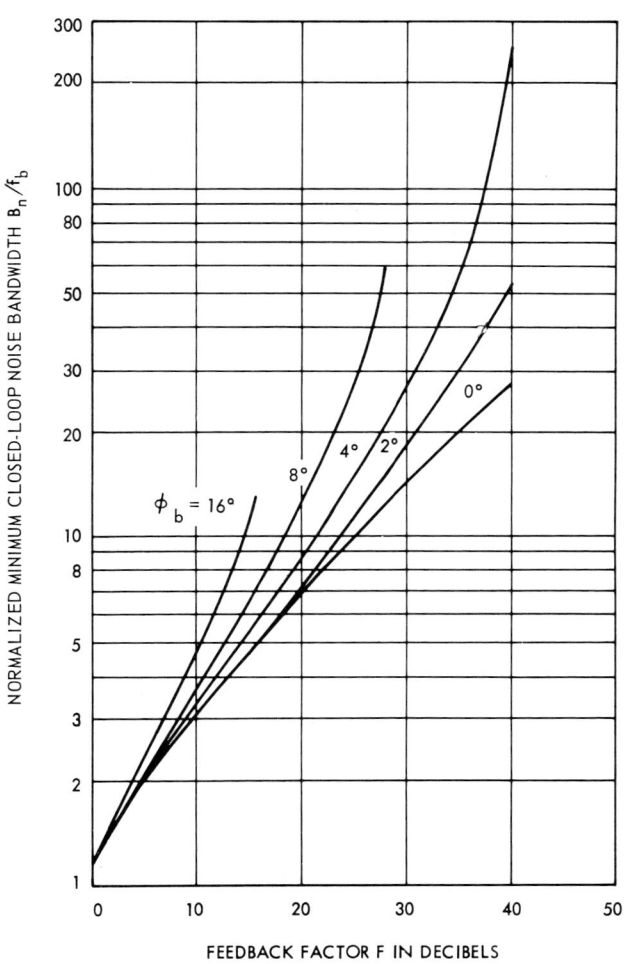

Fig. 7-24. Minimum closed-loop noise bandwidth as a function of F for various ϕ_b (from Enloe[1b]).

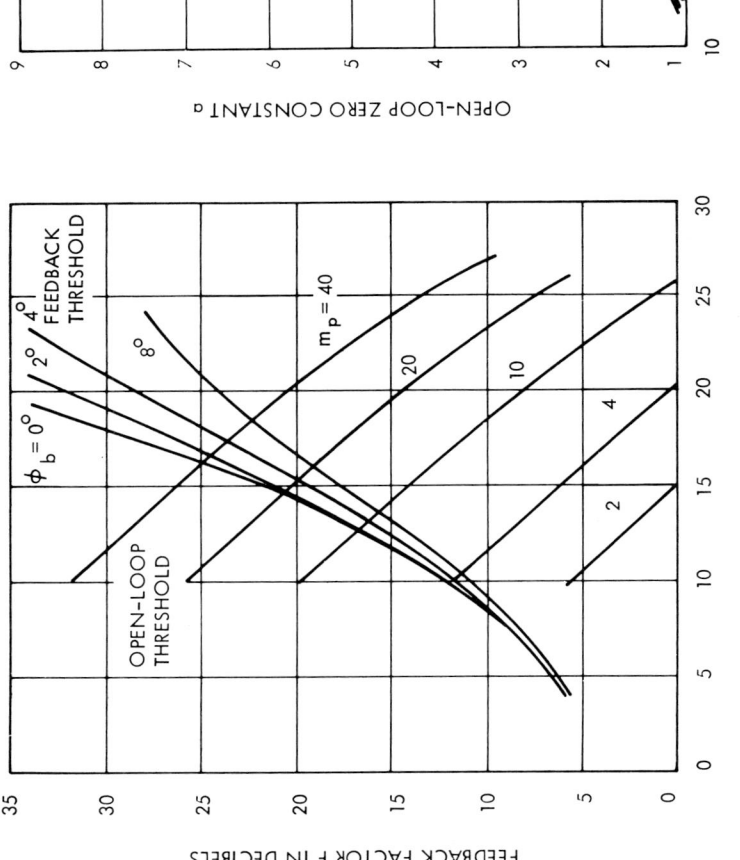

Fig. 7-25. Open-loop and feedback thresholds for an FMFB with delay (from Enloe[1b]).

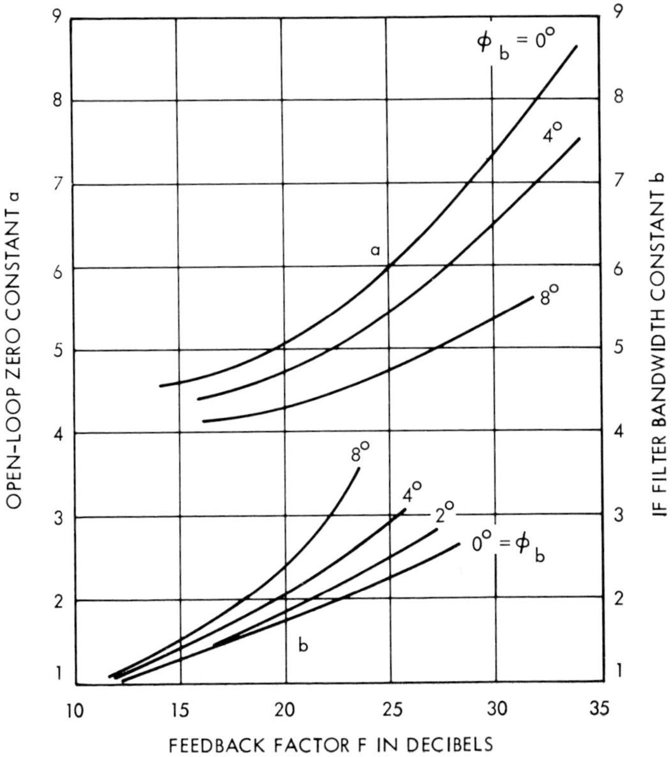

Fig. 7-26. Filter parameters as functions of feedback factor and delay (from Enloe[1b]).

output may be achieved by introducing an equivalent modulation component in the input signal. For example, specifying the input signal phase modulation to be

$$\phi_i(t) = \phi_{is}(t) + \mu \int e_n(t)\, dt \qquad (7\text{-}108)$$

where $\phi_{is}(t)$ is received signal phase modulation, μ is VCO sensitivity in radians per volt-second, and $e_n(t)$ is equivalent noise input generator that accounts for VCO noise (the equivalent generator is removed from the loop), fully accounts for the effects of VCO drift.

Because the influence of modulation on the threshold operation of the FMFB is not completely understood, the effect of VCO noise cannot be included in the minimum threshold design procedure. Furthermore, VCO noise, interpreted as an equivalent modulation input, does not resemble the modulation types that were considered earlier; FDM and speech signal spectra abruptly terminate at a top baseband frequency, whereas the spectra of VCO frequency jitter normally decay 6–12 dB/octave over all frequencies. Quantitatively, a rough estimate may be made on the seriousness of VCO jitter for a particular loop design by first forming the equivalent input generator indicated by Eq. (7-108), and then calculating the mean-square frequency error that results in the loop IF. If this component is 20 dB below the intended signal modulation component, little threshold degradation will result by implementing the loop with the particular VCO under consideration.

An estimate may be made on the asymptotic SNR as limited by the VCO jitter. By integrating the VCO frequency jitter spectrum over the baseband of interest, the relative jitter or noise power may be determined with respect to the signal power. This is the limiting output SNR for the loop. Note that, as in the case of VCO distortion, the noise contribution to the FMFB output due to VCO jitter may be reduced by the feedback factor of the loop when taking the loop output after the VCO (through a linear LD) rather than before the VCO.

7.4. Step-by-Step Design Procedure and Examples

This discussion consolidates the design data developed in the chapter by first reviewing the design and performance charts and equations. Examples are then considered which illustrate the means to "walk-through" the charts and establish loop design and system performance on a step-by-step basis, with all calculations carried through to completion. The discussion is organized so as to be maximally self-contained and require least backtracking. Unfortunately, some of the distortion calculations, which form an important part of proper FMFB design, are too involved to permit their inclusion here;

therefore the reader is requested to refer to the distortion section where detailed examples are discussed; no summary of these procedures is given here. The examples given here are concerned primarily with threshold and the "linear" SNR-CNR region above threshold. Figures 7-21–7-23 summarize the FMFB design and performance characteristics as developed in Section 7.3. The designs are based essentially on the Enloe procedure,[1] from which the basic material of Section 7.3 was drawn. The functional block diagram of the FMFB is given in Fig. 7-27. Before considering the design in detail, certain general points are covered.

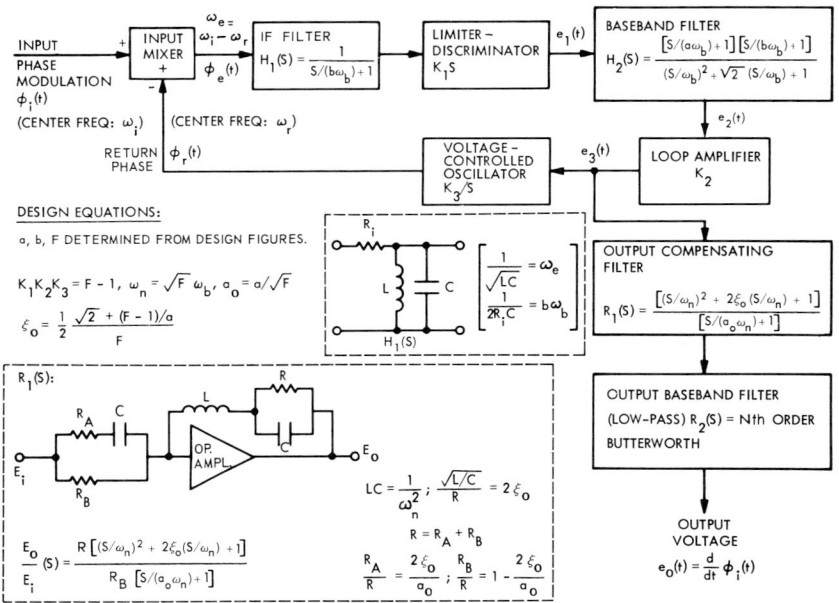

Fig. 7-27. FMFB functional block diagram.

A. Introduction and Summary

The first phase in FMFB design is to establish the required loop feedback factor F (Fig. 7-21). This is usually based on the modulation characteristics of the signal, which, in turn, is derived from system considerations such as a required loop output SNR per input CNR (Figs. 7-20 and 7-23). The remaining loop parameters (zero constants a and b, found in Fig. 7-22) are also found in terms of system modulation characteristics.

The baseband filter for the loop may be derived by reference to Fig. 7-28, which contains a realization of the required filter transfer characteristic.

7.4. Step-by-Step Design Procedure and Examples

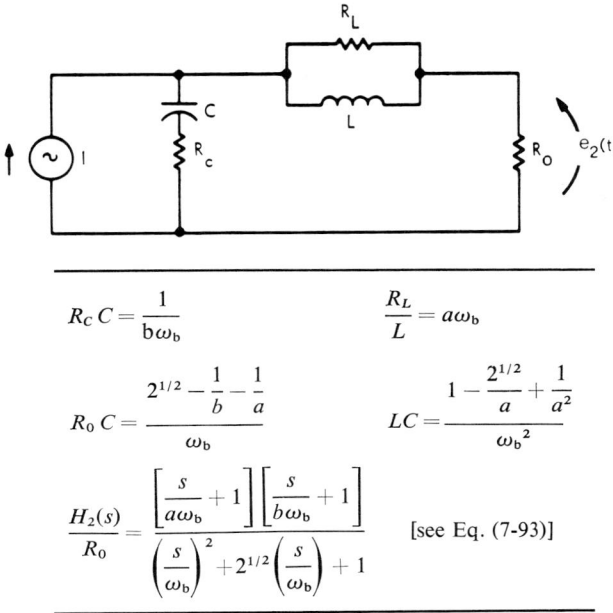

$$R_c C = \frac{1}{b\omega_b}$$

$$R_o C = \frac{2^{1/2} - \frac{1}{b} - \frac{1}{a}}{\omega_b}$$

$$\frac{R_L}{L} = a\omega_b$$

$$LC = \frac{1 - \frac{2^{1/2}}{a} + \frac{1}{a^2}}{\omega_b{}^2}$$

$$\frac{H_2(s)}{R_o} = \frac{\left[\dfrac{s}{a\omega_b}+1\right]\left[\dfrac{s}{b\omega_b}+1\right]}{\left(\dfrac{s}{\omega_b}\right)^2 + 2^{1/2}\left(\dfrac{s}{\omega_b}\right)+1} \quad \text{[see Eq. (7-93)]}$$

Fig. 7-28. Network with transfer function $H_2(s)$ (from Enloe[1b]).

From the factors a, b, and F, the internal IF bandwidth and center frequency may be chosen, together with the required LD bandwidth and sensitivity. The VCO center frequency, together with sensitivity and linearity, follows from these factors, together with fixed interface considerations. With VCO and LD detector sensitivities known, the required baseband amplifier gain is computed to realize the required loop gain K or feedback factor F.

Not all specifications for the loop elements can be uniquely determined. Some factors are a function of engineering judgment and available technology. The following rules of thumb, based upon common transistor circuit characteristics, serve as guides and are not to be considered inviolable.

A. Internal IF Center Frequency (ω_e)

In order to restrict carrier frequency components and harmonics (due to imperfect LD balance) from propagating around the loop, causing undesirable effects (amplifier saturation and spurious VCO modulation), the IF center frequency should be 30–100 times the noise bandwidth of the loop. LD linearity and bandwidth will benefit from the choice of larger IF center frequencies. However, IF center frequencies greater than 100 times the noise bandwidth of the loop generally require difficult-to-obtain circuit "Q's" in the implementation of the single-pole IF. Also, component and IF center frequency drift problems develop at too high an IF center frequency.

B. LD Sensitivity

Sufficient LD bandwidth is required to insure that the LD baseband response does not interfere with the required open-loop transfer characteristic. Also, Davis[8] has found that an insufficiently wide discriminator contributes to the threshold effect through dynamic nonlinearities. Furthermore, by wide banding the LD, detection linearity is achieved, thereby eliminating LD signal distortion from being a critical factor in FMFB performance. With wide bandwidth comes low sensitivity, which must be made up by the VCO or the baseband amplifier gain to realize the required loop gain K. Typically, an LD peak-to-peak separation (or bandwidth) of 5 to 15 times the closed-loop noise bandwidth is desirable, and as the peak output from a discriminator is restricted in voltage, 10 V for example, we might expect typical discriminator sensitivities of $2/B_n$ V/Hz. The limiter limiting level is usually designed to be 10–20 dB below the noise rms value when the signal is removed, i.e., with only noise present at the FMFB input.

C. VCO Center Frequency and Sensitivity

It is desirable, in order to attain high VCO sensitivity and to restrict carrier frequency components and their harmonics from propagating around the loop, for the VCO as well as the input center frequency to be 10–100 times the FMFB internal IF center frequency. Note that VCO linearity and sensitivity depend on center frequency, as described in Section 3.5. The VCO linearity is specified in terms of its peak-to-peak deviation, and more conservatively (especially for smaller modulation indices), in terms of Carson's rule bandwidth. Note that the VCO deviation is smaller than that of the signal by the factor K/F.

D. Amplifier Gain and Dynamic Range

The amplifier may be AC coupled,† with flat response required down to the low-frequency limit of the signal modulation. Care must be taken, however, in the time constants and the number of capacitively coupled stages, in order to limit excess phase shift in the low-frequency region. Normally, amplifier gain is restricted to be less than 100. Dynamic range requirements may be relaxed by placing the loop filter ahead of the baseband amplifier so that out-of-band noise is eliminated from the amplifier input. The dynamic range is normally specified equal to five times the peak-to-peak signal avail-

---

† This is in contrast to the PLL requirement where the baseband amplifier must pass DC.

7.4. Step-by-Step Design Procedure and Examples

able from the discriminator multiplied by the amplifier gain, which is equal to 5 × (VCO peak-to-peak deviation/VCO sensitivity).

The final step in the design is to establish the filters required within and following the loop. The baseband filter utilized to achieve the desired FMFB open-loop response is of the form shown in Fig. 7-28, with the parameters related to the loop design parameters by the equations below the figure.

A unique solution for the filter does not exist, as there are five unknowns and three fixed parameters (a, b, ω_b). Normally, a trial-and-error procedure is used to find the most convenient group of components. The postdemodulator filter should contain a section which compensates for the closed-loop response of the FMFB. This may be achieved by the filter shown in Fig. 7-27, with the interrelationship of the parameters given in the figure.

The key figures necessary to carry out a basic FMFB design are Figs. 7-21–7-23. Refinements in the design, or system calculations, may be obtained from other figures appearing in the chapter, and are not repeated here. The key figures are summarized as follows.

A. FIGURE 7-21

The FMFB feedback factor may be determined for various types of modulation covered in this chapter in terms of rms modulation index (defined as the rms frequency deviation/top baseband frequency). As an example, consider $\sigma = 10$ for FDM-FM; the design feedback factor F is 19.9 dB.

B. FIGURE 7-20

This curve presents the threshold CNR referred to twice base-bandwidth as a function of system rms modulation index, for the various modulation cases considered. The curve may be used in two ways. First, given the modulation index, the threshold $(CNR)_{AM}$ may be determined. As an example, for $\sigma = 1$, the threshold $(CNR)_{AM}$ for FDM-FM in the top channel is 11.8 dB. Second, if the desired threshold $(CNR)_{AM}$ is given, then the system modulation index may be determined.

C. FIGURE 7-22

This curve is similar to Fig. 7-21, except that its purpose is to provide the zero constants a and b necessary to complete the FMFB design. As an example, for speech modulation with $\sigma = 3$, the zero constants are $a = 5.25$, $b = 1.05$.

D. FIGURE 7-23

This curve is similar to Fig. 7-20 except that its purpose is to provide the threshold SNR or worst-channel NPR (for FDM-FM), with the 1-dB threshold degradation factor included, as a function of system rms modulation index for the various modulation cases considered. As an example, for TT at top baseband frequency with $\sigma = 2$, the threshold SNR is 28.4 dB.

B. Design Examples

This section consolidates the design data developed in the chapter by examining typical design problems on a step-by-step basis. The first design problem deals with FMFB design for single-channel speech applications.

Design Problem I

For an FMFB-FM demodulator and single-channel speech input: (1) evaluate threshold performance, both SNR and CNR, for a 1-kHz test tone; and (2) specify the design parameters.

A. GIVEN

Baseband	300–4000 Hz
Peak signal deviation	10 kHz
Predetection bandwidth	35 kHz

B. SOLUTION

Step 1: System modulation index

$$m_p = 10 \text{ kHz}/4 \text{ kHz} = 2.5$$

$$m_{rms}(\text{TT}) = 2.5/2^{1/2} = 1.77$$

Step 2: Threshold CNR. From Fig. 7-20,

$$(\text{CNR}_{AM})_{TH} = 11.7 \text{ dB} \qquad (m_{rms} = 1.77)$$

Referring this threshold CNR to the 35-kHz IF bandwidth, we have

$$(\text{CNR}_p)_{TH} = (\text{CNR}_{AM})_{TH}(2f_b/B_p) = 14.8(8 \text{ kHz}/35 \text{ kHz}) = 3.4 = 5.3 \text{ dB}$$

This represents a *4.7-dB improvement* with respect to the normally encountered 10-dB threshold for the LD.

Step 3: Threshold SNR. From Fig. 7-23 we have, as the TT SNR at threshold

$$(\text{SNR}_{TT})_{TH} = 23.5 \text{ dB}\dagger$$

Step 4: Loop design and calculation of loop parameters. From Figs. 7-21 and 7-22, the basic loop parameters are determined as

$F = 15 \text{ dB} = 5.6$ numeric $\qquad$ (TT, $f_b = 4f_T$ curve)
$K = 4.6$
$a = 4.6$
$b = 0.7$
$\omega_b = 2\pi(4k)$ rad/sec

† A penalty of 0.7 dB was taken for presumed nonideal postdemodulation filtering (see Fig. 7-3).

7.4. Step-by-Step Design Procedure and Examples

Note that K represents the product of the LD sensitivity, amplifier gain, and VCO sensitivity. The critical loop design parameters are reduced to the following circuit design specifications by referring to the general comments made earlier in this section.

A. INTERNAL IF

3-dB bandwidth	5.6 kHz	[Eq. (7-96)]
Center frequency	200 kHz	

B. DISCRIMINATOR

Center frequency	200 kHz
Peak-to-peak bandwidth	100 kHz
Sensitivity	0.2 V/kHz

C. VCO

Type	Varactor controlled
Center frequency	2 MHz
1% linear range	±10 kHz
Sensitivity	25 kHz/V
Output level	5 V peak-to-peak

D. INPUT

Center frequency	1.8 MHz
Level	1 volt peak-to-peak (signal only)

E. DC AMPLIFIER

Gain $K/K_1 K_3 = 4.6/5 = 0.92$

Since the gain requirement is less than one, a passive attenuator network rather than an amplifier is needed.

Design Problem II

For FM signals with FDM basebands, determine: (1) worst-channel threshold NPR, TT SNR, and threshold CNR; and (2) VCO linearity to achieve worst-channel high-level NPR = 45 dB.

A. GIVEN

Modulation	120-channel FDM-FM (4-kHz bandwidth/channel)
Baseband	60-554 kHz, with 10-dB peak-to-rms amplitude ratio
Peak frequency deviation	5.5 MHz

B. SOLUTION

Step 1: System modulation index and predemodulation bandwidth. Applying the 10-dB peak factor, the system rms modulation index σ is

$$\sigma = 5.5/(10^{1/2})(0.554) = 3.14$$

and the Carson's rule predemodulator bandwidth B_p is

$$B_p(3 \text{ dB}) = 2(\Delta f_p + f_b) = 12 \quad \text{MHz}$$

Step 2: Threshold CNR. From Fig. 7-20, the threshold $(CNR)_{AM}$ for the top channel (in this case, the worst FDM channel) is $(CNR)_{AM} = 15.9$ dB $= 39$, and when referred to a Carson's rule predemodulator bandwidth, results in an input CNR of

$$(CNR_p)_{TH} = (2f_b/B_p)(CNR_{AM})_{TH} = (1.1/12)(39) = 3.6 = 5.55 \quad \text{dB}$$

Step 3: Threshold NPR and TT SNR. From Fig. 7-23 the NPR at threshold for the top channel is $NPR_{TH} = 28$ dB.

We note the relation between psophometrically weighted TT SNR and NPR,[15]

$$SNR_{TT} = NPR + BWR - L + 2.5 \quad \text{dB}$$

where SNR_{TT} is channel TT SNR, NPR is noise power ratio, BWR is base bandwidth to channel-bandwidth ratio and L is multiplex noise loading ratio.† Referring to the CCIR recommended loading formula for FDM,

$$L \quad (\text{dB}) = -15 + 10 \log_{10} N$$

where N is the number of FDM channels, we have

$$L = -15 + 10 \log_{10}(120) = 5.8 \quad \text{dB}$$

$$BWR = 10 \log_{10} \frac{f_b - f_a}{B} = 10 \log_{10} \frac{0.554 - 0.06}{0.004} = 20.9 \quad \text{dB}$$

where B is the channel bandwidth in Hz. This results in

$$SNR_{TT} = 28 + 20.9 - 5.8 + 2.5 \quad \text{dB} = 45.6 \quad \text{dB} \qquad \text{(at threshold)}$$

Step 4: FMFB Design. From Fig. 7-21, the FMFB design feedback factor is $F = 13.6$ dB $= 4.8$, with the result that the required open-loop gain constant is $K = F - 1 = 3.8$.

From Fig. 7-22, the FMFB zero constants, a and b, are

$$a = 4.6, \qquad b = 2.05$$

† This is the decibel increase of baseband noise loading power over a 1-mW TT at the 0-dBm0 point of the system.[15]

7.4. Step-by-Step Design Procedure and Examples

Step 5: VCO linearity. The required worst-channel NPR at high input CNR is given as 45 dB. The minimum VCO linearity may be derived by application of Eq. (6-43) and reference to Fig. 6-12. We have

$$\Delta = \{10^5/[\text{NPR } W_{dn}(f)]\}^{1/2} \% \qquad (6\text{-}43)$$

Note from Fig. 6-12 that the bottom channels experience the largest values of $W_{dn}(f)$ and, thus, the lowest NPR for a given Δ. For the bottom channels, we have $W_{dn}(f) \cong 1$ and with NPR specified as 45 dB

$$\Delta = (10^5/3.16 \times 10^4)^{1/2} = 1.78\%$$

indicating that the VCO must be within $\pm 1.78\%$ of linear deviation over the required peak-to-peak deviation. The FMFB VCO deviates approximately $(F-1)/F$ times the input deviation; for this case, the stated linearity must exist over a peak-to-peak deviation of $(3.8/4.8)(11)$ MHz $= 8.7$ MHz.

Step 6: Circuit design. The loop design parameters may be reduced to the following circuit design specifications by referring to the general comments made earlier in the section.

A. INTERNAL IF

Type	Single-pole filter
3-dB bandwidth	2.27 MHz [Eq. (7-96)]
Center frequency	100 MHz

B. DISCRIMINATOR

Center frequency	100 MHz
Peak-to-peak bandwidth	40 MHz
Detection sensitivity	0.5 V/MHz

C. VCO

Type	Varactor controlled
Center frequency	600 MHz
1.78% linear range	± 4.35 MHz
Sensitivity	10 MHz/V
Output level	2 V peak-to-peak (10 dBm across 50 Ω)

D. INPUT

Center frequency	500 MHz
Level (signal only)	0.35 V peak-to-peak

E. BASEBAND AMPLIFIER

Gain	$K/K_1 K_3 = 3.8/5 = 0.76$

Since the gain requirement is less than unity, a passive attenuator network rather than an amplifier is needed.

References

1a. L. H. Enloe, Decreasing the threshold in FM by frequency feedback. *Proc. IRE* **50**, 18 (1962); also *in* "Selected Papers on Frequency Modulation" (J. Klapper, ed.). Dover, New York, 1970.
1b. L. H. Enloe, The synthesis of frequency feedback demodulators. *Proc. Nat. Electron. Conf.* **18**, 477 (1962).
2. J. B. Izatt, The distortion produced when frequency-modulated signals pass through certain networks. *Proc. IEE (Brit.)* **110**, No. 1 (1963).
3. E. Bedrosian and S. O. Rice, Distortion and cross-talk of linearly filtered, angle-modulated signals. *Proc. IEEE* **56**, No. 1, 2–13 (1968); also *in* "Selected Papers on Frequency Modulation" (J. Klapper, ed.). Dover, New York, 1970.
4. K. G. Fancourt and J. K. Skwirzynski, Design of a simple linear frequency discriminator. *Marconi Rev.* **19**, No. 121, 61–77 (1956).
5. A. J. Giger and J. G. Chaffee, The FM demodulator with negative feedback. *Bell Syst. Tech. J.* **42**, No. 4, Pt. 1, 1109 (1963).
6. S. O. Rice, Noise in FM receivers. *In* "Selected Papers on Frequency Modulation" (J. Klapper, ed.). Dover, New York, 1970.
6a. S. O. Rice, Properties of a sine wave plus random noise. *Bell Syst. Tech. J.* **27**, 109–157 (1948); also *in* "Selected Papers on Frequency Modulation" (J. Klapper, ed.). Dover, New York, 1970.
7. F. H. Lefrak, Advanced modulator-demodulator system. RCA Rep. CR-67-565-3, 3rd Quart. Rep. prepared for Commun. Satellite Corp. January 7, 1967.
8. B. R. Davis, Factors affecting the threshold of feedback FM detectors. *IEEE Trans.* **SET-10**, No. 3, 90 (1964).
9. E. J. Baghdady, FM demodulation with frequency-compressive feedback. *Proc. Nat. Telemetering Conf., Washington, D.C., 1962*, Vol. 2.
10. P. Frutiger, Noise in FM receivers with negative frequency feedback. *Proc. IEEE* **54**, No. 11, 1506–1520 (1966); Correction. *Proc. IEEE* **55**, 1674 (1967).
11. J. A. Develet, Statistical design and performance of high-sensitivity frequency-feedback receivers. *IEEE Trans.* **MIL-7**, 281 (1963).
12. D. L. Schilling and J. Billig, A Comparison of the Threshold Performance of the Frequency Demodulator Using Feedback and the Phase Locked Loop. Res. Rep. PIBMRI-1207-64. Polytech. Inst. of Brooklyn, New York, February 28, 1964; also, expanded paper presented at *Int. Space Electron. Symp., Miami Beach, Florida, 1965*.
13. F. H. Lefrak, H. Moore, A. Newton, and L. Ozolins, The frequency modulation feedback system for the lunar-orbiter demodulator. *RCA Rev.* **27**, No. 4, 563 (1966).
14. S. Kobayashi and S. Saito, Optimal design for frequency-compression demodulator. *Electron. and Commun. Jap.* **49**, 59 (1966).
15. Int. Radio Consultative Comm. (CCIR), Recommendation #393. *Docu. Plenary Assembly, 10th, 1963*, **4**. Int. Telecommun. Union, Geneva, 1963.

CHAPTER

8

Design of Compound and Multiple Loops for Low-Threshold Demodulation

As described in the previous chapter, the FMFB demodulator threshold is determined by both the feedback and open-loop detector thresholds. The design strategy is to implement the loop so that both thresholds occur simultaneously, without either of them dominating or determining overall loop performance. Improvement of either threshold mechanism without deterioration of the other will result in an improved demodulation system: Improvement of one will allow a new choice of loop design parameters (through the Enloe design procedure) that will result in a lower demodulation threshold. In this light, instead of the LD, we may incorporate the PLL,[1-5] FMFB, OR ERPLD† as the internal detector circuit (see Fig. 8-1) to provide the means of improving the open-loop threshold. We shall refer to such a system as the *compound-loop* demodulator and designate it as FMFB-PLL, FMFB-FMFB, or FMFB-ERPLD, where the second term refers to the internal demodulator. The "second-generation" demodulation systems, FMFB-PLL, FMFB-FMFB, and ERPLD, evaluated in this chapter operate with approximately the same demodulation efficiency. The FMFB-ERPLD represents a "third-generation" device, providing further improved threshold performance (see Fig. 3-8).

† The ERPLD (extended-range phase-locked demodulator) is an improved PLL circuit which exhibits lower threshold than does the "conventional" PLL. This circuit, discussed in detail in Section 8.4, is also representative of the *multiple-loop* demodulators.

226   8. Design of Compound and Multiple Loops

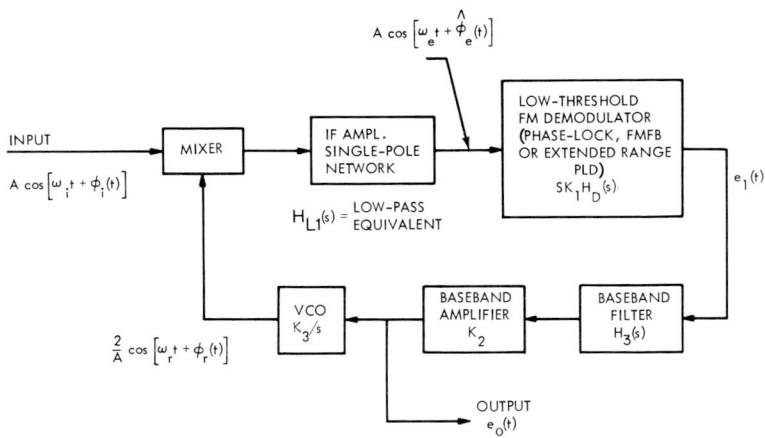

Fig. 8-1. Block diagram of compound-loop demodulator.

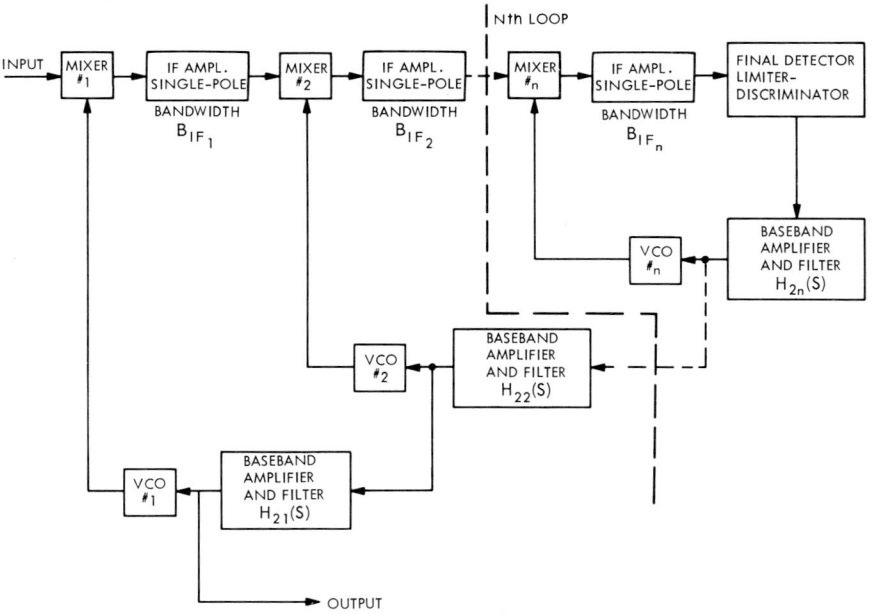

Fig. 8-2. Distributed feedback FM demodulator.

8.1. General Design Considerations

Further threshold reduction is available through a circuit expansion in the FMFB-FMFB system. Note that the inner loop has as the final detector an LD which itself can be replaced by an FMFB loop. The process, as depicted in Fig. 8-2, by successive repetition "telescopes" into the multiple-loop system which we call the *distributed-feedback* FM demodulator. To what degree this system can be expanded and further reduce threshold is not altogether clear, as a process of diminishing returns exists. Furthermore, from information theory arguments, an upper bound on FM detection may be derived (Chapter 3 and Appendix C). This suggests a performance limit or minimum attainable threshold.

This chapter presents the design considerations for compound- and multiple-loop demodulators, with details for the design and performance of the FMFB-PLL, FMFB-FMFB, ERPLD, and the FMFB-ERPLD. The material relies heavily on that presented in the preceding chapter.

8.1. General Design Considerations

To independently control the open- and closed-loop threshold mechanisms, we apply the Enloe design procedure to the compound-loop demodulator. Briefly reviewing, we note that the Enloe design for the second-order loop is predicated on synthesizing a specific open-loop transfer characteristic $G(s)$

$$G(s) = \frac{K[(s/a\omega_b) + 1]}{[(s/\omega_b)^2 + 2^{1/2}(s/\omega_b) + 1]} \quad (7\text{-}94) \quad (8\text{-}1)$$

The compound-loop demodulator internal detector network, in this case a low-threshold FM demodulator, is characterized by its own transfer function that in turn must be compensated for by the loop filter $H_3(s)$ to realize the required second-order compound loop $G(s)$. Note that both the second-order PLL and FMFB have closed-loop responses of the form†

$$H_D(s) = \frac{s/(a_0\omega_n) + 1}{(s/\omega_n)^2 + 2z(s/\omega_n) + 1} \quad (5\text{-}34) \quad (7\text{-}10) \quad (8\text{-}2)$$

for which (from Chapters 6 and 7)

$$\left.\begin{array}{l} a_0 = 1 \\ z = \xi = \tfrac{1}{2} \end{array}\right\} \quad \text{PLL} \quad (6\text{-}68) \quad (8\text{-}3)$$

---

† The detection sensitivity (gain K_1), as well as the characteristic differentiation (the s factor), is taken apart from $H_D(s)$ in the transfer characteristic of the demodulation stage (see Fig. 8-1).

$$\left.\begin{array}{l} a_0 = a/(1+K)^{1/2} = a/F^{1/2} \\ z = \zeta_0 = \dfrac{2^{1/2} + (F-1)/a}{2F^{1/2}} \\ \omega_n = F^{1/2}\omega_b \end{array}\right\} \quad \text{FMFB} \quad (7\text{-}96) \quad (8\text{-}4)$$

with the normalized parameters being determined† by the input signal characteristics (Sections 6.3 and 7.3).

In order to design a compound loop with the specified $G(s)$, it is necessary to cancel both the pole of the internal IF and the two poles and the zero of $H_D(s)$. This is accomplished (as in the conventional FMFB) by inserting appropriate zeros and poles in the baseband filter $H_3(s)$. Thus, from Eqs. (8-1) (the gain constant K is not part of the baseband filter), (8-2), and (4-61), the required $H_3(s)$ is

$$H_3(s) = \frac{G(s)}{KH_{L1}(s)H_D(s)}$$

$$= \frac{[s/(a\omega_b) + 1][s/(b\omega_b) + 1]}{(s/\omega_b)^2 + 2^{1/2}(s/\omega_b) + 1} \cdot \frac{(s/\omega_n)^2 + 2z(s/\omega_n) + 1}{[s/(a_0\omega_n)] + 1} \quad (8\text{-}5)$$

Several methods are available to synthesize $H_3(s)$; a simple and straightforward way is to retain the indicated partition and implement the filter with two sections coupled by an isolation stage. By this procedure, the first section $H_3'(s)$ is

$$H_3'(s) = \frac{[s/(a\omega_b) + 1][s/(b\omega_b) + 1]}{(s/\omega_b)^2 + 2^{1/2}(s/\omega_b) + 1} \quad (8\text{-}6)$$

which is precisely the Enloe filter whose synthesis is discussed in Section 7.4 (Figure 7-28). For convenience, the filter and a summary of the design equations is repeated in Fig. 8-3. The indicated lead network is required for the case of small b's which normally lead to negative $R_0 C$ values. The design then proceeds by selecting arbitrarily a larger b in the design of $H_3'(s)$, which we call b', thereby producing a dummy zero at $b'\omega_b$. The purpose of the lead network then is to cancel the dummy zero by providing a pole at $b'\omega_b$, and to insert a zero at the correct frequency $b\omega_b$.

The second section $H_3''(s)$ required to complete $H_3(s)$ is

$$H_3''(s) = \frac{(s/\omega_n)^2 + 2z(s/\omega_n) + 1}{[s/(a_0\omega_n)] + 1} \quad (8\text{-}7)$$

† For the FMFB, feedback factor F ($F = 1 + K$) and zero constant a are determined explicitly; a_0 and ω_n then follow.

8.1. General Design Considerations

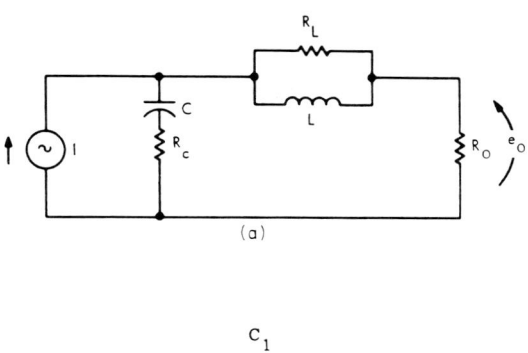

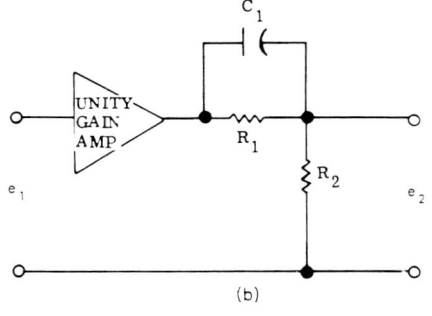

$$R_c C = \frac{1}{b\omega_b} \qquad \frac{R_L}{L} = a\omega_b$$

$$R_0 C = \frac{2^{1/2} - \frac{1}{b} - \frac{1}{a}}{\omega_b} \qquad LC = \frac{1 - \frac{2^{1/2}}{a} + \frac{1}{a^2}}{\omega_b^2}$$

$$\frac{H_3'(s)}{R_0} = \frac{E_0(s)}{I(s)} = \frac{\left(\frac{s}{a\omega_b} + 1\right)\left(\frac{s}{b\omega_b} + 1\right)}{\left(\frac{s}{\omega_b}\right)^2 + 2^{1/2}\left(\frac{s}{\omega_b}\right) + 1}$$

Fig. 8-3. (a) Network with transfer function $H_3'(s)$ [or $H_2(s)$]; (b) Lead network. [Cascade lead network with $H_2(s)$ to realize overall filter for $b < a/(2^{1/2}a - 1)$.]

For $a_0 = 1$ and $\zeta = \frac{1}{2}$, which is the case for the FMFB-PLL, $H_3''(s)$ is fully represented by the admittance of the RLC network given in (a) of Fig. 8-4, or the impedance of the RLC network shown in (b) of Fig. 8-4; therefore, either the "impedance inversion" technique[6] shown in (a) of Fig. 8-4, or the method given in (b) of the figure, may be applied to realize $H_3''(s)$ (see also Fig. 6-26). In the case of the FMFB-FMFB, we have $a_0 > 1$ and $\zeta_0 > \frac{1}{2}$, which requires the

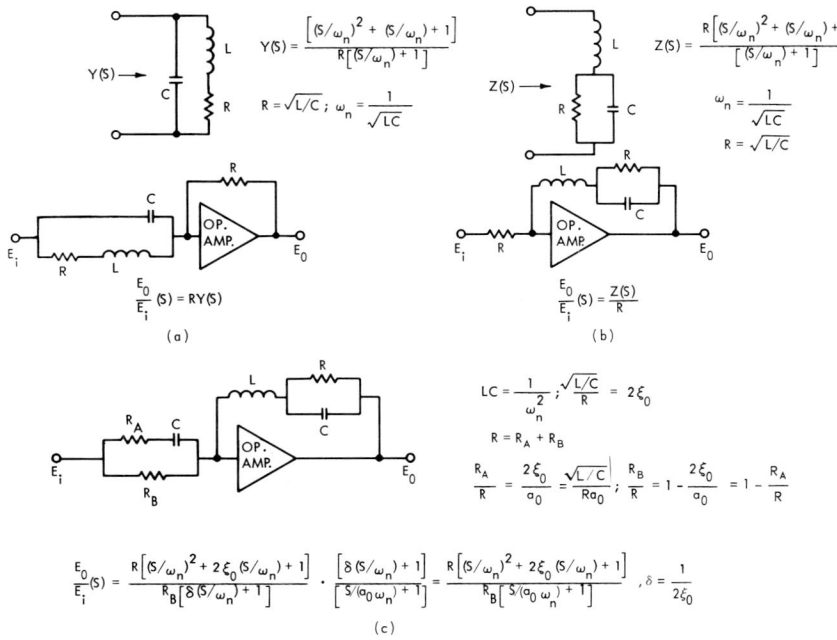

Fig. 8-4. Filter realizations for $H_3''(s)$ required in compound loop baseband. (a) Admittance. (b) Impedance. (c) Lag impedance.

addition of the lag impedance, as indicated in (c) of Fig. 8-4 to implement $H_3''(s)$. Observe that $H_3''(s)$ is implemented by a method in which there is a pole-zero cancellation (see also Fig. 7-27).

The subsequent design and loop performance are illustrated using a 1-kHz TT in a 4-kHz baseband, with carrier peak frequency deviation equal to that of the intended voice signal. TT modulation is a representative test condition for single-channel basebands, and, as was shown in Chapter 6, is an appropriate substitute for the voice signal, leading to essentially equivalent loop design parameters. Although the loop design parameters are considered to be the same for both modulations, threshold will differ when switching from TT to speech modulation, with TT yielding the higher threshold.† The extension to other basebands is straightforward, using the same methods but different detailed data from the last two chapters. Also, we do not indicate in this chapter the postdemodulation filter to compensate for the compound-loop response and to perform the lowpass filtering, as it is identical to that treated in the preceding chapters.

We now proceed to consider particular compound loop designs.

† For the PLL this difference is 1.2 dB (Section 6.4).

8.2. The FMFB-PLL Compound Loop

This section develops a design of the second-order FMFB-PLL compound-loop demodulator suitable for FM voice signals. The functional block diagram of this demodulator is shown in Fig. 8-5, with the linear equivalent model shown in Fig. 8-6. The design proceeds by finding the intersection of the open-loop and feedback threshold curves, as indicated in Fig. 8-7. The open-loop threshold curves are those of a PLL given by Eq. (6-105) for a TT signal with $\omega_T/\omega_b = \frac{1}{4}$

$$(CNR_{AM})_{TH} = 17.5(\omega_T/\omega_b)^{1/2}(m_p')^{1/2} = 10.4(\sigma')^{1/2} \qquad (8\text{-}8)$$

where m_p' and σ' are the peak and rms modulation indices, respectively, at the PLL input. (Note $\sigma' = m_p'/2^{1/2}$.) Since the PLL is demodulating the compressed input frequency deviation we have

$$\sigma' = \sigma/F \qquad (8\text{-}9)$$

where σ is rms modulation index at the compound-loop input. The open-loop threshold characteristic in terms of input modulation index σ and loop feedback factor F is then

$$(CNR_{AM})_{TH} = 10.4(\sigma/F)^{1/2} \qquad (8\text{-}10)$$

and represents the family of open-loop threshold curves in Fig. 8-7. The feedback threshold curve is obtained from Fig. 7-17. The intersection of open-loop and feedback threshold curves determines F, and the overall demodulator threshold CNR, CNR_{TH} (FMFB-PLL). The system design curves describing FMFB-PLL threshold performance are given in Fig. 8-8. These curves, derived from Fig. 8-7 and Eq. (6-3),† aid in the determination of system modulation index and carrier power required to achieve a specified performance level in output SNR. A comparison of demodulation efficiency in terms of conventional LD and other demodulation techniques discussed herein is available in Chapter 3 (Fig. 3-8).

The loop design parameters, feedback factor F, and zero constants a and b, are presented in Fig. 8-9 as functions of input modulation index. The design feedback factor F is taken from Fig. 8-7 and determines, with Fig. 7-16, the zero constant a. The single-pole IF bandwidth and compensating zero constant b may be obtained from the Carson's rule design equation based on a 1-kHz TT. From Eq. (7-103),

---

† Equation (6-3) gives the linear demodulation SNR, while Fig. 8-8 incorporates the 1-dB degradation at threshold.

232   8. Design of Compound and Multiple Loops

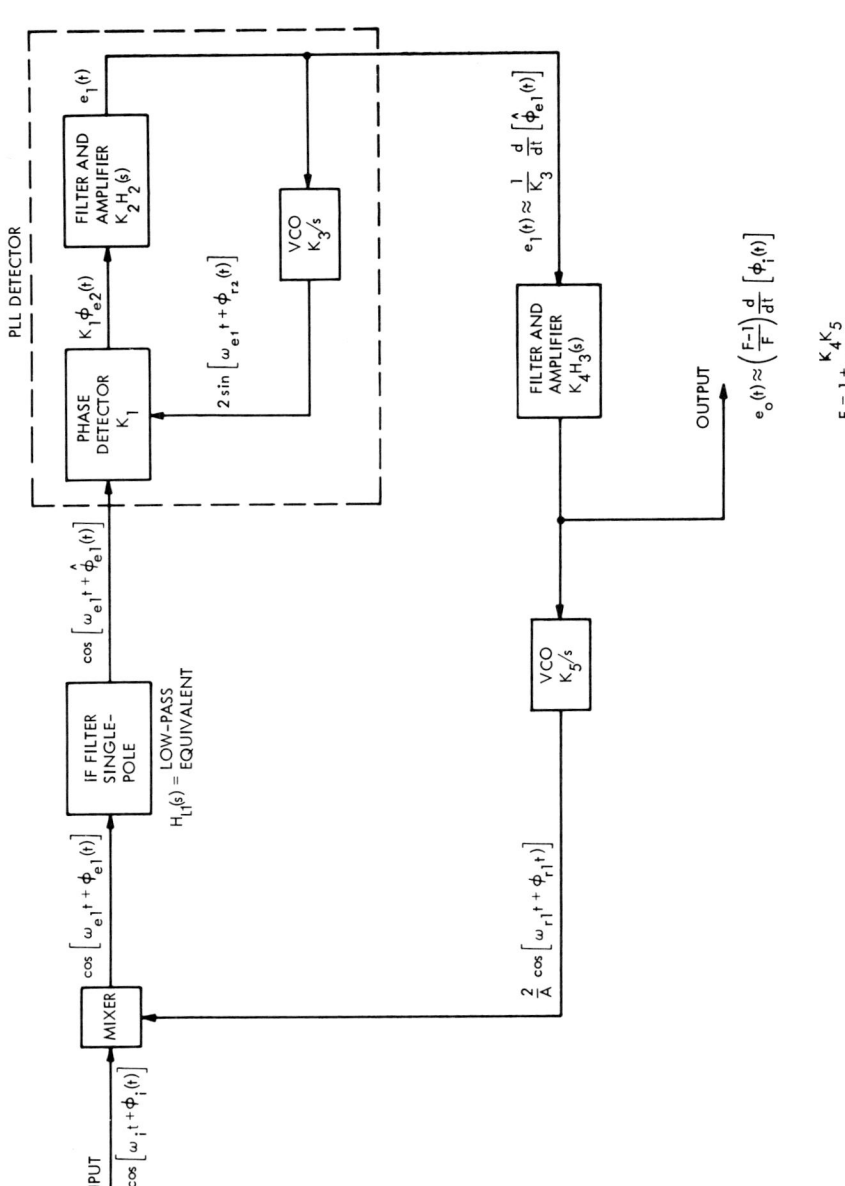

Fig. 8-5. Block diagram of FMFB-PLL demodulator.

8.2. The FMFB-PLL Compound Loop

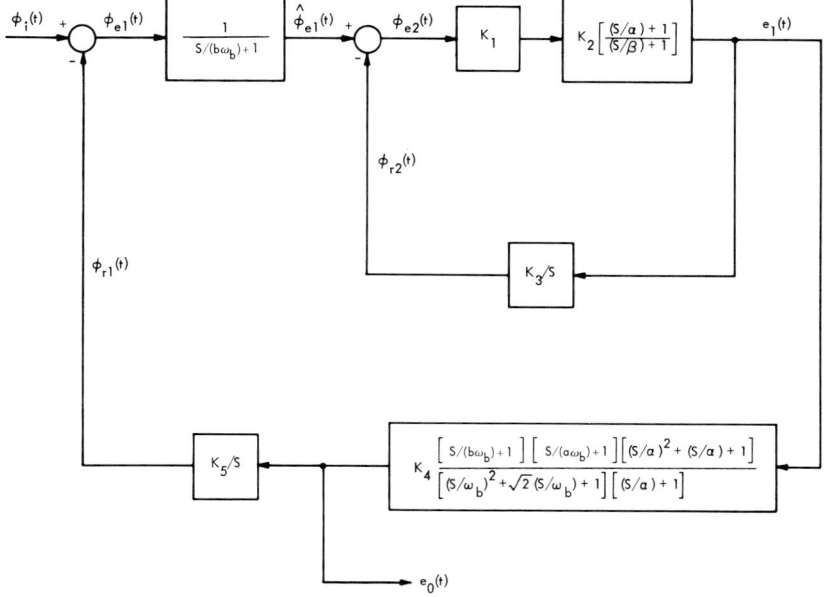

Fig. 8-6. Linear model of FMFB-PLL demodulator.

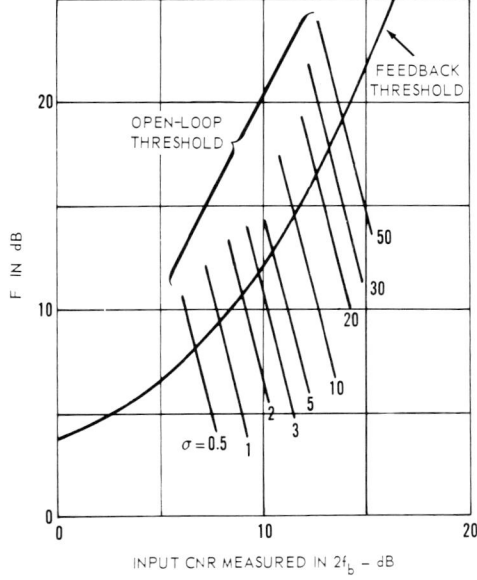

Fig. 8-7. Threshold determination for compound FMFB-PLL demodulator (test-tone at $f_b/4$).

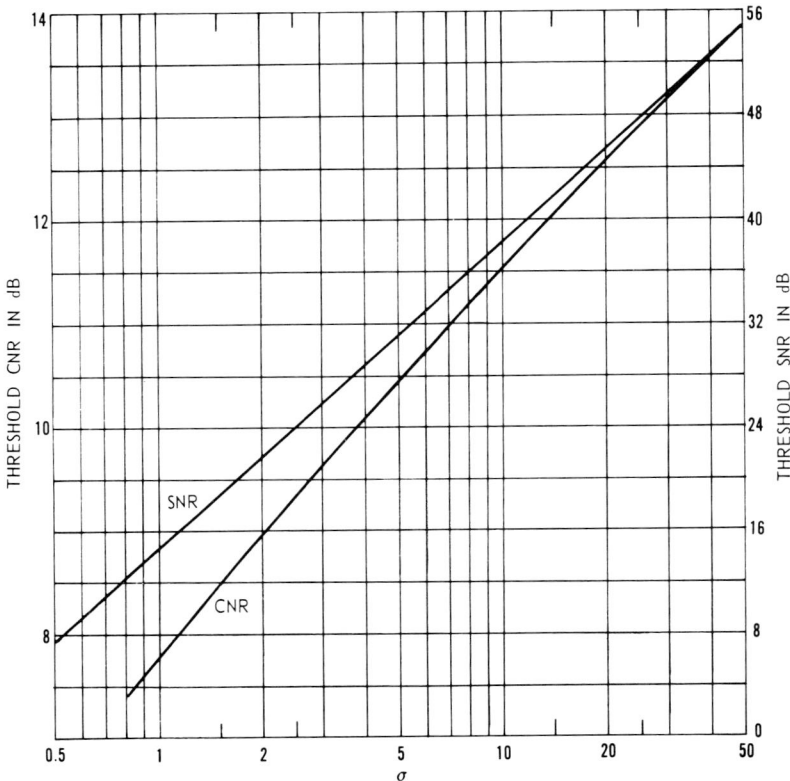

Fig. 8-8. FMFB-PLL compound-loop threshold CNR (measured in twice base bandwidth) and SNR (measured in base bandwidth) as a function of input rms modulation index (test-tone at $f_b/4$).

$$B_{IF}(3 \text{ dB}) = \frac{f_b}{2}\left[1 + \frac{4\Delta f_p}{F \cdot f_b}\right] = \frac{f_b}{2}\left(1 + \frac{4(2)^{1/2}\sigma}{F}\right) \quad (8\text{-}11)$$

and

$$b = \frac{B_{IF}(3 \text{ dB})}{2f_b} = \frac{1}{4}\left[1 + \frac{5.65}{F}\sigma\right] \quad (8\text{-}12)$$

which is presented in Fig. 8-9.

The basic loop parameters are now available to carry forth a particular loop design. The following example illustrates the application of the derived performance and design curves.

8.2. The FMFB-PLL Compound Loop

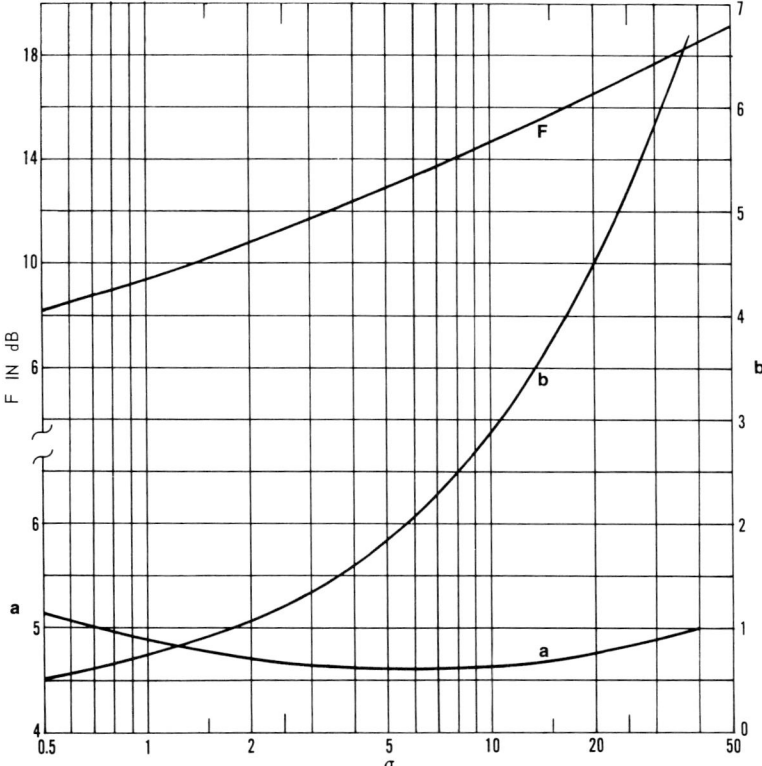

Fig. 8-9. FMFB-PLL outer-loop design parameters as a function of rms modulation index (test-tone at $f_b/4$).

A. Design Problem: Single-Channel Signal

For an FMFB-PLL compound-loop demodulator: (1) evaluate SNR and CNR threshold performance for a 1-kHz test tone (TT), and (2) specify design and element parameters.

A. GIVEN

Baseband	300–4000 Hz
Peak signal deviation	10 kHz
Receiver predetection bandwidth	35 kHz

B. SOLUTION

Step 1: System modulation index.
$$m_p = 10 \text{ kHz}/4 \text{ kHz} = 2.5$$
$$\sigma(\text{TT}) = 2.5/2^{1/2} = 1.75$$

8. Design of Compound and Multiple Loops

Step 2: Threshold CNR. From Fig. 8-8

$$(CNR_{AM})_{TH} = 8.75 \quad dB$$

Referring this threshold CNR to the 35-kHz receiver predetection bandwidth, we have

$$(CNR_p)_{TH} = (CNR_{AM})_{TH}(2f_b/B_{IF})$$
$$= 8.75 + 10 \log_{10}(8 \text{ kHz}/35 \text{ kHz}) \quad dB = 2.35 \quad dB$$

This represents a *2.65-dB improvement* with respect to the PLL, and a *2.95-dB improvement* with respect to the FMFB.*

Step 3: Threshold SNR. From Fig. 8-8 we have as the TT SNR at threshold,

$$(SNR_{TT})_{TH} = 20.4 \quad dB\dagger$$

Step 4: Calculation of outer-loop parameters. From Fig. 8-9, the basic loop parameters are determined as

$$F_{OL} = 10.5 dB = 3.36\ddagger$$
$$a_{OL} = 4.72$$
$$b_{OL} = 1.0$$
$$K_{OL} = F_{OL} - 1 = 2.36$$

Note that K represents the product of detector sensitivity, amplifier gain, and outer-loop VCO sensitivity. Since the detector is a PLL circuit, its detection sensitivity (volt-seconds per radian) is the reciprocal of the PLL-VCO sensitivity, or $1/K_3$.

Step 5: Calculation of inner-loop parameters. The PLL design parameters are established by first noting that the PLL input signal modulation index is

$$\sigma' = \sigma/F = 1.75/3.36 = 0.52$$

where σ' is rms modulation index at PLL input and σ is rms modulation index at FMFB-PLL input.

---

* See Design Problems in Chapters 6 and 7.

† The penalty due to nonideal postdemodulator filtering may be obtained from the correction curves found in Chapter 7 (Fig. 7-3). For this problem, with no closed-loop response compensation, the output SNR is degraded by 1.3 dB when using a 6-pole Butterworth postdemodulator filter.

‡ The subscript OL refers to outer-loop parameters; an IL subscript is used to designate inner-loop parameters.

8.2. The FMFB-PLL Compound Loop

From Fig. 6-17 directly, or Eq. (6-104), the PLL design noise bandwidth based on TT modulation is

$$B_n = 2.8 f_b m_p^{1/2} = 2.8(4\text{kHz})[2^{1/2}(0.52)]^{1/2} = 9.6 \quad \text{kHz}$$

The open-loop design of the high-gain second-order type-one PLL realizes the following gain constant and filter characteristics for minimum noise bandwidth (Figs. 8-5, 8-6, and Chapter 6),

$$K_{IL} = K_1 K_2 K_3$$

$$H_2(s) = \frac{s/a_{IL} + 1}{s/b_{IL} + 1}$$

with $a_{IL} = 2B_n$ [Eq. (6-131)] $= 2(9.6)(10^3) = 19.2 \text{ krad/sec} = 3.1$ kHz. The bottom of the baseband is 300 Hz and, therefore, maximum gain across the baseband is guaranteed by specifying b_{IL} to be no greater than 300 Hz; thus

$$b_{IL} = 300 \text{Hz} = 1.88 \text{ krad/sec}$$

Finally, the loop gain constant is established by [Eq. (6-132)]

$$K = \frac{4B_n^2}{b} = \frac{(4)(9.6)^2(10^6)}{(1.88)(10^3)} = 1.96 \times 10^5 \quad \text{sec}^{-1}$$

The loop design parameters may now be reduced to the following circuit design specifications by referring to the general comments made in Sections 6.4 and 7.4.

Step 6: Loop component specifications

A. INNER LOOP (PLL)

1. VCO

Type	Varactor controlled
Center frequency	300 kHz
1% linear range	± 3 kHz
Sensitivity	7 kHz/V
Output level	5 V peak-to-peak

2. Input

Center frequency	300 kHz
Level	1 V peak-to-peak
Phase detector sensitivity	0.25 V/rad

3. DC amplifier

Gain (K_2)	$\dfrac{K_{IL}}{K_1 K_3} = \dfrac{(1.96)(10^5)}{(0.25)(2\pi)(7)(10^3)} = 17.9$
Output dynamic range	9 V peak-to-peak
3-dB bandwidth	DC to 150 kHz [$\theta_b \cong 1.5°$, using Eq. (6-130)]

4. Filter (see Section 6.4)

R_1 17.7 kΩ
R_2 1.73 kΩ
C 0.03 μF

B. OUTER LOOP

1. IF

Center frequency   300 kHz
3-dB bandwidth $2b_{OL} f_b = 8$ kHz
Gain   10 dB†

2. VCO

Type   Varactor controlled
Center frequency   2.1 MHz
1% linear range ± 10 kHz
Sensitivity   25 kHz/V
Output level   5 V peak-to-peak

3. Input

Center frequency   1.8 MHz
Level (signal only)   1 V peak-to-peak

4. DC amplifier

$$\text{Gain } (K_4) \quad \frac{K_{OL} K_3}{K_5} = \frac{(2.36)(7)(10^3)}{(25)(10^3)} = 0.66$$

Since the gain requirement is less than unity, a passive attenuator network rather than an amplifier is needed.

5. Filter

a. $H_3'(s)$ (see Fig. 8-3)

$R_c C$ $(3.98)(10^{-5})$
$R_o C$ $(8.05)(10^{-6})$
LC $(1.18)(10^{-9})$
R_L/L $(1.18)(10^5)$
Let C 0.02 μF (convenient value)
then L 59 mH
R_L 6,962 Ω (6.9 k)
R_o 402.5 Ω (400 Ω)
R_c 1,990 Ω (2 k)

† The IF gain is made to compensate for the conversion loss in the outer-loop input mixer which is assumed here to be 10 dB.

b. $H_3''(s)$ [see Fig. 8-4b (compensates for PLL response)]

LC $(2.72)(10^{-9})$ [for ω_n see Eq. (6-67)]
Let C $0.05\ \mu F$
then L 54.5 mH
R $1040\ \Omega\ (1\ k)$

8.3. The FMFB-FMFB Compound Loop

This section develops a design of the *second-order* FMFB-FMFB compound-loop demodulator suitable for FM voice signals. The functional block diagram of this demodulator is shown in Fig. 8-10, with the linear equivalent model shown in Fig. 8-11. As in Section 8.2, the design proceeds by finding the intersection of the open-loop and feedback threshold curves, as indicated in Fig. 8-12. The open-loop threshold (being that of the inner loop) is approximated by equating the FMFB threshold characteristic curve, Fig. 7-20 (TT $f_T = f_b/4$ curve) to the empirical equation

$$(\text{CNR}_{\text{AM}})_{\text{TH}} = (A^2/2)/2\eta f_b = A_1[\sigma']^{A_2} \tag{8-13}$$

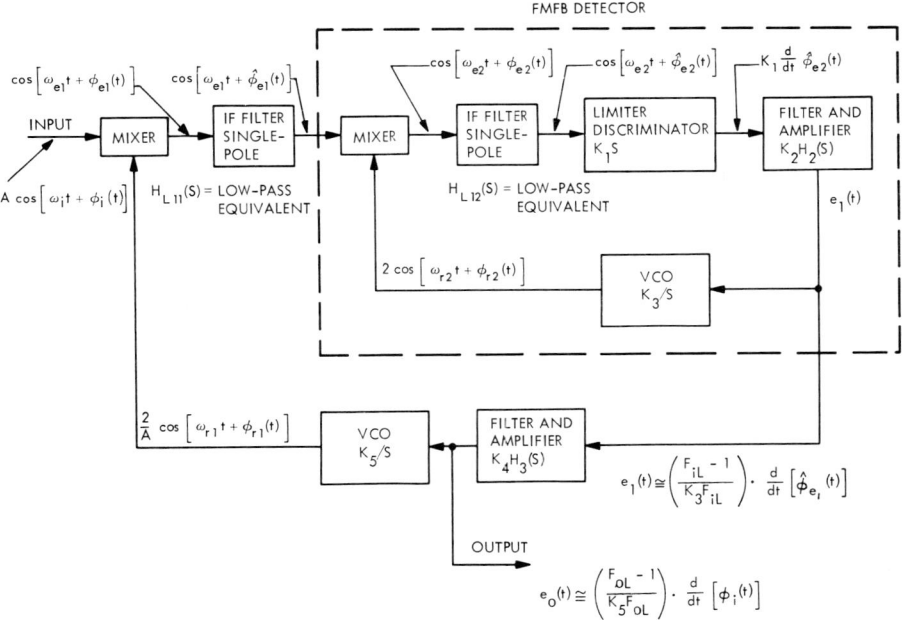

Fig. 8-10. Block diagram of FMFB-FMFB compound-loop demodulator.

240 8. Design of Compound and Multiple Loops

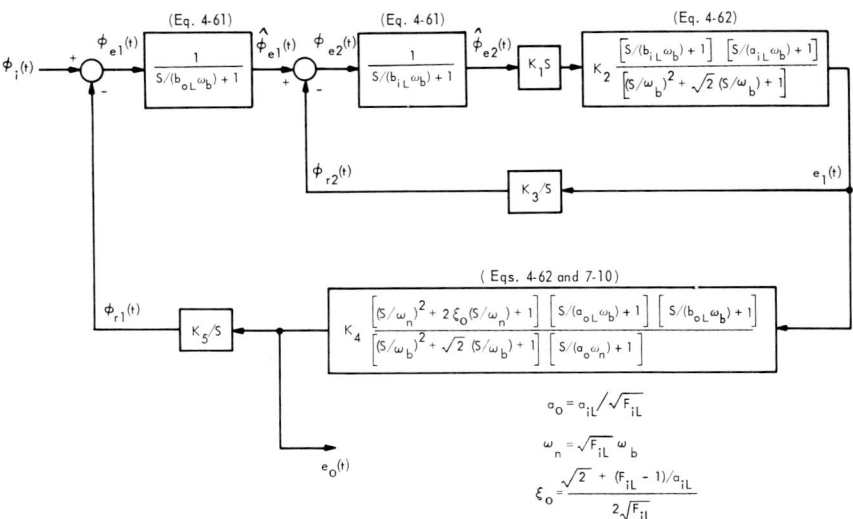

Fig. 8-11. Linear model of second-order FMFB-FMFB compound-loop demodulator.

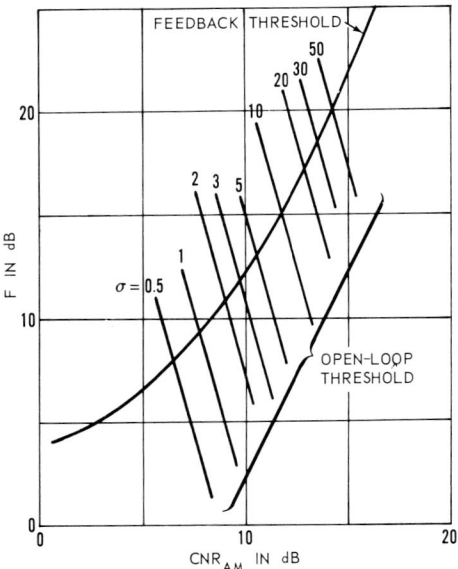

Fig. 8-12. Threshold determination for compound FMFB-FMFB demodulator (test-tone at $f_b/4$).

8.3. The FMFB-FMFB Compound Loop

where A_1 and A_2 are constants to be determined and σ' is rms modulation index at FMFB input. Over the range $0.2 \leq \sigma' \leq 70$ the constants

$$A_1 = 10.8 \quad \text{and} \quad A_2 = 0.56 \tag{8-14}$$

provide an accurate fit between the empirical equation (8-13), and the derived threshold characteristic (Fig. 7-20).

As the internal FMFB loop is demodulating the compressed input frequency deviation, we have

$$\sigma' = \sigma/F_{\text{OL}} \tag{8-15}$$

where σ is rms modulation index at the compound-loop input. The open-loop threshold characteristic in terms of input modulation index σ and loop feedback factor F is then

$$(\text{CNR}_{\text{AM}})_{\text{TH}} = 10.8[\sigma/F_{\text{OL}}]^{0.56} \tag{8-16}$$

which represents the family of open-loop threshold curves in Fig. 8-12. The feedback threshold curve is obtained from Fig. 7-17. The intersection of open-loop and feedback threshold curves determines the loop design feedback factor F and the overall demodulator threshold CNR, CNR_{TH}(FMFB-FMFB). The system design curves describing FMFB-FMFB threshold performance are given in Fig. 8-13. These curves, derived from Fig. 8-12 and Eq. (6-3),† aid in the determination of system modulation index and carrier power required to achieve a specified performance level in output SNR. A comparison of demodulation efficiency in terms of conventional LD and other demodulation techniques discussed herein is available in Chapter 3 (Fig. 3-8).

The outer loop design parameters, feedback factor F, and zero constants a and b, are presented in Fig. 8-14 as functions of input modulation index. The design feedback factor F is taken from Fig. 8-12 and determines, with Fig. 7-16, zero constant a. The single-pole IF bandwidth and compensating zero constant b may be obtained from the Carson's rule design equation based on a 1-kHz TT. From Eq. (7-103),

$$B_{\text{IF}} = 2f_{\text{T}}[1 + (\Delta f_{\text{p}}/F_{\text{OL}} f_{\text{T}})] \tag{8-17}$$

and from Eq. (8-12)

$$b_{\text{OL}} = B_{\text{IF}}(3 \text{ dB})/2f_{\text{b}} = \tfrac{1}{4}[1 + (5.65\sigma/F_{\text{OL}})] \quad (8\text{-}12) \tag{8-18}$$

which is also presented in Fig. 8-14.

† Equation (6-3) gives the linear demodulation SNR, while Fig. 8-13 incorporates the 1-dB degradation at threshold.

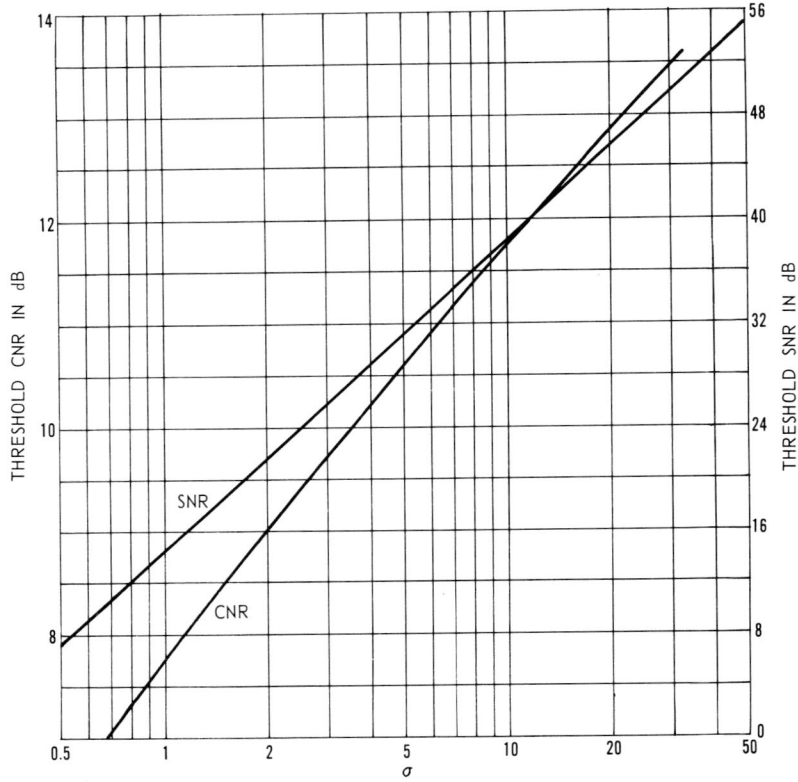

Fig. 8-13. FMFB-FMFB compound loop threshold CNR and SNR as a function of input rms modulation index (test-tone at $f_b/4$). (CNR measured in twice base bandwidth, SNR measured in base bandwidth.)

Consider now the design of the inner loop, which itself is an FMFB circuit. The inner-loop input modulation index is σ/F_{OL} which, for $\sigma = 5$ as an example is, by Fig. 8-14

$$\sigma' = \sigma/F_{OL} = 5/4.46 = 1.12; \qquad F_{OL} = 13 \quad \text{dB} \qquad (8\text{-}19)$$

The inner-loop design, based on $\sigma' = 1.12$, from Fig. 7-21, calls for an inner-loop feedback factor $F_{IL} = 13$ dB. The equality between inner- and outer-loop feedback factors is not a coincidence; it is a result of the design requirement for coincidence of open-loop and feedback thresholds in the outer loop. For equal open-loop and feedback thresholds, the inner- and outer-loop noise bandwidths must be equal, with the result, by Eq. (7-87) and Fig. 7-16, that the inner- and outer-loop feedback factors must be equal. This same line of reasoning may be extended to the design of the distributed-feedback loop.

8.3. The FMFB-FMFB Compound Loop

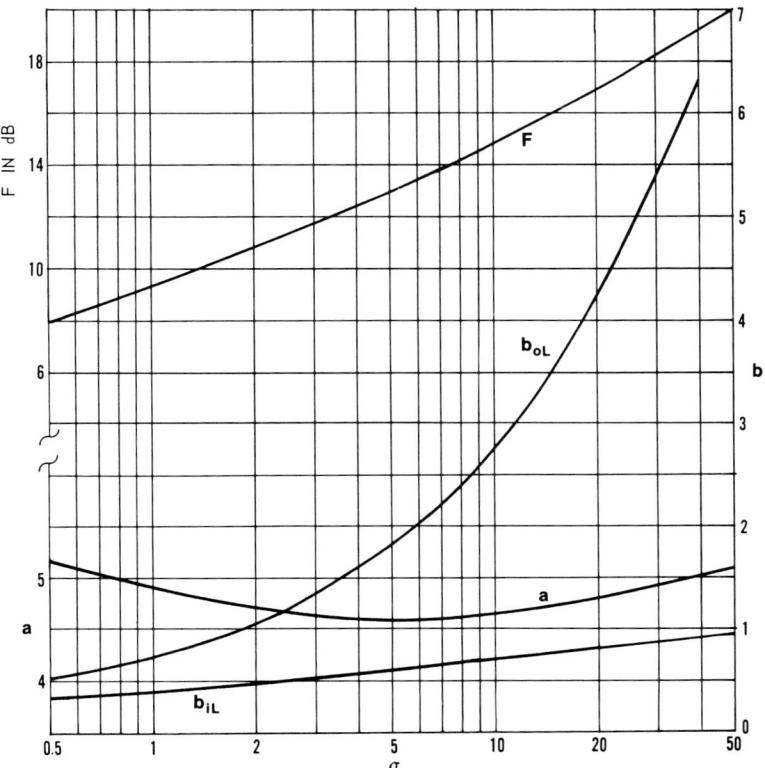

Fig. 8-14. FMFB-FMFB design parameters as a function of input rms modulation index (test-tone at $f_b/4$).

Here each successive loop is required to have the same closed-loop noise bandwidth, so that both open-loop and feedback thresholds for each loop are coincident. Thus, no single loop dominates and determines the demodulation threshold. The end result is that each FMFB loop in a distributed system must have the same feedback factor F.

Once F is specified, zero constant a is known [refer to Eq. (7-98)], with the result that the a's are equal for each FMFB loop in a distributed-feedback system, and in the inner and outer loop of the FMFB-FMFB system. Zero constant b dependent on the loop input modulation index, however, is different for each loop. Applying Eq. (8-18) to the inner loop of the FMFB-FMFB system yields

$$b_{iL} = \tfrac{1}{4}[1 + (5.65\sigma/F^2)] \qquad (8\text{-}20)$$

which is plotted in Fig. 8-14.

The basic loop parameters are now available to carry forth a particular loop design. The following example illustrates the application of the derived performance and design curves.

A. Design Problem: Single-Channel Signal

For an FMFB-FMFB compound-loop demodulator: (1) evaluate SNR and CNR threshold performance, for a 1-kHz test tone (TT), and (2) specify design and element parameters.

A. GIVEN

Baseband	300–4000 Hz
Peak signal deviation	28 kHz
Receiver predetection bandwidth	80 kHz

B. SOLUTION

Step 1: System modulation index

$$m_p = 28 \text{ kHz}/4 \text{ kHz} = 7$$
$$\sigma(TT) = 7/2^{1/2} = 4.95$$

Step 2: Threshold CNR. From Fig. 8-13

$$(CNR_{AM})_{TH} = 10.6 \text{ dB} = 11.45 \quad (\text{for} \quad \sigma = 4.95)$$

Referring this threshold CNR to the 80-kHz receiver predetection bandwidth, we have

$$(CNR_p)_{TH} = (CNR_{AM})_{TH}(2f_b/B_p) = 11.45(8 \text{ kHz}/80 \text{ kHz}) = 1.145 = 0.6 \text{ dB}$$

Step 3: Threshold SNR. From Fig. 8-13, we have as the TT SNR at threshold,

$$(SNR_{TT})_{TH} = 31.2 \text{ dB}†$$

Step 4: Calculation of outer-loop parameters. From Fig. 8-14, the basic loop parameters are determined as

$$F_{OL} = 13 \text{ dB} = 4.47$$
$$a_{OL} = 4.6$$
$$b_{OL} = 1.82$$
$$K_{OL} = F_{OL} - 1 = 3.47$$

† The penalty due to nonideal postdemodulator filtering may be obtained from the correction curves found in Chapter 7. For this problem, with no closed-loop response compensation, the output SNR is degraded by 1.2 dB when utilizing a 6-pole Butterworth postdemodulator filter.

8.3. The FMFB-FMFB Compound Loop

Note that K represents the product of detector sensitivity, amplifier gain, and outer-loop VCO sensitivity. Since the detector is an FMFB circuit, its detection sensitivity (volt-seconds per radian) is equal to $(F_{IL} - 1)/(K_3 F_{IL})$ (see Fig. 8-10), where K_3 is the inner-loop VCO sensitivity in radians per volt-second and F_{IL} is the inner-loop feedback factor.

Step 5: Calculation of inner-loop parameters. The inner-loop parameters, as discussed previously, are the same as the outer-loop parameters, except for zero constant b. From Fig. 8-14 we have

$$F_{IL} = 13 \text{dB} = 4.47$$

$$a_{IL} = 4.6$$

$$b_{IL} = 0.6$$

$$K_{IL} = F_{IL} - 1 = 3.47$$

The loop design parameters may now be reduced to the following circuit design specifications by referring to the general comments made in Sections 6.4 and 7.4, with some exceptions necessitated by the greater complexity of the compound loop.

Step 6: Loop component specifications

A. INNER LOOP (FMFB)

1. Internal IF

Center frequency	150 kHz
3-dB bandwidth	4.8 kHz

2. Discriminator

Center frequency	150 kHz
Peak-to-peak bandwidth	80 kHz
Sensitivity	0.2 V/kHz

3. VCO

Type	Varactor controlled
Center frequency	650 kHz
1% linear range	±4.9 kHz
Sensitivity (K_3)	6.5 kHz/V
Output level	5 V peak-to-peak

4. Input (also outer-loop IF output)

Center frequency	500 kHz
Level (signal only)	1 V peak-to-peak

5. Baseband amplifier

Gain (K_2) $K_{1L}/K_1 K_3 = 2.67$
Output dynamic range ± 4 V
3-dB bandwidth 115 kHz ($2°$ at f_b)

6. Filter $H_2(s)$ (See Figs. 8-10 and 8-3)

$R_C C$ $(6.6)(10^{-5})$
$R_0 C$ $(-1.9)(10^{-5})$
LC $(1.17)(10^{-9})$
R_L/L $(1.15)(10^5)$

Note that the $R_0 C$ product is negative, prohibiting the realization of $H_2(s)$, as shown in Fig. 8-3 and, therefore, requiring the modification for small b indicated therein. Arbitrarily setting $b'_{1L} = 2$, which sets $R_0 C = (2.8)(10^{-5})$, we may derive the element values as follows: For convenience choose

C 0.023 μF
then L 51 mH
R_L 5,900 Ω
R_0 1,200 Ω
and R_C 860 Ω
Choosing C_1 0.01 μF (convenient value)
we have R_1 6,660 Ω (6.7 k)
R_2 2,860 Ω
and μ 2,780 μmho (filter driver transconductance for unity gain)

B. OUTER LOOP (FMFB)

1. IF

Center frequency   500 kHz
3-dB bandwidth   14.6 kHz
Gain   10 dB†

2. VCO

Type   Varactor controlled
Center frequency   2 MHz
1% linear range ± 22 kHz
Sensitivity (K_5) 20 kHz/V
Output level   5 V peak-to-peak

3. Input

Center frequency   2.5 MHz
Level   1 V peak-to-peak

† Gain is required to recover conversion loss in outer-loop input mixer; it is assumed here to be 10 dB.

4. Baseband amplifier

Gain (K_4) $\quad K_4 = K_{OL} \Big/ \left[K_5 \left(\dfrac{F_{IL} - 1}{K_3 F_{IL}} \right) \right] = 3.47 \Big/ \left[20 \left(\dfrac{3.47}{(6.5)(4.47)} \right) \right] = 1.45$

Output dynamic range $\quad \pm 5.5$ V
3-dB bandwidth $\quad$ DC to 500 kHz

5. Baseband filter

a. $H_3'(s)$ (see Figs. 8-10 and 8-3)

$R_C C$	$(2.2)(10^{-5})$
$R_O C$	$(2.58)(10^{-5})$
LC	$(1.17)(10^{-9})$
R_L/L	$(1.15)(10^5)$
Let C	$0.023\ \mu F$
then L	51 mH
R_L	5.8 kΩ
R_O	1,120 Ω (1.2 k)
R_C	955 Ω

b. $H_3''(s)$ [see Fig. 8-4c and Eq. (8-4)]

$$LC = \omega_n^{-2} = (F \cdot \omega_b^2)^{-1} = [4.47(2\pi)^2(4)^2(10^6)]^{-1} = 3.57(10^{-10})$$

$$\dfrac{(L/C)^{1/2}}{R} = 2\zeta_0 = \dfrac{2^{1/2} + (F-1)/a}{F^{1/2}} = \dfrac{2^{1/2} + (3.47/4.6)}{(4.47)^{1/2}} = 1.03$$

$$\dfrac{R_A}{R} = \dfrac{2\zeta_0}{a_0} = \dfrac{2^{1/2} + (F-1)/a}{a} = \dfrac{2^{1/2} + (3.47/4.6)}{4.6} = 0.47$$

$$\dfrac{R_B}{R} = 1 - \dfrac{2\zeta_0}{a_0} = 0.528$$

Let C	0.01 μF
then L	35.7 mH (36 mH)
R	1,835 Ω (1.8 k)
R_A	865 Ω
R_B	965 Ω

8.4. Extended-Range Phase-Locked Demodulator (ERPLD) and the FMFB-ERPLD

As described in Chapter 5, there are two distinct sources of cycle slippings in a PLL: the ThI which are generated in the same manner as in the discriminator, and the LLI which are due to the periodic response of the phase detector. A minimum threshold design provides for a minimum rate in the sum of these cycle slippings, referred to also as the "minimum spike rate."

The ThI are caused by input noise and interference, and appear at the demodulator output when the PLL "follows" (and demodulates) the cycle slippings that appear in the input resultant IF wave. Since the tracking accuracy of the PLL improves with increasing loop bandwidth, the rate of ThI appearing at the loop output will also increase with bandwidth.

The LLI, on the other hand, require that the PLL lose lock with respect to the signal, and they will occur due to a combination of noise and signal-transients which the PLL is not able to follow. There are two basic parameters that determine the LLI rate for a given signal and noise environment: the PLL frequency response characteristic, and the detector response characteristic. For a given PLL embodiment (namely second-order with a given damping factor) the frequency response characteristic is fully described by the bandwidth. Since modulation tracking deteriorates with decreasing loop bandwidth, the LLI will tend to increase. This, when taken with the behavior of the ThI, implies an optimum loop bandwidth which results in a minimum spike rate. It is possible to specify a phase detector characteristic which would reduce the rate of LLI without unduly altering the rate of the ThI, resulting thereby in a more favorable tradeoff between LLI and ThI. A helpful qualitative though simplified physical picture is as follows.

A conventional phase detector (of sinewave characteristic) has a $\pm 90°$ monotonic range. If the phase error exceeds $\pm 90°$, the probability is high that a loss of lock will take place, due to the change in sign of the phase detector sensitivity which throws the loop into a regenerative mode, as described in Chapter 5. It has, therefore, been reasoned that a phase detector having a monotonic range greater than $\pm 90°$ would yield a lower rate of LLI for the same phase error statistics and power. The bandwidth of the PLL can now be reduced to reduce the rate of the ThI, until a new optimum bandwidth is reached where the sum of the rates of the ThI and LLI is minimum and lower than in the conventional PLL. This, in fact, is borne out by calculations on the first-order PLL where exact results are available. It is interesting to note that not only the monotonic range of the phase detector characteristic but also its shape is a critical factor. Furthermore, calculations indicate that a linear characteristic does not necessarily produce a minimum spike rate, as described presently.

It is recalled from Chapter 6 that our optimum design strategy for the PLL involves the minimization at threshold, of the total phase error, which consists of both the noise-induced and modulation-induced components, and that threshold occurs when the total phase error has a mean-square-value of 0.25. The threshold reduction of the extended- (monotonic) range phase detector reflects itself in a mean-square phase error at threshold which is greater than 0.25. From Eq. (6-79) we find the additional threshold extension from the relationship between optimum noise bandwidths for two second-

8.4. Extended-Range Phase-Locked Demodulator

order PLLs ($\xi = \frac{1}{2}$) with different levels of the mean-square phase error at threshold, namely,

$$(\alpha_1/\alpha_2)^{1/4} = B_{n1}/B_{n2} = [\overline{\phi_2^2(t)}/\overline{\phi_1^2(t)}]^{1/4} \qquad (8\text{-}21)$$

where B_n is closed-loop noise bandwidth (hertz), α is loop input CNR (referred to bandwidth B_n) at threshold (for the conventional loop, $\alpha = 4 = 6$ dB), $\overline{\phi^2(t)}$ is *total* mean-square phase error jitter at threshold (for the conventional loop, $\overline{\phi^2(t)} = \frac{1}{4}$), and the subscripts 1 and 2 designate respective loops.

A means of examining range extension and threshold reduction is contained in the loss-of-lock analysis presented by Viterbi[7] for the first-order loop; in terms of the Viterbi analysis, several hypothetical phase detector characteristic functions are examined[4] with the results interpreted in terms of second-order PLL performance. The method of analysis begins by considering the PLL model shown in Fig. 8-15, where an arbitrary phase detector characteristic is represented by the functional element $q(\phi_e)$. The model is developed by substituting $q(\phi_e)$ for $\sin \phi_e$ in the basic PLL differential equation (Section 5.2).

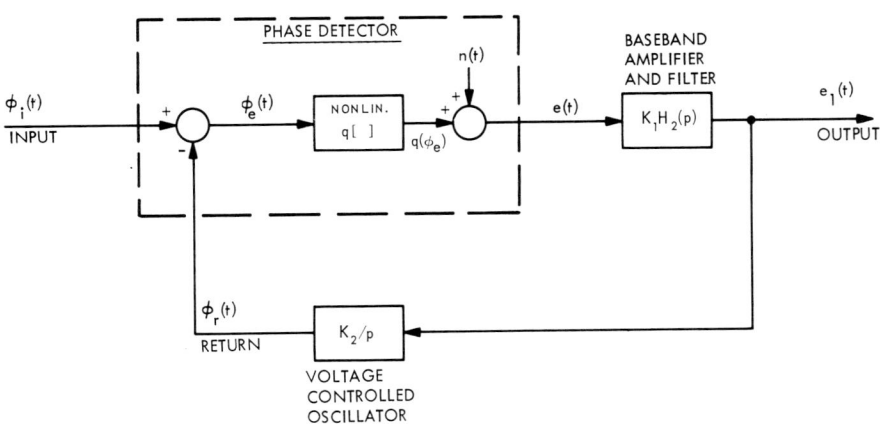

Describing equation	Actual components
$\dot{\phi}_e(t) = \dot{\phi}_i(t) - K_1 K_2 H_2(p) \cdot [q(\phi_e) + n(t)]$	Input: $A \cos[\omega_i t + \phi_i(t)] + N(t)$ Return: $-(2/A) \sin[\omega_i t + \phi_r(t)]$ $N(t) = x(t) \cos \omega_i t + y(t) \sin \omega_i t$ $n(t) = [x(t)/A] \sin[\phi_r(t)] + [y(t)/A] \cos[\phi_r(t)]$ Note: If input noise $N(t)$ is white with density η, then $n(t)$ is white with density $2\eta/A^2$.

Fig. 8-15. PLL synchronous model with generalized phase detector. (For comparison, see Fig. 5-2.)

The noise is introduced by way of the equivalent noise generator $n(t)$, which is derived from the product of input noise and VCO signal. For the conventional loop, $n(t)$ can be shown to be white and Gaussian with one-sided density $2\eta/A^2$ when the input noise $N(t)$ is white and Gaussian with density η (the signal carrier power is $A^2/2$). In terms of this equivalent noise generator and unmodulated carrier signal, Viterbi calculates, for an arbitrary $q(\phi_e)$ function, the mean time $T(\phi_e)$ for the phase error (ϕ_e) to progress from an initial zero value to ϕ_e rad. A loss of lock, or slipped cycle, is evident when the phase error has progressed to $\phi_e = 2\pi$ rad and, therefore, $T(2\pi)$ is the mean time to loss of lock, whereas $1/[T(2\pi)]$ is the mean loss-of-lock rate. The loss-of-lock rates for two ideal phase detector cases (see Fig. 8-16) have been calculated; when applied to the threshold impulse noise model (in terms of the second-order loop) threshold reductions of 4 dB (for the sawtooth characteristic) and 7 dB (for the truncated sine characteristic), with respect to the conventional second-order loop, are predicted.[4] Before discussing methods of loop implementation, a review of those factors that form the basis of the indicated improvement is in order.

The change from the conventional system here is the substitution of other phase detector responses $q(\phi_e)$ *while keeping the equivalent noise generator fixed*. In other words, it is assumed that $q(\phi_e)$ is implemented such that the equivalent noise generator retains its basic characteristic, i.e., it is Gaussian in statistics and remains spectrally *a linear translation* of the input noise to low frequency. These requirements are incorporated in the equivalent model (Fig. 8-15), and are used in the subsequent analysis. In certain instances, the physical attainment of a specific $q(\phi_e)$ may not be possible without altering the equivalent noise generator such that no improvement, or an actual deterioration in PLL performance with respect to the conventional PLL, is experienced.

Within the framework of the Viterbi model, the question arises as to what is the optimum phase detector characteristic. An analysis[8] based on minimizing the area under the tail of the $|\phi_e|$ probability density function, and holding the equivalent noise generator $n(t)$ fixed, results in the square-wave phase detector characteristic (see Fig. 8-16) as the optimum. This response characteristic might be obtained through carrier signal waveshaping discussed presently.

Several approaches may be taken to realize an extended-range phase detector. They fall under the following classifications: (1) carrier signal waveshaping, (2) postdetection linear synthesis, and (3) postdetection nonlinear synthesis. Carrier signal waveshaping, although of limited practicality, is discussed first, as it provides some basis for the other two techniques. It is based on choosing the input and reference carrier signal shapes to the multi-

8.4. Extended-Range Phase-Locked Demodulator

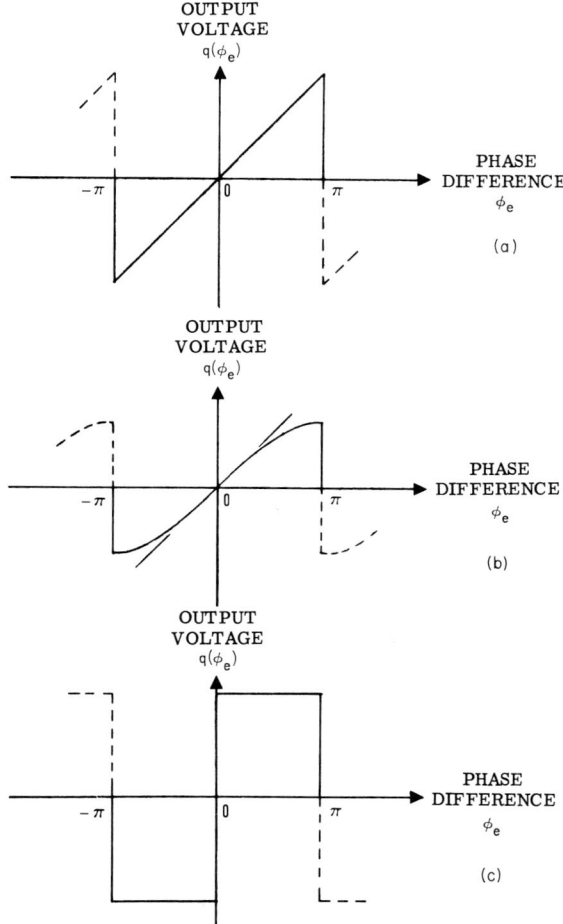

Fig. 8-16. Extended-range phase-detector responses. (a) Sawtooth. (b) Truncated sine. (c) Square wave.

plier type of phase detector in such a way that the low-frequency (or DC) output component has the desired response characteristic as a function of input carrier relative phase. An example of this technique is the hypothetical case of making one carrier a sawtooth and the other an impulse train. When these carriers are applied to a multiplier-type phase detector, they yield the sawtooth phase detector response characteristic shown in Fig. 8-16. The choice of signal and reference for a particular set of shaped signals is important. Although the signal response is independent of the choice, the equivalent noise generator† is a function of it. For example, consider the sawtooth carrier as the signal and the impulse train carrier as the reference. In this instance, the equivalent noise generator (formed from the product of additive input noise and reference carrier signal) has an unbounded spectral density, whereas in the reverse situation it has a bounded spectral density.‡ Waveshaping of the reference signal is easily accomplished, since it is a locally generated signal. The received signal, however, unless waveshaped at the transmitter prior to corruption by noise, cannot be shaped without at the same time also altering the noise. Furthermore, transmitter waveshaping requires prohibitive channel transmission bandwidths (to pass the required harmonic spectral zones), resulting in a generally impractical system. As a final example of this technique, consider the problem of realizing the square-wave phase detector characteristic (Fig. 8-16). Consider the choice of a square-wave for the VCO output signal and an impulse train carrier as the input signal. The multiplying type phase detector will produce the desired square-wave response in conjunction with linear spectral translation of the input noise. Note that the equivalent noise generator remains the same for both sinusoidal or square-wave VCO signals of equal average power[8] for white noise at the PLL input. The system is limited in a practical sense by the fact that the impulse train carrier signal requires infinite average power and bandwidth for transmission.

By examining the range extension mechanism in the waveshaping technique already discussed and then restricting the signal carrier shape to that of a sinusoid, postdetection synthesis procedures can be evolved which prove to be of greater practical value. We may represent shaped and unmodulated input and reference signals to the conventional PLL (see Fig. 5-1) by the following Fourier series

$$a(t) = \sum_{m=1}^{\infty} a_m \cos[m\omega_i t] \qquad (8\text{-}22)$$

$$f(t) = \sum_{n=1}^{\infty} b_n \sin[n\omega_i t - n\phi_e] \qquad (8\text{-}23)$$

where ϕ_e is relative phase shift between signal and reference.

The low-frequency components emanating from the multiplying type of phase detector in the absence of noise are, therefore,

$$e = a(t)f(t) = \sum_{n=1}^{\infty} a_n b_n \sin(n\phi_e) \qquad (8\text{-}24)$$

† This is the noise generator that is introduced in the PLL model (see Fig. 8-15) to represent input additive noise.
‡ This may be understood by noting that the impulse train reference signal translates harmonic zones of the input noise uniformly to low frequency; whereas the sawtooth reference signal translates harmonic zones with an attenuation in power of $(1/n^2)$, where n is harmonic number, to low frequency.

8.4. Extended-Range Phase-Locked Demodulator

It should be apparent, from Eq. (8-24), that the resultant phase detector response, (e vs ϕ_e), is a function of the superposition of Fourier terms that result from the intermodulation between spectral zones of the signal and reference of the same harmonic number. The proper choice of the $a_n b_n$ product (controlled by the shape of the carrier signals) results in various extended-range phase-detector responses. Direct synthesis of the Fourier series in Eq. (8-24) without carrier signal waveshaping brings us to the postdetection linear synthesis method, which is based on generating, *from postdetection signals*, the individual sinusoidal terms that comprise the Fourier series. The terms are combined by linear addition to produce an extended-range phase detector response. Two proposed systems that follow this approach are shown in Figs. 8-17 and 8-18.[9] The first approach is a feedback system where the error voltage, designed to be a linear function of the phase error, is fed back to introduce the phase error, and multiples of the phase error, to auxiliary phase detection loops. The results are phase-detector outputs proportional to $\sin(n\phi_e)$, which may then be appropriately weighted in amplitude, and combined to form a linear function of ϕ_e. The second system, which is an open-loop system, relies on quadrature detection and trigonometric multiplication of $\sin(\phi_e)$ and $\cos(\phi_e)$ terms to generate the $\sin(n\phi_e)$ terms; these terms are then weighted in amplitude and combined to form an extended-range phase-detector response. Both of these systems are restricted in the practical sense to forming a finite sum of sinusoidal terms; the question therefore arises

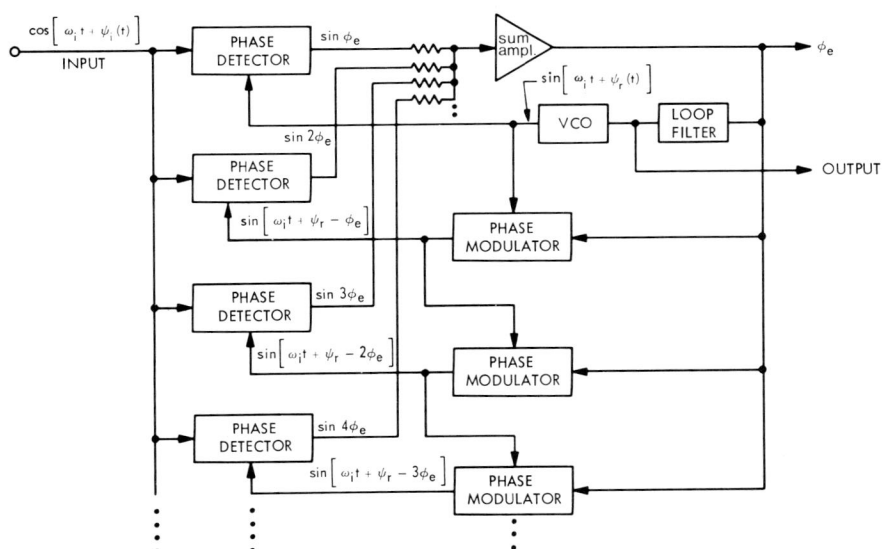

Fig. 8-17. Extended-range PLL using multiterm synthesis: Phase-error feedback method.

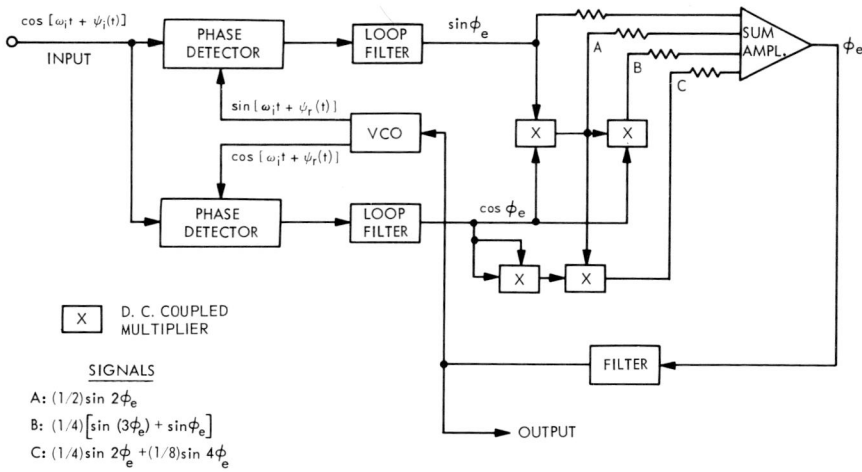

Fig. 8-18. Extended-range PLL using multiterm synthesis: Quadrature detection and trigonometric multiplication method.

as to what is the best choice of weighting coefficients ($a_n b_n$ products) for a finite series generation. For the approximation of the sawtooth phase-detector response, some analytical results were obtained by R. Sommers.[10] These are plotted in Fig. 8-19 as an aid in the computation of range extension as a function of the number of terms and the nature of the weighting coefficients (Fourier, Fejer,[11] or maximal linear). The figure presents the fraction of the period occupied by the monotonic response as a function of the highest harmonic number used in the synthesis. Both of these systems are not straightforward extended-range phase detectors when operating under noise conditions, because of the complex internal noise processing. Therefore, the threshold reductions reported earlier for the extended-range phase detectors, using Viterbi's analytical formulation, do not simply apply. The calculation of the equivalent noise generator, required to evaluate loop performance in terms of the PLL model, is a difficult analytical task, with the result that the threshold performance of these two systems is as yet unknown.

The third general technique, postdetection nonlinear synthesis, is best illustrated by reference to a practical system for which experimental results are available. The Tanlock system,[12] shown in Fig. 8-20, is based on generating the function

$$q(\phi_e) = \frac{(1 + K) \sin \phi_e}{1 + K \cos \phi_e}, \qquad 0 \leq K < 1 \qquad (8\text{-}25)$$

8.4. Extended-Range Phase-Locked Demodulator

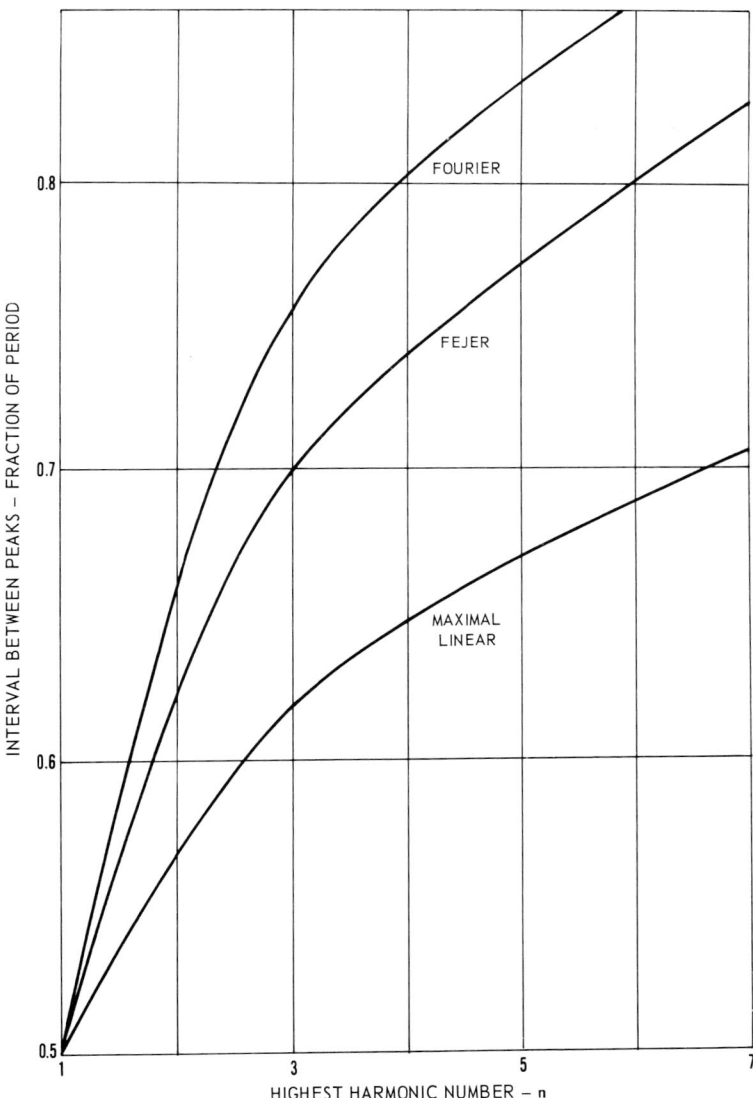

Fig. 8-19. Comparison of Fourier, Fejer, and maximal-linear approximations to the sawtooth.

256   8. Design of Compound and Multiple Loops

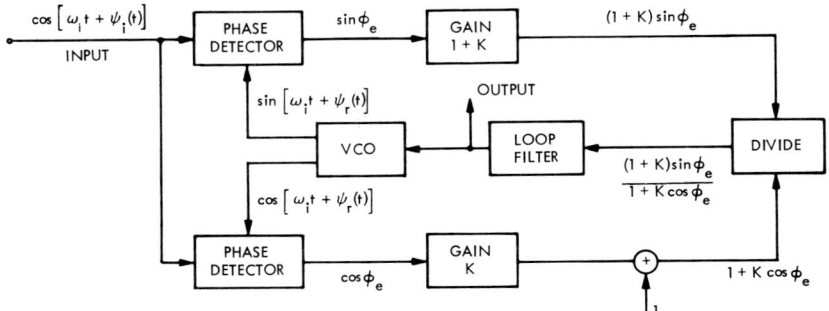

Fig. 8-20. Tanlock PLL.

which is plotted in Fig. 8-21 for various values of K. By using values of K approaching unity, monotonic range extensions approaching $\pm \pi$ rad are possible, but at a sacrifice of linearity. A reasonable compromise between range extension and linearity is obtained for $K = 0.7$, for which a threshold reduction of 4 dB, with respect to a conventional loop, is predicted.[13] Extensions of the Tanlock system are reported in the literature,[14] where performance comparisons (mainly in the absence of signal modulation) between Tanlock

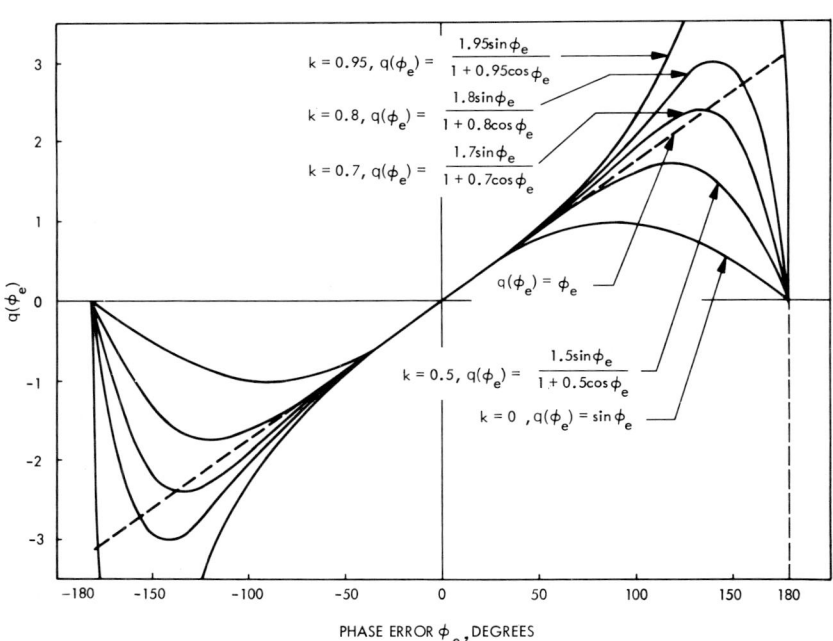

Fig. 8-21. Tanlock phase comparator characteristic curves (from Robinson[12]).

8.4. Extended-Range Phase-Locked Demodulator

and conventional PLL systems, of both first- and second-order and with loops incorporating the generalized Tanlock comparator,

$$q(\phi_e) = \frac{(1 + K)^n \sin \phi_e}{(1 + K \cos \phi_e)^n}, \qquad 0 \le K < 1 \tag{8-26}$$

are presented. Note that the parameter n provides an additional degree of freedom in forming the extended-range phase-detector response. Although there is limited experimental evidence, there appears to be disagreement[12,14] as to whether the Tanlock system actually provides threshold reduction with respect to the conventional PLL. Comprehensive experiments in the presence of modulation are required to resolve the issue.

We consider finally a phase detector implementation that claims monotonic range extension by application of phase feedback,[15] as shown in (a) of Fig. 8-22. When applied to the PLL we have the implementation (called ERPLD) shown in (b) of Fig. 8-22. It is convenient to consider the circuit in (a) of Fig. 8-22 as an effective phase detector, with the applied phase difference being $\psi = \phi_i(t) - \phi_{r1}(t)$, i.e., the phase difference between the input signal and the VCO output as in the conventional PLL. The characteristic equation of this phase feedback phase detector in the absence of noise may be written as

$$e = \sin[\psi - K_3 e] \tag{8-27}$$

where e is the detector output and K_3 is the sensitivity of the phase modulator. When plotted as a function of ψ, the detector exhibits the characteristic curves shown in Fig. 8-23, which offer monotonic range extension and improved linearity. Another interesting feature of these curves is that they become multivalued in e for values of $K_3 > 1$. The maximum monotonic range extension for a single-valued characteristic is achieved when $de/d\psi = \infty$ for $\psi = \pi$, which occurs for $K_3 = 1$, and results in a monotonic range of

$$-(\pi/2 + 1) \le \psi \le (\pi/2 + 1) \tag{8-28}$$

radians. The slope of the curves (sensitivity) near the origin is

$$1/(1 + K_3) \quad \text{V/rad}$$

With ERPLD design based on achieving the same closed-loop response,

$$\frac{\Phi_{r1}}{\Phi_i}(j\omega)$$

and noise bandwidth B_n (taken at the VCO output), as for a conventional PLL, experimental data indicate[15] that the ERPLD exhibits a 3-dB lower

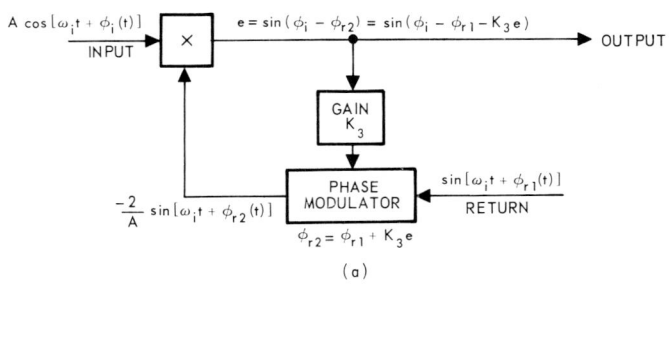
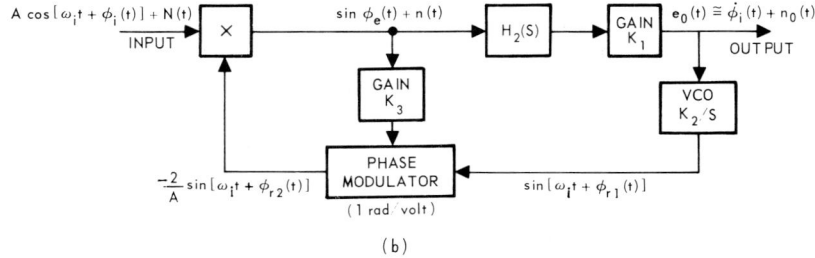
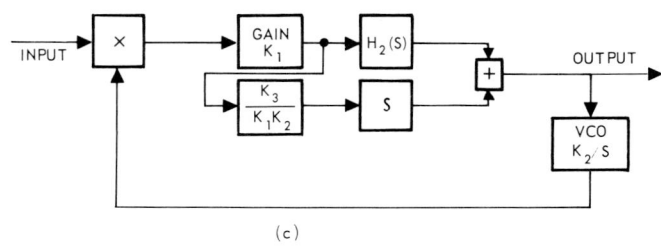

Fig. 8-22. ERPLD block diagrams. (a) Extended-range phase detector. (b) Extended-range phase-locked demodulator. (c) Equivalent filter realization of ERPLD.

threshold ($\Delta f_p = 10$ kHz, $f_T = 1$ kHz, $f_b = 4$ kHz, $B_p = 36$ kHz) than the conventional loop. Note that phase feedback reduces the phase detection sensitivity (slope of e vs ψ curve, Fig. 8-23), thus requiring a 6-dB increase in the main loop gain $K_1 K_2$ over that normally required for a conventional loop when choosing $K_3 = 1$.

Under noise conditions, however, we cannot simply insert the extended-range phase-detector response into the Viterbi formulation for the calculation of the cycle slipping rate because of the noise accompanying the fed-back phase. To reconcile analytically the exhibited experimental performance,

8.4. Extended-Range Phase-Locked Demodulator

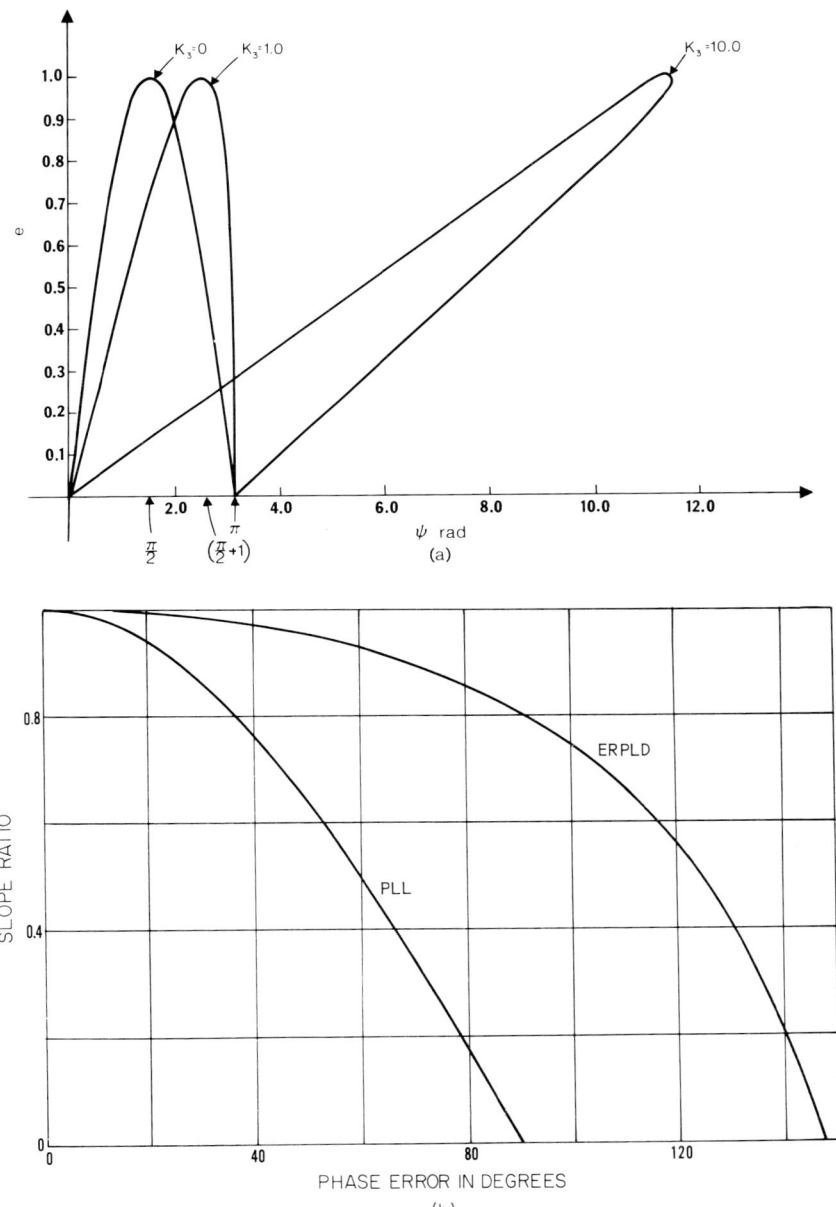

Fig. 8-23. Response functions and linearity of the extended-range phase detector by phase feedback. (a) Response functions. (b) Linearity in terms of the ratio of the slope of the response functions at a given phase error to that at the origin ($K_3 = 1$ for ERPLD; from Acampora and Newton[15]).

we examine the ERPLD from another point of view. Formulation of the loop differential equation (see Fig. 8-22b) results in

$$\dot{\phi}_e(t) = \dot{\phi}_i(t) - K_1 K_2 F'(p)[\sin \phi_e + n(t)] \quad (8\text{-}29)$$

where

$$\phi_e(t) = \phi_i(t) - \phi_{r2}(t) \quad (8\text{-}30)$$

$$F'(p) = \left[H_2(p) + \frac{K_3}{K_1 K_2} \cdot p \right] \quad (8\text{-}31)$$

The equivalent noise generator $n(t)$ is Gaussian and white with density $2\eta/A^2$ when the input noise $N(t)$, is Gaussian and white with density η ($A^2/2$ is signal carrier power), as derived for the conventional PLL. Equation (8-29) is identical in form to the describing equation for the conventional loop, with the difference being the filter function $F'(s)$. In the presence of input noise the ERPLD can therefore be represented by the conventional loop model, with the loop filter replaced by $F'(s)$.

Examination of Eq. (8-31) suggests alternate implementations of the ERPLD and an explanation of the observed performance improvement. The alternate implementation is called the "equivalent filter approach," where the ERPLD is realized in a conventional PLL configuration by substituting for the $H_2(s)$ filter, the $F'(s)$ filter. The basis for the equivalence, aside from Eq. (8-31), may be understood by examining the functional layout of the ERPLD (see Fig. 8-22b). The phase modulator function can be performed by a VCO preceded by a differentiator; when applied to the ERPLD, this results in the functional layout shown in (c) of Fig. 8-22. Note that the equivalent filter is easily identified, together with reduction of the system to the form of a conventional loop. Evaluation of the equivalent filter $F'(s)$, for the case where $H_2(s)$ is the lag filter, results in

$$F'(s) = \frac{s^2/\omega_n^2 + 2\xi(s/\omega_n) + 1}{(s/\beta) + 1} \quad (8\text{-}32)$$

$$H_2(s) = (s/\alpha + 1)/(s/\beta + 1) \quad (8\text{-}33)$$

and

$$\omega_n^2 = 1/\gamma\beta, \quad 2\xi/\omega_n = (1/\alpha) + (1/\gamma) \quad (8\text{-}34)$$

with

$$\gamma = K_1 K_2 / K_3 \quad (8\text{-}35)$$

indicating that the effect of phase feedback in a conventional second-order loop is to produce a second-order loop with complex zeros. The design of

8.4. Extended-Range Phase-Locked Demodulator

second-order loops with this generality is still unexplored, and this form of the PLL has as yet not been tried experimentally.

A direct linear analysis of the ERPLD, formulated about a conventional phase detector with the addition of an auxiliary phase feedback loop, follows the same procedure as in Chapters 5 and 6 for the PLL, and results in the following[16]:

1. The closed-loop response may be written in the normalized second-order form:

$$H(s) = \frac{\Phi_{r1}(s)}{\Phi_i(s)} = \frac{(s/a) + 1}{(s/\omega_n)^2 + 2\xi(s/\omega_n) + 1} \quad (8\text{-}36)$$

where

$$\omega_n^2 = Kb/\alpha_1, \quad \xi = 0.5(\alpha_1 b/K)^{1/2} + 0.5(Kb/\alpha_1)^{1/2}/a$$

$$\alpha_1 = 1 + \alpha_2, \quad \alpha_2 = K_3 \delta, \quad K = \delta K_1 K_2$$

and δ is the phase detector sensitivity which was assumed to be unity in the earlier analysis. The constants a and b are from the filter $H_2(s)$ which is taken to be of the same form as in the conventional second-order PLL.

2. The noise bandwidth of the loop is given by

$$B_n = [Kb/(4a\alpha_1)][(Kb/a\alpha_1) + a]/[(Kb/a\alpha_1) + b]$$
$$\cong 0.25[(Kb/a\alpha_1) + a] \quad (8\text{-}37)$$

[For B_n of conventional PLL, see Eq. (6-65)].

3. The mean-square value of the phase error (with $\phi_e = \phi_i - \phi_{r2}$) due to both signal and noise is

$$\overline{\phi_e^2(t)} = [y/(Kb)^2] + [B\eta/A^2] \quad (8\text{-}38)$$

where y is given by Eq. (6-74), the bandwidth B is defined by

$$B = (a/2\alpha_1^2) + [Kb(\alpha_1 + \alpha_2)/2\alpha_1^3 a] + (B_p \alpha_2^2/\alpha_1^2) \quad (8\text{-}39)$$

and B_p is the predetection bandwidth.

4. The carrier-to-noise ratio at threshold, with the noise measured in twice the base-bandwidth, is given by

$$(\text{CNR}_{\text{AM}})_{\text{TH}} = B/\{4f_b[v - y/(Kb)^2]\} \quad (8\text{-}40)$$

where v is the mean-square value of the phase error at threshold (taken as 0.25 for the conventional PLL).

It is clear that unlike the PLL, the threshold CNR of this loop is a function of the predetection filtering (through B_p); this was not evident from the extended-range phase-detector formulation.

5. Optimization of the loop in a manner similar to that of the conventional PLL in Chapter 6, results in

$$a(\text{optimum}) = [(1 + 2\alpha_2)Kb/\alpha_1]^{1/2} \qquad (8\text{-}41)$$

and

$$(Kb)^{5/2} = [5y(Kb)^{1/2}/v] + \{4y\alpha_2{}^2 B_p[\alpha_1/(\alpha_1 + \alpha_2)]^{1/2}/v\} \qquad (8\text{-}42)$$

A computer solution of the above equations indicates that $\alpha_2 = 1$, as used by Acampora and Newton,[15] is a reasonably optimum value for a large range of system parameters. The above equations accurately predict the available experimental results when v is taken as 0.25, as done earlier in the conventional PLL analysis. Threshold reduction in the Acampora–Newton system with respect to the conventional PLL is predicted[16] only for $(B_p/a) < 2.75$. Furthermore, this optimum design indicates a threshold

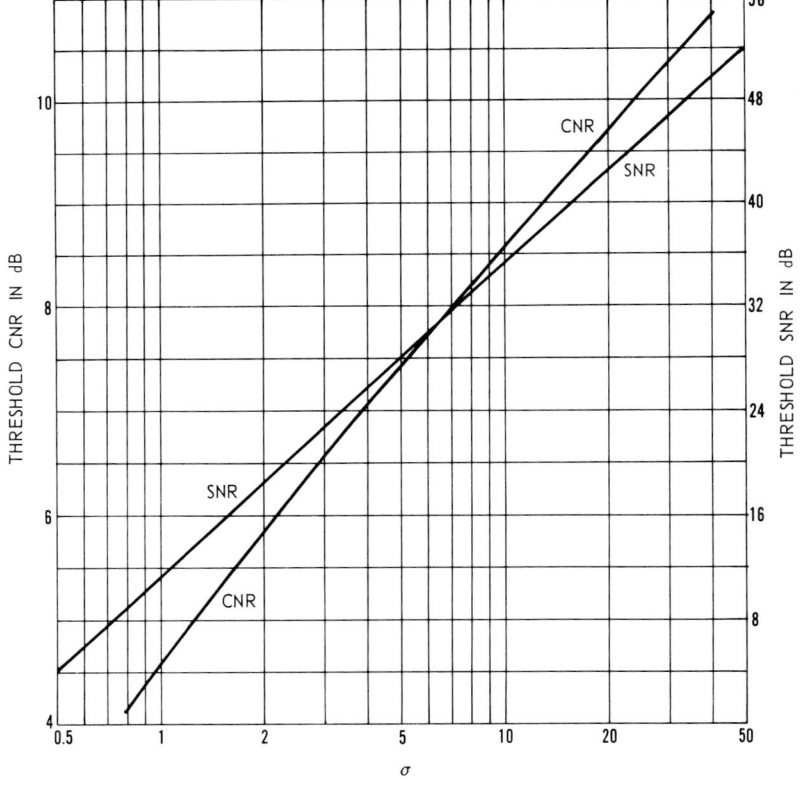

Fig. 8-24. FMFB-ERPLD compound loop threshold CNR and SNR as functions of input rms modulation index (test-tone at $f_b/4$; CNR measured in twice base bandwidth, SNR measured in base bandwidth).

8.4. Extended-Range Phase-Locked Demodulator

extension of 1.2 dB beyond that achieved by Acampora and Newton, thereby indicating a threshold CNR of about 4 dB less than that obtained with a conventional PLL for the system parameters stated above.

In summary, we see that attempts have been made to find extended-range phase detectors which would reduce threshold as calculated for the characteristics shown in Fig. 8-16. Some interesting and useful PLL configurations have been found along the way, and some of these (as the ERPLD) have indeed experimentally been shown to reduce threshold. However, under noise conditions and due to the fed-back noise, these cannot be considered purely on the basis of range extension, but on the basis of the actual circuitry and nonlinearities involved. For the most general nonlinear analysis, the reader is referred to a paper by Lindsey.[17]

To conclude this section we present the design curves (Figs. 8-24 and 8-25) applicable to an FMFB-ERPLD, where the internal detector, in general, is presupposed to have a threshold 4 dB lower than that of the conventional second-order PLL. The design follows almost exactly that of the FMFB-PLL.

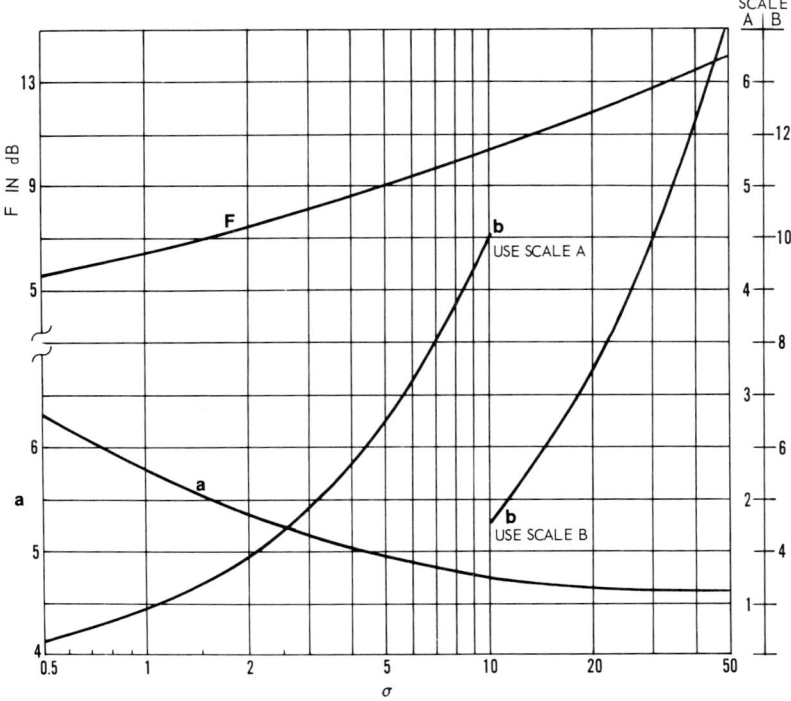

Fig. 8-25. FMFB-ERPLD design parameters as a function of input rms modulation index (test-tone at $f_b/4$).

References

1. R. M. Gagliardi, Transmitter power reduction with frequency tracking FM receivers. *IEEE Trans.* **SET-9**, 18 (1963).
2. M. Morita and S. Ito, High sensitivity receiving system for frequency modulated wave. External Publ. Nippon Electric Co. Ltd., Tokyo, Japan; *IRE Int. Conv. Rec.* **8**, Pt. 5, 228 (1960).
3. R. F. Stone, A transportable tracking and communication terminal for passive satellites. *Proc. Mil-E-Con, 7th, Washington, D.C., September 9–11, 1963*.
4. J. T. Frankle, Threshold performance of analog FM demodulators. *RCA Rev.* **27**, No. 4, 521 (1966).
5. H. Heinemann, A. Newton, and J. T. Frankle, Multiple-loop frequency-compressive feedback for angle-modulation detection. *RCA Rev.* **29**, No. 2, 252–269 (1968).
6. A. Acampora, Advanced FM demodulator. 3rd Quart. Progr. Rep., Contract DA28-043-AMC-02473(E). RCA Commun. Syst. Div., New York, June 1967.
7. A. J. Viterbi, Phase-locked loop dynamics in the presence of noise by Fokker-Planck techniques. *Proc. IEEE* **51**, 1737 (1963).
8. J. J. Stiffler, On the selection of signals for phase-locked loops. *IEEE Int. Conf. Commun. Digest of Papers, Minneapolis, Minnesota, June 12–14, 1967*.
9. Advanced ECCM Techniques (U). 1st Quart. Progr. Rep., Contract DA28-043-AMC-00167(E) (CONFIDENTIAL). RCA Commun. Syst. Div., New York, October 1964.
10. Advanced ECCM Techniques (U). 2nd Quart. Progr. Rep., Contract DA28-043-AMC-00167(E) (CONFIDENTIAL). RCA Commun. Syst. Div., New York, January 1965.
11. E. A. Guillemin, "The Mathematics of Circuit Analysis." Wiley, New York, 1949.
12. L. M. Robinson, Tanlock: A phase lock loop of extended tracking capability. *Proc. Nat. Conf. Military Electron., Los Angeles, California, February 1962*, pp. 396–421.
13. A. Acampora, Advanced FM Demodulator, Final Rep., Contract DA28-043-AMC-02473(E). RCA Commun. Syst. Div., August 1967.
14. J. J. Uhran and J. C. Lindenlaub, Effects of a class of phase comparators on the threshold and lock range of phase lock loop systems. *Int. Conf. Commun. 3rd, Minneapolis, Minnesota, June 1967*; Experimental results for phase-lock loop systems having a modified nth-order tanlock phase detector. *IEEE Trans.* **COM-16**, No. 6, 787–795 (1968).
15. A. Acampora and A. Newton, Use of phase subtraction to extend the range of a phase-locked demodulator. *RCA Rev.* **27**, No. 3, 577–599 (1966).
16. J. Klapper and G. M. Veiga, Threshold performance of a PLL with phase feedback. *Proc. Allerton Conf. on Circuit and System Theory, Monticello, Illinois, October 1971*.
17. W. C. Lindsey, Nonlinear analysis of generalized tracking systems. *Proc. IEEE* **57**, No. 10, 1705–1722 (1969).

CHAPTER
9

Digital FM and Other PLL Applications

The first part of this chapter presents principles of digital FM demodulation by first reviewing general principles and LD performance, and then presenting guidelines for PLL design and application. The remainder of the chapter gives a brief review of other PLL applications, with the purpose of stressing the varied applications of the phase-lock principle.

The analytical state-of-the-art in digital FM reception with angular-feedback demodulators has not yet progressed to the same level as threshold calculations for analog FM signals. However, there are a number of pressing factors for knowing these demodulators in terms of digital signals, among which are: (1) the existence of improved performance in some applications; (2) the improved performance, if it exists, generally appears for all received CNR's and not only below threshold; and (3) in a dual-purpose receiver designed for both analog and digital signals,† the low-threshold demodulator aimed primarily at analog signal demodulation will also generally be used to demodulate the digital signals. Current knowledge in this area, though incomplete, is developed in this chapter to provide useful theoretical insights and design guidelines.

We begin by placing the digital FM system in perspective with respect to other digital-modulation carrier systems. We then describe the link between the threshold mechanism in analog signaling and error generation for digital

---

† It is generally a simple matter to switch a PLL from one set of parameters to another, as it usually involves only the components of the lowpass filter.

signals. This leads to the calculation of error rates with discriminator or PLL detection, and to the optimum design of a PLL for binary FM signals. No analytical or experimental data are as yet available on the performance of an FMFB as a demodulator of digital FM.

9.1. Introduction to Digital FM Systems

Digital FM is a form of frequency shift keying (FSK) in which the signal wave is treated like an FM signal. For the transmitter, it means that the signal is generated by feeding the digital baseband to an FM modulator, rather than by switching between independent oscillators. The result is a wave with phase continuity at the switching instants, giving more confined spectra[1] and lower error rates.[2] Recent practice is to generate FSK in this manner unless modulator frequency instabilities are excessive. The use of an FM modulator provides additional flexibility in shaping the transitions in the digital baseband, i.e., it permits premodulation filtering.

In the receiver, digital FM is demodulated by an FM demodulator, such as the LD or any of the angular-feedback demodulators described in the earlier chapters. Other types of FSK demodulators, such as correlation[3] and twin-filter types,† may also be used for digital FM, but will not be considered herein.

A simplified block diagram of a digital FM system is shown in Fig. 9-1. A number of "transmitter" and "receiver" functional elements, which are basically concerned with signal amplification and frequency translation, are omitted, as they are not intended to change the intrinsic characteristic of the signal or the noise. Only the elements fundamental to the operation of a digital FM system are shown.

In the transmitter, the applied digital baseband signal is inscribed in the frequency of the FM modulator output. The baseband signal is often "premodulation" filtered, usually for the sake of reducing the frequency splatter of the transmitter output. The transmitted wave is corrupted in the "chan-

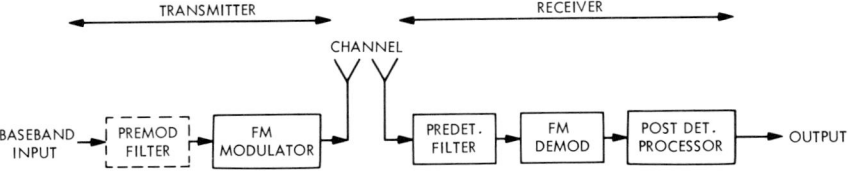

Fig. 9-1. Simplified block diagram of a digital FM system.

† The work of Stein and Jones[3a] is recommended for a lucid treatment of the FSK twin-filter approach.

nel," which may be the space through which it is radiated, or possibly a telephone system. The corruption may take the form of additive noise and/or fading conditions. The theoretical discussion is based mainly on additive Gaussian noise interference, both for analytical convenience and due to the fact that it represents sufficiently well an important class of situations.

The first critical receiver function is predetection filtering, by which as much noise as possible is eliminated without unduly distorting the FM wave. A rule of thumb (based on experimental evidence and some theoretical justification) suggests a predetection filter bandwidth approximately equal to the bit rate,[4-9] provided system constraints (such as center frequency uncertainties) permit this and the implied small deviation indexes. More details will be given later.

The FM demodulator may be an LD or any of the angular-feedback types. Known performance differences of the various demodulators are described later. Operation of the post-detection processor is generally a variation on one of the following: (1) The output of the demodulator is filtered, preferably by a delay-equalized filter, to remove excess noise without unduly distorting the digital wave. At the same time, the filter removes output components of the carrier frequency and its harmonics, and may also block any DC component stemming from signal and demodulator center-frequency uncertainties. The filtered wave is then sampled at the instant when the signal is expected to have a maximum amplitude (usually at the center of the delayed bit) and the digital decision is based on the value of the sample. (2) The output of the demodulator is filtered by an integrate-dump circuit which operates as follows: Integration takes place between the beginning and the end of each bit, and the integrator content is dumped at the end of each bit period to prepare for a new integration interval. The binary decision is based on sampling the integrator just prior to dumping.

The integrate-dump processor is a matched filter† for rectangular bits immersed in white Gaussian noise. The output noise of the FM demodulator is neither white nor Gaussian, and the bit transitions are not rectangular because of shaping introduced by premodulation filtering and/or by receiver filtering. The matched filter for this case is not yet known. Of the two popular processors described above, the integrate-dump circuit was found experimentally to give somewhat better performance[4]; this conclusion is also backed by some analytical considerations.[10] As will be seen, the integrate-dump processor has the added advantage of considerably simplifying the theory, and for these reasons it will be assumed in much of the development which follows.

† A matched filter is the *best possible* filter for a digital signal in additive, stationary, Gaussian noise, and the best possible *linear* filter if the additive noise does not have Gaussian statistics.

A. System Considerations

We now examine the required received signal power for a given message rate, permitted error rate, and interference environment. It is meant to give some perspective on the performance of the various digital angular transmission systems and their associated demodulators. Additive Gaussian noise as the sole source of interference and binary signaling are assumed.

It is appropriate to give the system performance as the probability of bit error (or bit error rate, BER) as a function of the energy ratio E/η, where E is the received energy per bit and η is the noise power per hertz of bandwidth. By giving the performance as a function of E/η, it is possible to compare the various systems on the basis of required energy expenditure per bit of information for a given white interference background. Figure 9-2 presents the performance of the salient systems.

The most efficient binary system for additive Gaussian noise interference is coherent PSK with phase shifts of $\pm 90°$.[3] Its demodulation is by correlation techniques and, thus, requires a perfect phase reference in the receiver. A system not dependent on the availability of a phase reference is differentially coherent PSK (DPSK) in which demodulation takes place by comparing the phase of the current bit against that of the preceding one.[11] It is observed that the additional E/η required for this system is a function of the BER, and that it approaches the performance of coherent PSK in very high reliability systems. Next in performance is binary FM (phase continuous, rectangular transitions) with the optimum deviation index of 0.7, and demodulated by a "maximum likelihood receiver" (MLR).[2,12] The performance of binary FM demodulated by an LD is next.[6,7,13] Performance with angular feedback demodulators is described presently. It is also observed that binary FM demodulated by an LD has about the same performance as coherently demodulated orthogonal FSK. Finally, the performance of noncoherent FSK (i.e., only envelope information is used) forms the right-most curve. Carrier on–off signaling is the same as FSK on an average power basis. (However, FSK has important advantages in a fading environment, see Stein and Jones[3a]). The performance of binary FM with PLL demodulation is about 0.5 dB to the left of the LD curve, as described later. Figure 9-2 indicates also the minimum E/η required for errorless transmission, derived from Shannon's theory.[15]

Our consideration in this section is with binary FM of rectangular transitions, the performance limit of which is given by curve 3. It has recently been shown[9] that binary FM can perform almost as well as DPSK (curve 2), if the transmitter wave is optimally filtered before transmission. However, the following sections will assume no rf filtering in the transmitter, consistent with current communication practice.

9.1. Introduction to Digital FM Systems

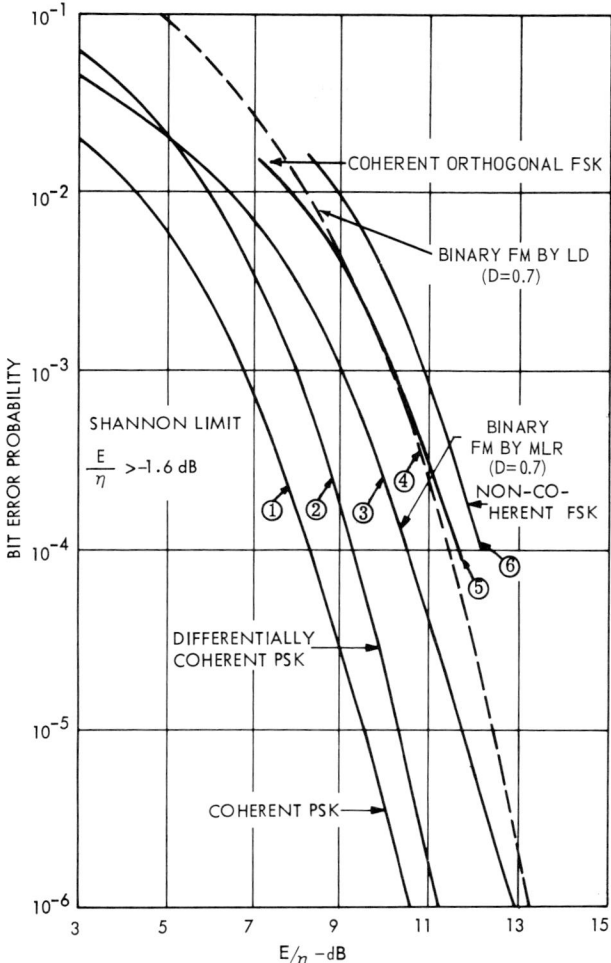

Fig. 9-2. Performance of binary systems.

B. The Error Mechanism

The instantaneous output of an FM demodulator is proportional to the instantaneous frequency deviation of the wave at the input to the demodulator. This wave is filtered by the relatively narrowband predetection filter of the receiver and is comprised of both signal and noise components. The noise is minimized by the predetection filtering, but at the same time the signal is also shaped by the filter and, generally, in a way that increases its susceptibility to noise. The increased noise susceptibility is due to: (1) smearing of the bit energy into adjacent time-slots, generally referred to as "intersymbol interference," and (2) the creation of amplitude modulation on the FM wave which creates regions of decreased CNR (ρ). We will ignore these complications for the present.†

If an integrator is placed at the output of the FM demodulator, then its instantaneous output is proportional to the instantaneous phase deviation of the wave at the input to the demodulator. In the integrate-dump circuit, a new integration period begins with each bit. Thus, at the end of the bit, the output is proportional to the difference in phase between the beginning and end of the bit. The following important conclusion is reached: *The binary decision is based on the polarity of the difference between the final and initial phases of the resultant; the resultant is the sum of the signal and noise waves at the demodulator input.*

Transforming the decision into that of phase difference greatly simplifies the analytical problems at hand, and yields more readily physical insights. The phase difference accumulated between the beginning and the end of the bit has been contributed by two factors: signal, and noise (and distortion). If the phase difference contributed by the noise cancels that of the signal, an error in decision occurs.

What is the phase difference contributed by the signal? For any shaped bit it is given by θ_s,

$$\theta_s = 2\pi \int_0^T (f - f_0) \, dt \qquad (9\text{-}1)$$

where $(f - f_0)$ is the instantaneous frequency deviation and T is the bit duration. For rectangular transitions, it reduces to

$$\theta_{sr} = 2\pi \, \Delta f \, T \qquad (9\text{-}2)$$

where Δf is the peak frequency deviation in hertz from the center frequency. The output of the integrate-dump circuit is, for this case, a series of positive and negative-going ramps with the peak values given by Eq. (9-2). This is illustrated in Fig. 9-3.

It is convenient to define a deviation index D as

$$D = 2 \, \Delta f / B_R \qquad (9\text{-}3)$$

where B_R is the bit rate and equals the reciprocal of T. Observe that D becomes the conventional modulation index of analog signaling if the bits are

---

† The inclusion of these effects in the calculation of error rates is generally a difficult analytical problem (see Bennett and Salz[7]). In practice, the engineer often compensates for them by adding a penalty factor, e.g. 1 dB (see Meyerhoff and Mazer[5]).

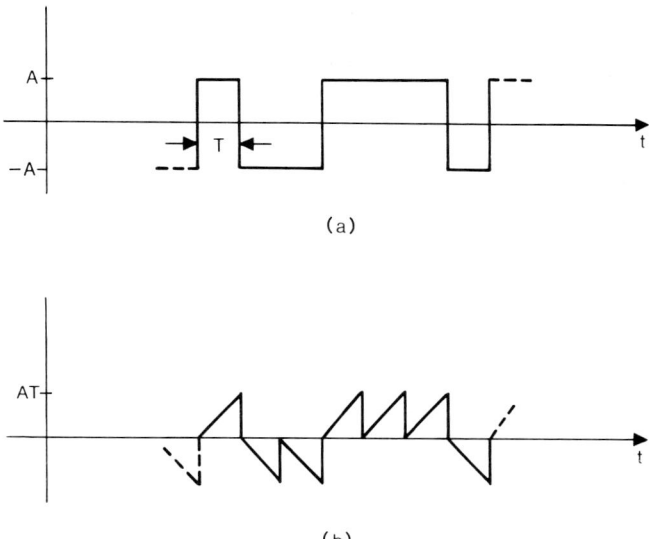

Fig. 9-3. Baseband signal and integrate-dump output. (a) Example of digital baseband. (b) Output of integrate-dump circuit for the above wave.

half sinewaves and a repetitive mark-space message is sent. In terms of the deviation index,

$$\theta_{sr} = \pi D \qquad (9\text{-}4)$$

Figure 9-5 illustrates the value of this angle for various deviation indexes. Note that a complete cycle is introduced within a bit period by a $D = 2$.

What is the phase difference introduced by the noise? An illustration was given in Figs. 3-5 and 3-7. The noise and the carrier form a resultant, and the angle of the resultant with respect to the carrier is the angle introduced by the noise. The noise amplitude is small most of the time, and so is the noise angle. However, once in a while the noise is sufficiently large to make the resultant encircle the origin and thereby introduce a 2π phase shift. The small noise now undulates about the new reference position. Also, for any reasonable ratio of signal deviation to receiver bandwidth, the encirclements are preponderantly in the direction to oppose the signal.[13] This is also the case with the cycles slipped due to loss of lock in the angular feedback demodulators.[14] As an illustration, Fig. 9-4 shows oscillograms of a quasi-random digital baseband at a CNR of 7 dB, demodulated by a PLL type of demodulator (the ERPLD discussed in Chapter 8). Part (a) of the figure is a single trace in which one spike (or encirclement) is evident. Part (b) of the figure is a long exposure photograph in which a number of spikes are caught.

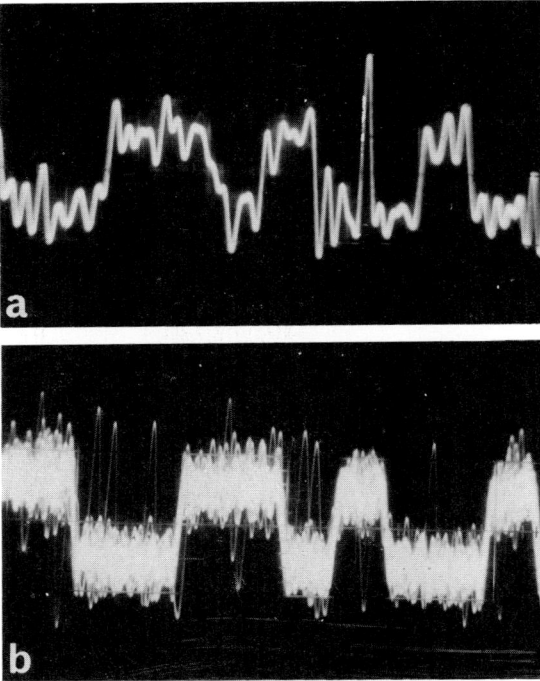

Fig. 9-4. Spikes in the output of an FM demodulator (ERPLD). (a) Single-shot oscillogram. (b) Long-exposure photograph.

We will now provide a very interesting and important link between the threshold mechanism in analog FM and the error mechanism in digital FM.[13] It is recalled that the phase steps in (a) of Fig. 3-7, which generally result in output noise spikes (Fig. 3-7b) are the main cause of threshold in analog FM. It will be described here how these phase shifts of 2π, or "encirclements," or "cycle slippings," are also the main cause of errors in digital FM, with one essential difference: When the CNR is sufficiently high to place the demodulator above threshold, the cycle slippings occur very rarely. In analog FM demodulation, rare spikes contributing very little to the average output noise power value can be, and are, ignored. The situation is different in digital FM, where the rare events are of critical importance. It turns out that no matter what ρ is, or how much above threshold the demodulator is, the origin encirclements produced by the noise are the main cause of errors in digital FM for all except very small deviation indexes.

It is observed from Fig. 9-5 that the signal contributes an origin encirclement for each increase in D by a value of two. For noise to cause an error, it must contribute an equal or greater angle, but opposite in phase. These

9.1. Introduction to Digital FM Systems

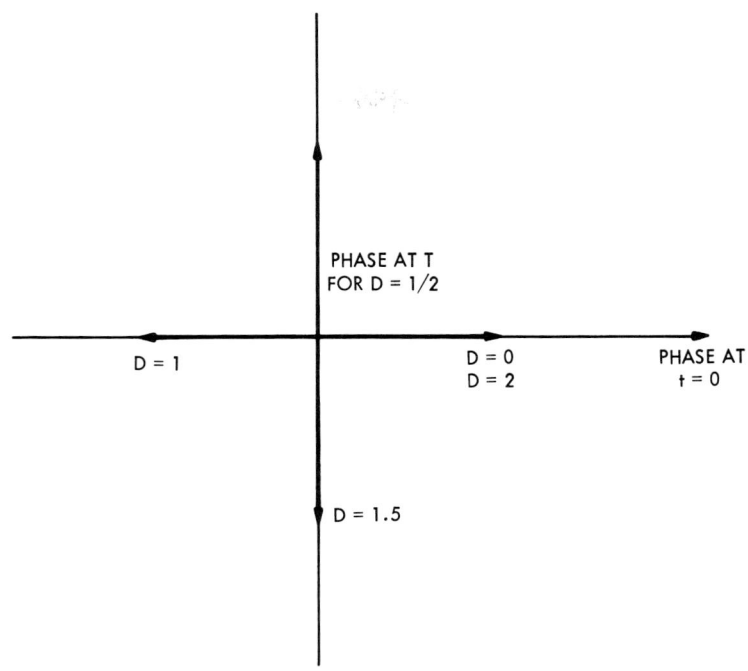

Fig. 9-5. Signal angle at end of bit for various deviation indexes (rectangular bits; $\theta_s = \pi D$).

considerations lead to the necessary and sufficient conditions for error generation summarized in Fig. 9-6. The noise must contribute at least one encirclement for an error to occur to a digital signal with $D > 2$. At least two origin encirclements are required for $D > 4$, etc. Thus, it is unquestionable that the threshold mechanism is the cause of errors for $D > 2$. However, there is analytical evidence that the encirclements are the primary cause of errors for all deviation indices greater than 0.5.[16] Experimental data available between $D = 0.7$ and $D = 10$ (essentially the range of practical interest for D) agree well with this assertion.

Why is the spike noise† so effective in causing errors? The nonencirclement noise has undulations in both directions, and the probability of a large undulation at a sampling instant is small. The encirclements, however, introduce a large phase shift usually in the direction to oppose the signal deviation and, generally, no return of this phase step takes place. Now, if we compare the effect of spike (or encirclement) noise against that of the doublet type, we

---

† The terms "spike noise," "encirclement noise," "2π phase shifts," "cycle skippings," and "cycle-slippings" are used here interchangeably, although the noise is not necessarily of spike form.

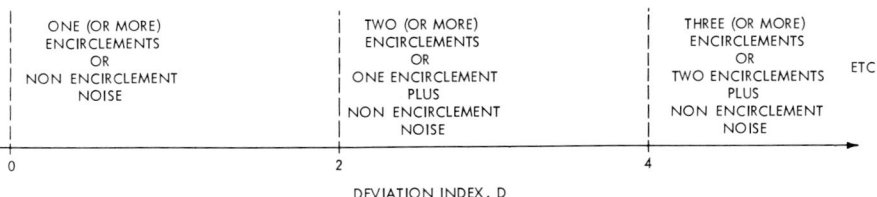

Fig. 9-6. Causes of errors.

find the following: (1) In analog FM demodulation, the encirclement noise is much more destructive because of its high low-frequency content (flat power spectrum); and (2) in digital FM demodulation, the encirclement noise causes errors because of the large opposing area (phase difference) it represents.

It is a difficult mathematical problem to include, on a quantitative basis, the nonencirclement noise in predicting BER. Fortunately, for any except very small D's it suffices to deal with the encirclement noise only. The following sections will make the assumption that the encirclements are the only cause of errors, while keeping in mind that the nonencirclement noise will cause some modification (smoothing) of the results. This will permit a quantitative application of the theory. Intersymbol interference can be interpreted as a reduction in the phase difference introduced by the signal. However, the analytical complexity is not warranted at this time, and the effect of intersymbol interference will not be included in a quantitative manner.

9.2. Binary Error Rates with Limiter-Discriminator† Demodulation

We will now apply the theory developed earlier to predicting the probability of error P_e for binary FM with an LD as a demodulator and additive Gaussian noise interference. The effect of the following parameters will be examined: CNR at the LD input (ρ), E/η, D, and signal offset from center frequency. The encirclement noise will be considered the sole cause of errors.

It is clear that for an unmodulated carrier centered in the noise power spectrum, the probability of positive or negative encirclements is identical. However, it may be shown that for a carrier substantially deviated from the

---

† As stated in Chapter 3, the limiter-discriminator, or LD, is used here as a generic term for all known noncompressive types of FM demodulators that are insensitive to AM, and includes the Foster-Seeley, pulse-counting, delay line, zero crossing, and the ratio detector types.

9.2. Binary Error Rates with LD Demodulation

center frequency, the encirclements in the direction aiding the frequency deviation may generally be ignored. This is illustrated in Fig. 9-7 by a plot of aiding $(N+)$ and opposing $(N-)$ encirclement rates as a function of the carrier deviation from the center frequency.[13] The curve for the opposing spikes is a plot of Eq. (3-30), while the rate of aiding spikes is a plot of[17]

$$\frac{N+}{\Delta f} = 0.5 \left\{ \left[\left(\frac{r}{\Delta f}\right)^2 + 1 \right]^{1/2} \operatorname{erfc}\left[\rho + \rho\left(\frac{\Delta f}{r}\right)^2\right]^{1/2} - e^{-\rho} \operatorname{erfc}\left(\frac{\Delta f}{r} \rho^{1/2}\right) \right\} \quad (9\text{-}5)$$

where erfc is the complementary error function available in tabulated form.

It is interesting to observe that the rate of opposing spikes is not a function of the shape of the predetection filter (i.e., the shape of the power spectral

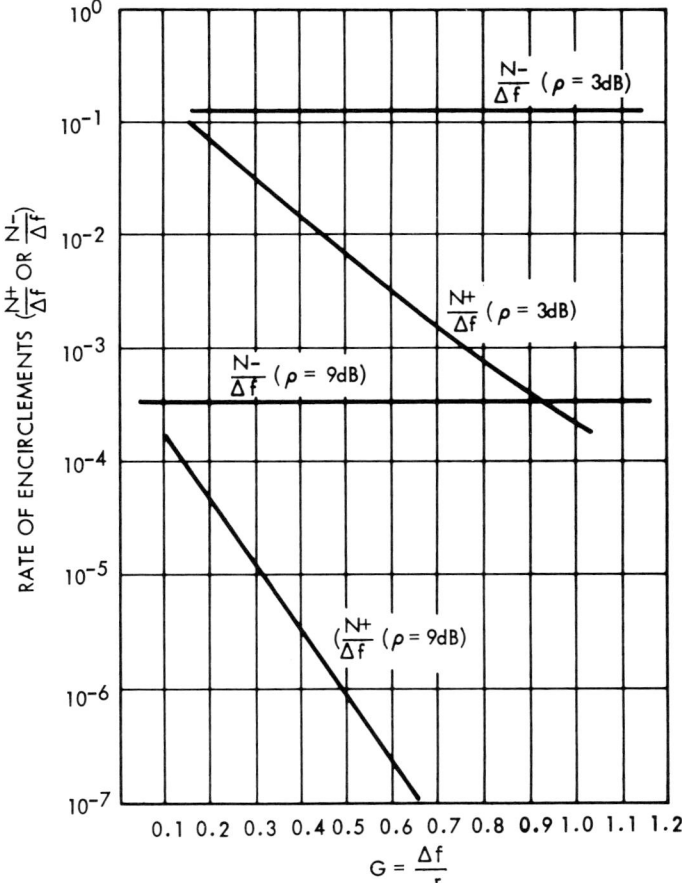

Fig. 9-7. Comparative rates of encirclements (from Klapper[13]).

density of the noise at the LD input), only of ρ, as long as the rate of aiding spikes is negligible.

The ordinate is normalized with respect to the deviation (in hertz), and the abscissa is normalized with respect to the radius of gyration of the noise spectrum r defined by Eq. (3-29). The parameter is ρ, the CNR at the LD input. It is observed that, for normalized deviations greater than about 0.2, the aiding encirclements may be ignored. Since, in the usual application, this condition is met, only the opposing encirclements will be considered in the following development.†

Another condition which has been utilized in theory,[17] and verified experimentally,[18] is that the encirclements follow a Poisson distribution. A Poisson distribution is one found in describing discrete random phenomena, and may be considered the counterpart of the Gaussian distribution found in continuous processes. It is given by[19]

$$P(k) = e^{-NT}(NT)^k/k! \tag{9-6}$$

where N is the average number of events per second, T is the time interval under consideration, and k is the number of events in time interval T. In terms of our problem, if N is the average number of cycle slippings per second and T is the bit length, then $P(k)$ is the probability of exactly k cycle slippings in a bit. Under usual operating conditions, the probability of "k or more" encirclements in a bit is essentially the same as the probability of "k encirclements in a bit." In terms of the Poisson distribution this is satisfied if

$$NT/(k+1) \ll 1 \tag{9-7}$$

Additional assumptions are that the signal dwells sufficiently long at each frequency so that an analysis based on a steady-state condition applies; and that the encirclements are rapid, so that their full value appears in a single bit.

The number of encirclements required to cause an error is k, and from Fig. 9-6 it is given by

$$k = D/2 \quad \text{(next higher integer)} \tag{9-8}$$

The rate of encirclements per second is given by Eq. (3-30)

$$N_- = (N_+) + \Delta f \, e^{-\rho} \cong \Delta f \, e^{-\rho} \tag{3-30}$$

Let I be defined as the average number of spikes (or encirclements) per bit. From Eq. (9-3)

$$I = NT = (D/2)e^{-\rho} \tag{9-9}$$

---

† Some analytical difficulty arises with a noise spectrum shaped by a single tuned circuit, since its r is unbounded. However, such a spectrum is unrealistic in a practical receiver since higher-order poles will eventually appear.

9.2. Binary Error Rates with LD Demodulation

Equation (9-9) may be generalized to the case where the center frequency of the signal is not the same as the center frequency of the noise, as follows,[13]

$$I = (D/2)(1 \pm \delta)e^{-\rho} \tag{9-10}$$

where

$$\delta = d/\Delta f \tag{9-11}$$

and d is the offset in the center frequency in hertz. The "plus" applies to the symbol furthest from the center frequency, and vice-versa, the "minus" applies to the symbol closest to the center frequency. Equation (9-10) assumes that the DC at the output of the FM demodulator produced by the frequency offset is blocked and the only effect is the altered rate of cycle slipping due to a new effective Δf in Eq. (3-22). In practice, a signal frequency shift may also be accompanied by a change in effective ρ due to a nonflat passband characteristic of the filter.

Using the formula for the Poisson distribution, Eq. (9-6), and Eqs. (9-8) and (9-9), we obtain for the probability of error

$$P_e = I^k e^{-I}/k! \tag{9-12}$$

Equation (9-12) is the result we have been looking for, namely, BER† as a function of the system parameters with LD demodulation. The following examples illustrate its use and present some further insight.

Example 1: BER for constant CNR

A. Problem

Suppose we are constrained to a given CNR at the LD input (ρ) due to a fixed predetection filter bandwidth, or for some other reason. What is the tradeoff between BER or P_e and the rate of information transmittal?

B. Solution

For a given CNR we would generally specify a deviation as large as possible, consistent with the predetection filter bandwidth. A solution to our problem is a plot of P_e as a function of D, with ρ as a parameter, obtained by using Eq. (9-9) in Eq. (9-12). Such curves are given in Fig. 9-8 for $\rho = 3, 6$, and 9 dB. As expected, P_e generally decreases with a decreasing information rate (increasing D), since the energy expended per bit increases with D. Figure 9-8 also gives a plot of the average number of encirclements per bit

---

† Bit error rate (BER), error rate, and probability of error (P_e) are used here interchangeably.

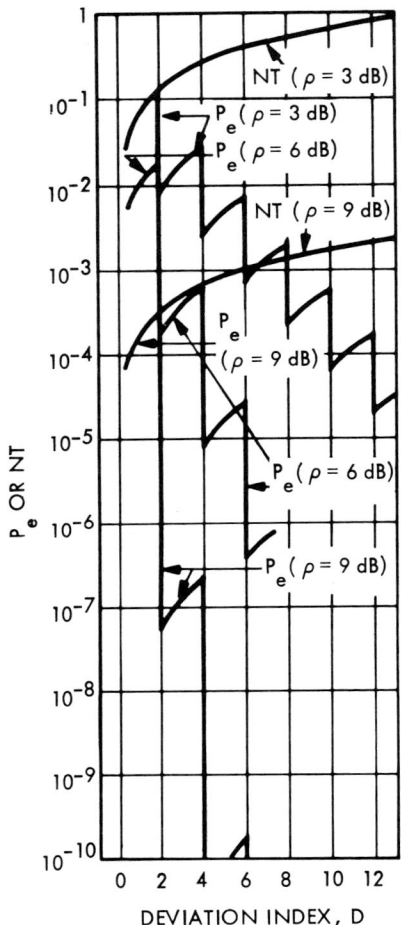

Fig. 9-8. Probability of error and spikes per bit as a function of deviation index (from Klapper[13]).

(NT). For a fixed Δf, the bit length increases with D; therefore, NT also increases. For $D < 2$, P_e and NT are about the same, since a single spike causes an error.

The interesting part in Fig. 9-8 is the "staircase" nature of the curves, which indicates that a number of relative minima may be available to the designer. This behavior is due to the consideration of only an integral number of encirclements per bit, and the neglect of the nonencirclement component of the noise. For example, at $D = 2$ we jump from the probability of one encirclement per bit to the probability of two (see Fig. 9-6), resulting in a

9.2. Binary Error Rates with LD Demodulation

large jump in BER. In practice, it has been found that Eq. (9-12) gives a reasonably good prediction of BER for any D greater than about 0.7. However, depending upon the bandwidth of the predetection filter and other parameters, the smearing of the encirclements into adjacent bits and the presence of the nonencirclement noise generally soften the staircase behavior.[8]

Example 2: Error rates for a constant energy ratio

A. PROBLEM

Find the performance of an LD receiver as a function of D, with E/η as a parameter. This will permit a comparison against other binary systems in Fig. 9-2.

B. SOLUTION

There is a need to transform ρ in Eq. (9-9) to E/η. The relation is

$$E/\eta = k_e \rho B_n T \qquad (9\text{-}13)$$

where B_n is the predetection noise bandwidth, and k_e compensates for modulation losses in the filter

$$k_e = \text{unmodulated } \rho/\text{modulated } \rho \geq 1 \qquad (9\text{-}14)$$

The reason for the inclusion of k_e is because E/η is nominally measured at the input to the receiver, and ρ is defined here as being measured at the input to the LD. In this way, any modulation losses in the filter are charged as a penalty to the receiver, to permit a fair comparison against the other digital receivers.

Observe in Eq. (9-13) that the conversion to E/η is heavily dependent on the filter type and bandwidth. Recall that in analog FM the receiver bandwidth is generally chosen according to Carson's rule [Eq. (3-25)] but that in binary FM the receiver bandwidth is chosen equal to, or a little larger than the bit rate for small D, and a little larger than the peak-to-peak deviation for large D. For large D, the two rules are about the same. For small D, the bit rate bandwidth is considerably smaller than that dictated by Carson's rule. Increasing the bandwidth from that of the "bit rate" to Carson's rule for small D decreases the filter losses to $k_e \cong 1$ and increases the admitted noise power; the two effects are somewhat compensatory. To permit a transformation applicable for a large range of parameters, we will use Carson's rule for the noise bandwidth as an approximation. The video bandwidth is taken as one half the bit rate, consistent with Nyquist's signaling rate,[20] thus

$$E/\eta = \rho B_{CR} T \qquad (9\text{-}15)$$

P_e is plotted as a function of D in Fig. 9-9 using Eqs. (9-12), (9-9), and (9-15), with E/η as a parameter.

Figure 9-9 reveals further interesting properties, in addition to the staircase behavior discussed earlier: (1) The tendency of the P_e vs D curve for a constant E/η is toward greater error rates for larger D. On the surface, this appears to be totally at odds with our picture of FM for analog signals where bandwidth may be traded for carrier power. While it has been known for some time that a D of 0.7 is optimum,[2] the physical reasoning behind it has only become clear with the establishment of the link between the threshold and error

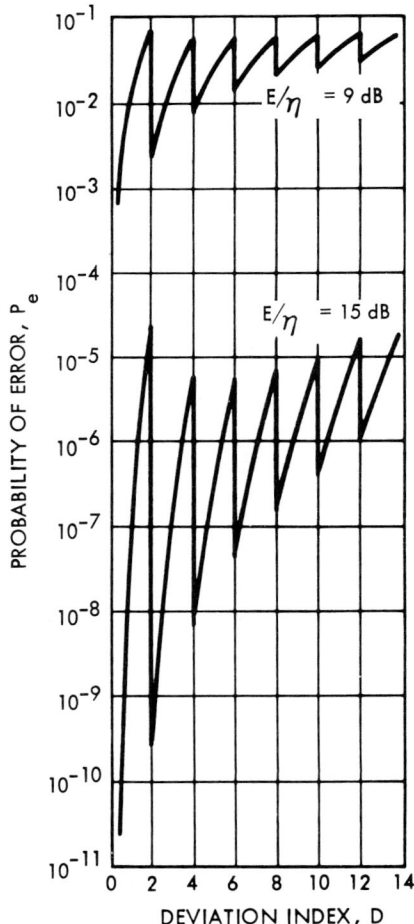

Fig. 9-9. Probability of error versus deviation index with E/η as a parameter (from Klapper[13]).

mechanisms discussed above. Thus, it may be stated that the increase in P_e with D is equivalent to the increase in the required CNR (measured in $2f_b$) at threshold with an increase in modulation index. (2) It would appear that P_e decreases indefinitely with D. This is obviously not so, and is due to the neglect of the nonencirclement noise the effect of which increases with decreasing D. Thus, the minimum P_e is at a D of about 0.7.

Example 3: Effect of center frequency offset

A. PROBLEM

Find the effect on P_e by an offset of signal center frequency, possibly due to the Doppler effect or to transceiver oscillator instabilities.

B. SOLUTION

It is assumed that the output of the demodulator is AC coupled, thus blocking the DC output due to the offset in center frequency. The remaining effect is the change in the probability of an encirclement due to the offset and due to a possible change in ρ because of the different positioning of the signal within the predetection filter. Assuming a sufficiently wide rectangular filter, there is no change in ρ; thus, P_e with frequency offset is again given by Eq. (9-12), except that I is given by Eq. (9-10). The plot of P_e versus center frequency shift is given in Fig. 9-10 for two values of D and a constant ρ at the LD input. Observe the following: (1) Total P_e increases with the frequency offset only moderately, and the increase is greater at lower bit error rates; (2) The brunt of errors occurs in the symbol which has been moved away from the center frequency, the error rate for the other symbol drops rapidly; (3) For the common filters, the effective ρ for the symbol which has been shifted away from the center frequency would be decreased due to the selectivity, aggravating the effects described in (1) and (2). DC coupling at the demodulator output would have a compensating effect, providing greater noise immunity for the symbol which has the greater probability of encirclements.

The example has shown some of the considerations associated with the type of coupling at the demodulator output. We continue by considering this further in greater detail. One reason for the use of AC coupling given earlier was to accommodate frequency instabilities in the transmitter (mainly in the modulator, since the other oscillators can have full crystal control stability), Doppler frequency shifts in the channel, and frequency instabilities in the receiver associated with its local oscillators. These instabilities can be quite serious, and in the past have forced many systems to operate with a substantially larger deviation index than the optimum of 0.7 for minimum error rates.

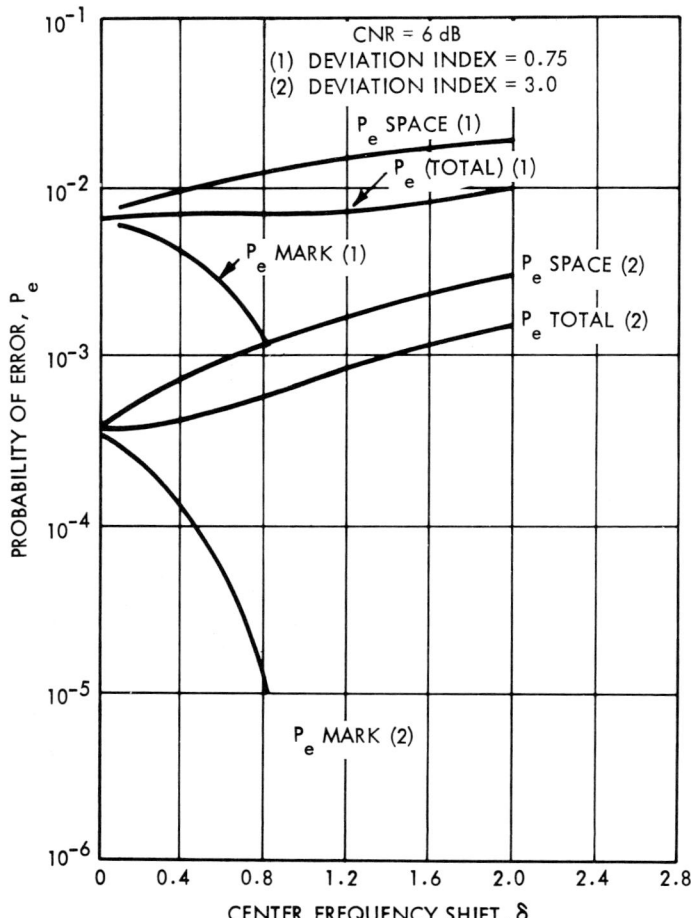

Fig. 9-10. Probability of error versus center frequency shift (from Klapper[13]).

For, if the frequency uncertainty is of the same order or greater than the peak frequency deviation and there is DC coupling, difficulties are encountered in making a correct binary decision at the demodulator output. Recently, transmitter and receiver oscillators have been greatly stabilized partly through the use of PLL circuits for frequency synchronization and frequency modulation.

There is, however, another very serious DC stability problem due to the demodulator itself, and interestingly enough it is associated with all known FM demodulators, including the PLL and FMFB. The FM wave at the input to the demodulator is at a center frequency which is very large compared to the frequency deviation. The effect of this center frequency must be canceled so that the demodulator output is proportional to the frequency deviation from center only. In an LD this is generally done by using a balanced circuit in which the relatively large DC output due to the center frequency is canceled out. It is immediately evident that small changes in the discriminator balance will result in appreciable DC outputs. In the PLL, the effect of the center frequency of the input wave is nullified by the center frequency of the VCO internal to the PLL. Again, any variations in the center frequency of the VCO will result in a DC output. The effect of these center frequency uncertainties may be substantial, and may require either a system with larger than optimum deviation indices, or AC coupling at the output of the demodulator.

The question remains, if AC coupling is such a simple remedy, why not use it? To answer the question, we must study the nature of the digital baseband signal. Consider a non return-to-zero (NRZ) baseband. If the signal were simply a repetitive mark-space sequence, i.e., a square wave, then there would indeed be no problem, as may be readily ascertained by studying either the time domain or frequency domain representation of the wave. But, a real digital wave is random in its nature, and problems arise during instances when there is a long string of the same symbol, because of the finite "droop" experienced in the output of highpass circuits. In order to reduce the droop, one would increase the time constant of the coupling network; however, this results in a slower recovery from the droop which may lead to errors while the system recovers from a long sequence of the same symbol. The problem is further aggravated in systems where "no message" is represented by a sequence of the same symbol.

There is no unique solution to the problem. If an acceptable time constant cannot be found and the system cannot be sufficiently stabilized in frequency, the designer may have to resort to special coding or baseband modulation (e.g., diphase). This generally requires more bandwidth, and/or transmitter power in order to assure a negligible low-frequency content in the baseband, thereby facilitating the use of AC coupling.

9.3. Binary FM Demodulation with Angular-Feedback Demodulators

The theory described above tends to indicate that low-threshold demodulators should also decrease the probability of error in binary FM. The reasoning is as follows: Threshold in analog FM demodulation is primarily due to the spikes or encirclement noise. Because the low-threshold demodulator has a lower rate of encirclements, it should also have a decreased P_e. However, it has been found in practice that the angular-feedback demodulators† do not present the same magnitude of improvement in digital FM as in analog FM. The main reason is believed to be the following: Binary FM permits a relatively narrow predetection filter since waveform fidelity is not the performance criterion, but rather, error rate. At the same time, it typically requires a wider PLL bandwidth for proper operation due to the sharp transitions characteristic of data. Both of these considerations reduce the noise filtering benefits of the PLL. The design theory of a PLL for binary FM is not yet as well developed as for analog FM. The material which follows is based mainly on experimental investigations.

To begin with, the question arises what type of PLL should be used to demodulate binary FM? Interestingly enough, experimental investigations[8] indicate that a first-order type is preferable over the second-order type. Some analytical justification will be given later. Actually, the preferred type was found to be a "modified" first-order loop, shown in Fig. 9-11. The pole at s_2 is made as low as possible, consistent with the practical constraints of the elements. The zero at s_1 is fixed once the loop bandwidth is specified, as

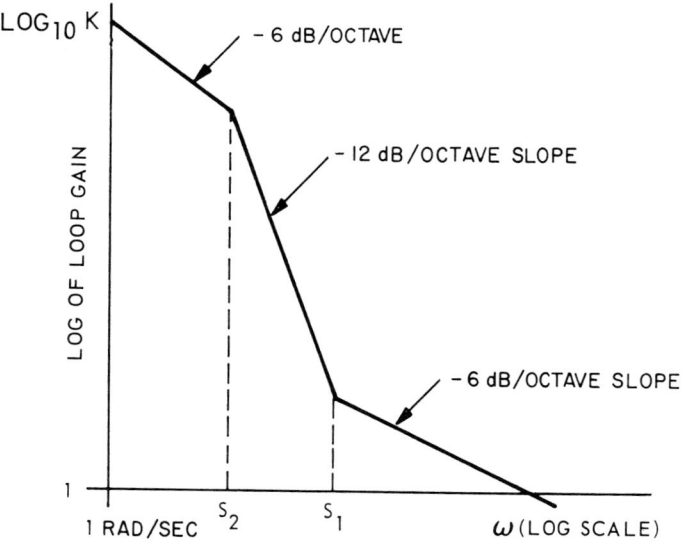

Fig. 9-11. Asymptotic open-loop response of modified first-order PLL (from Klapper et al.[8]).

† Only the PLL and the ERPLD type of demodulators were investigated, and only on binary (not multilevel) FM.

described presently. The modified first-order loop acts essentially like a first-order loop as far as the data are concerned, but it permits the inclusion of a much larger DC gain which reduces the static phase error due to average frequency offsets, such as those due to instabilities of the VCO.

The primary concern in design is the minimization of the total spike rate, as described earlier for the LD, and consistent with a reasonable intersymbol interference. It turns out that as bandwidth is reduced the PLL generally "breaks down" in spike rate before intersymbol interference becomes unreasonable, and since the latter can also be partially helped by post-PLL compensating networks, the design is aimed at spike-rate minimization.

As was done for the LD, we approximate the problem by assuming that the signal dwells sufficiently long at a MARK or SPACE frequency, so that we may treat it as a steady-state carrier offset from center frequency.

With the PLL frequency response shape fixed, the only parameter remaining variable is the PLL bandwidth. Experimental investigations[14] show that there is indeed a bandwidth which minimizes the spike rate for a given frequency offset. This is shown in Fig. 9-12, where the rate of opposing spikes is plotted as a function of the loop noise bandwidth B_n, with the frequency offset Δf and P_s/η as parameters, where P_s is the average input carrier power and η is the input noise power spectral density.

The existence of an optimum bandwidth has been conjectured from theory in earlier chapters, inasmuch as the ThI rate increases with increasing bandwidth while the LLI rate decreases. This points out further that to the right of the minimum we essentially deal with ThI and to the left with LLI.

As in the case of the LD, the rate of aiding spikes is negligible for reasonable frequency deviations, as is shown in Fig. 9-13. Figure 9-14 indicates that a simple approximate empirical formula for the spike rate is

$$N- \cong 0.6 B_n e^{-0.8\alpha}, \quad \text{spikes/sec} \qquad (9\text{-}16)$$

where α is the CNR in the loop bandwidth B_n. Furthermore, an examination of Fig. 9-12 leads to an approximate formula for the PLL noise bandwidth for a minimum spike rate

$$(B_n)_{opt} \cong 3.5\,\Delta f \quad \text{Hz} \qquad (9\text{-}17)$$

The available experimental results lead to the design and performance of a PLL for binary FM which may be summarized as follows:

(1) The PLL open-loop response shape is the modified first-order type, shown in Fig. 9-11.

(2) The optimum bandwidth of the loop for large deviation indices (those for which the carrier dwells for a relatively long time at one frequency) is

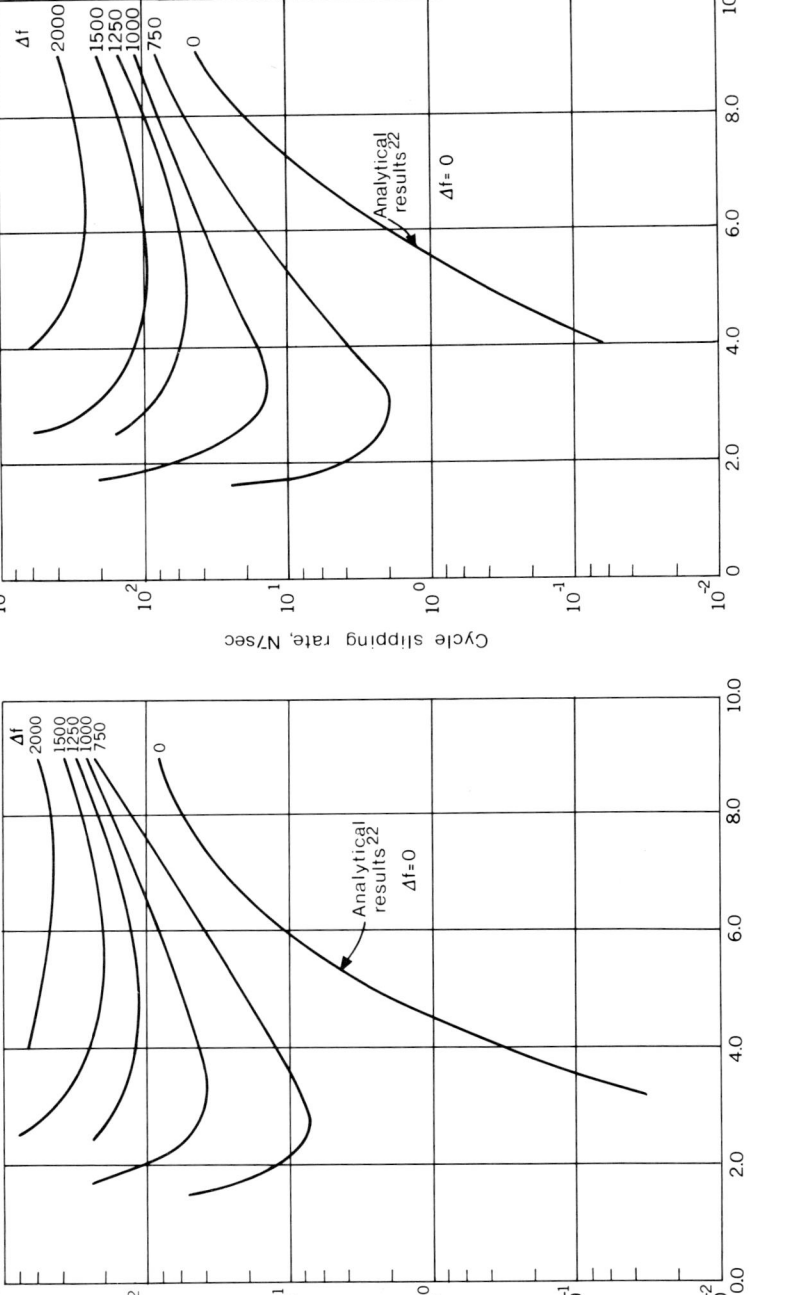

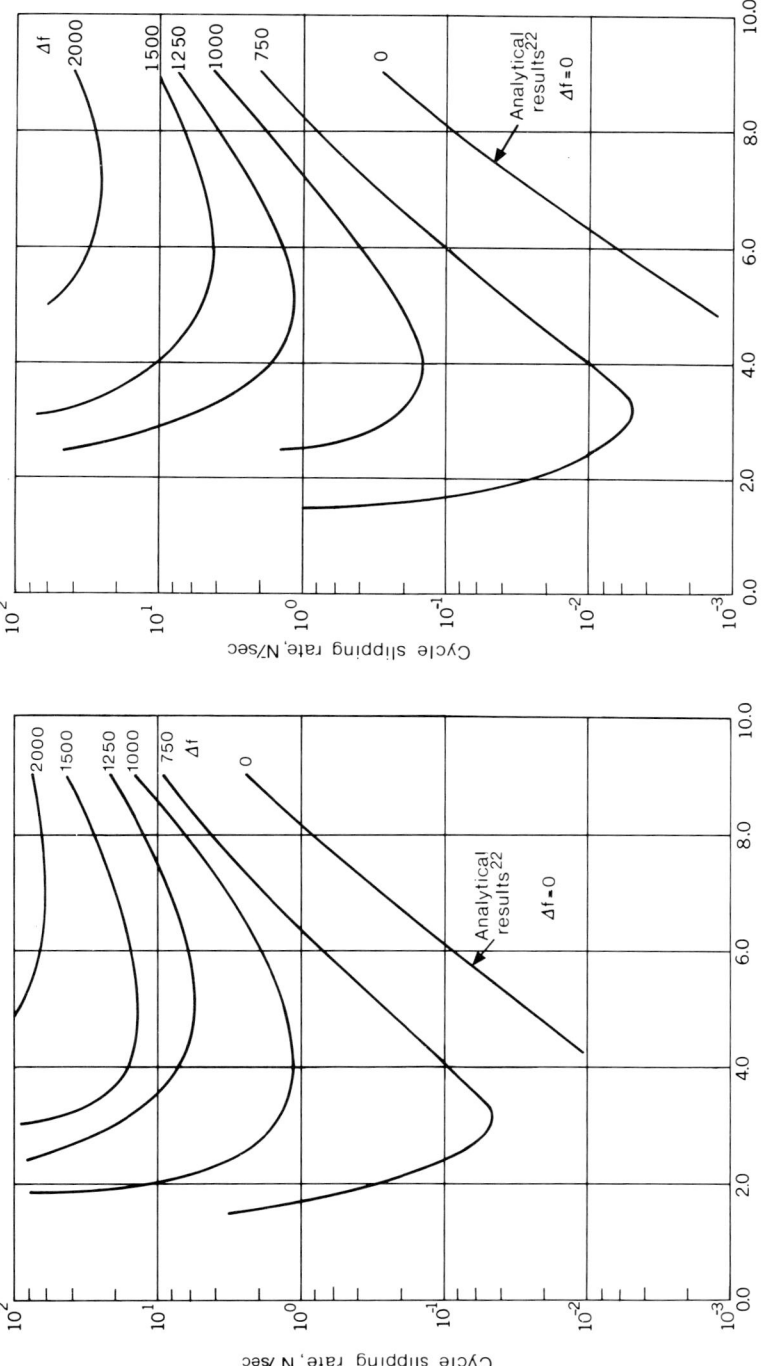

Fig. 9-12. Cycle slipping rate as a function of frequency offset. (a) $P_s/\eta = 17.3$ kHz; (b) $P_s/\eta = 21.8$ kHz; (c) $P_s/\eta = 34.6$ kHz; (d) $P_s/\eta = 43.5$ kHz (from Klapper and Creutz[14]).

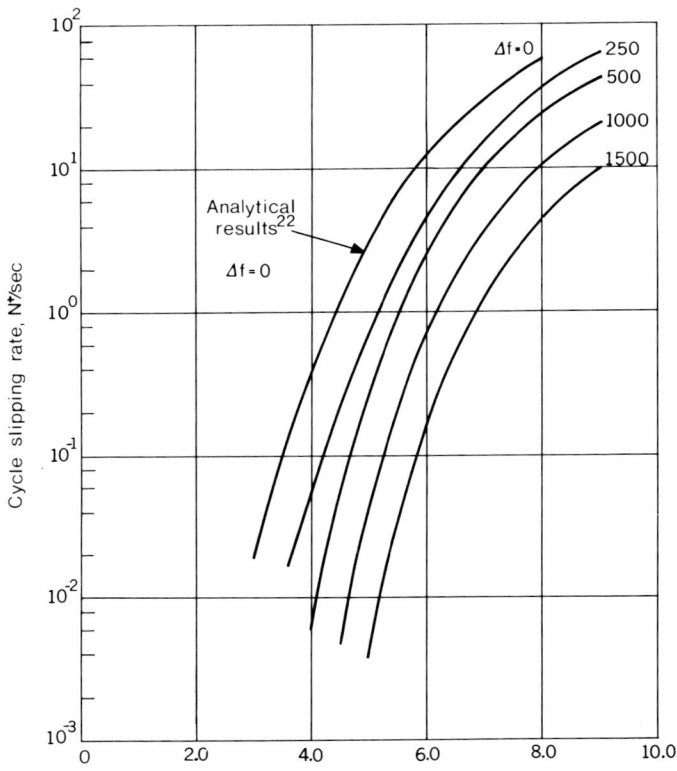

Fig. 9-13. Rate of aiding cycle slippings (from Klapper and Creutz[14]).

given simply by Eq. (9-17). To accommodate also small deviation indices, we add the proviso that the loop bandwidth should not be less than[8]

$$B_n \geq \tfrac{1}{2}\pi B_R \quad \text{Hz} \qquad (9\text{-}18)$$

where B_R is bit rate in bits per second.

(3) Inasmuch as the spike distribution of a first-order loop is Poisson[21], the probability of error is predicted simply from Eqs. (9-12) and (9-10), with the spike rate N obtained from Fig. 9-12 or the approximate empirical formula (9-16). The prediction is subject, of course, to modifications due to the neglect of the nonspike noise, as in the LD. The result gives the characteristic staircase of P_e as a function D, similar to Figs. 9-8 and 9-9.

9.3. Angular Feedback Demodulators

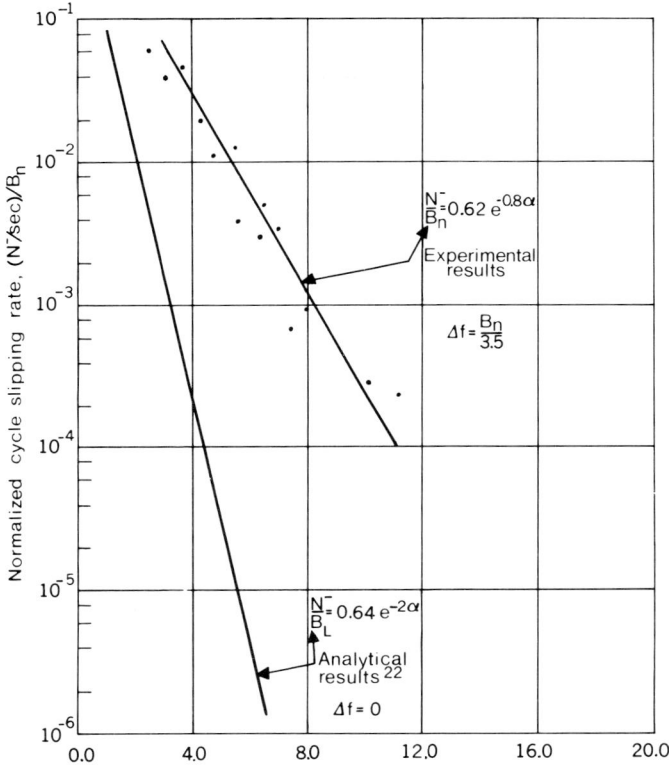

Fig. 9-14. Normalized cycle slipping rate (from Klapper and Creutz[14]).

(4) If the predetection bandwidth can be chosen on the basis of modulation only, and the optimum $D = 0.7$ is used, then the PLL requires about 0.5 dB less E/η than the LD. This improvement was measured for common predetection filters such as the Bessel type (see Fig. 9-2). Experimental measurements show some improvement in performance of the PLL with respect to the LD for $D < 2$ only, and the maximum improvement at $D = 0.7$.

(5) For an appropriately designed PLL, the bandwidth of the predetection filter may be considerably wider than optimum without a significant degradation of performance (see Fig. 9-15). In the case of an LD, the performance degrades directly with increasing bandwidth (ρ is proportional to bandwidth). This property of the PLL is useful in cases where the predetection bandwidth is forced to be wide, e.g., due to frequency uncertainties or compatibility with other signals.

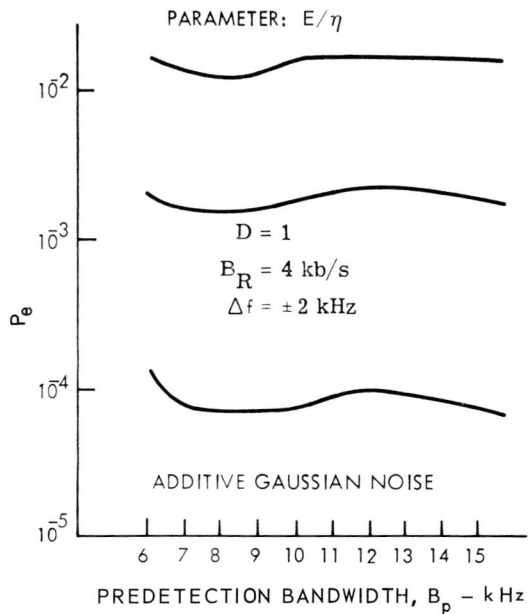

Fig. 9-15. ERPLD performance versus predetection bandwidth for binary FM (from Klapper et al.[8]).

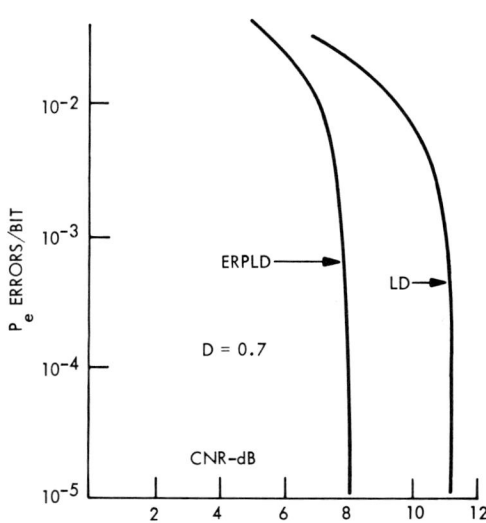

Fig. 9-16. Comparative performance with impulsive noise (from Klapper et al.[8]).

(6) The PLL may be substantially superior to the LD for certain non-Gaussian interferences. Figure 9-16 illustrates a 3-dB improvement for impulsive interference at $D = 0.7$. The improvement was found to drop off at larger D's in a manner similar to that of Gaussian noise.

A. Design Problem

A. PROBLEM

Design an optimum PLL demodulator for binary FM (rectangular transitions), faced with additive Gaussian noise interference. The information rate is 5.7 kb/sec, and there are no appreciable signal center-frequency uncertainties. What is the required E/η for $P_e = 10^{-3}$?

B. SOLUTION

Step 1: Since there are no frequency uncertainties, $D = 0.7$ is chosen. From Eq. (9-3), the peak deviation is ± 2 kHz.

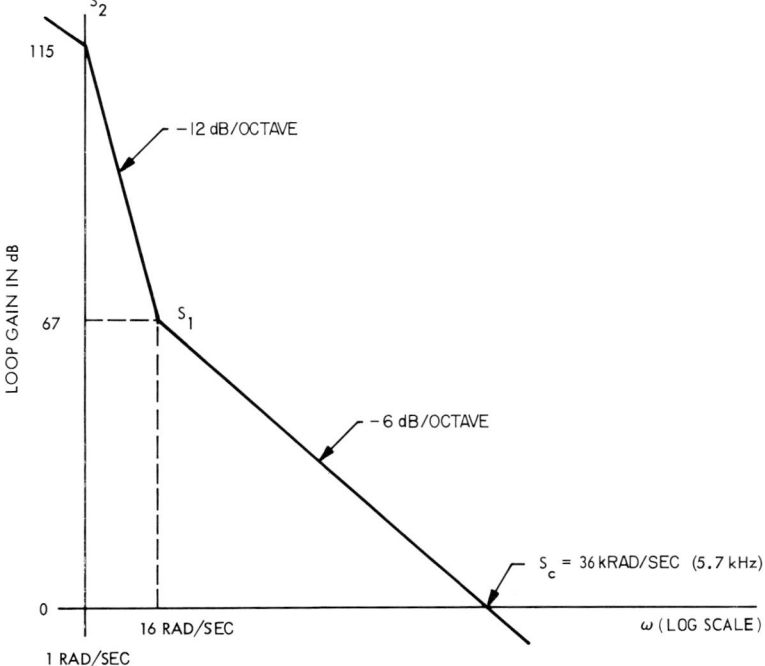

Fig. 9-17. Open-loop response: Design example (from Klapper et al.[8]).

Step 2. Predetection filter: The bandwidth of the predetection filter is chosen a little greater than the bit rate, say 6 kHz. A desirable filter is the equal-delay "Bessel" type; a 5–6-stage approximation will do.

Step 3. The phase-locked loop. The desired open-loop response is shown in Fig. 9-17. The 0-dB gain frequency (in hertz) equals the bit rate. The pole at s_2 (as low as practical) was chosen as 1 rad/sec, and with a DC gain (a practical value) of 115 dB, the zero at s_1 is at 16 rad/sec. The phase detector sensitivity (practical value) is 0.25 V/rad. The VCO sensitivity (practical value at 11-MHz center frequency) is 225 kHz/V. The loop filter component values are shown in Fig. 9-18.

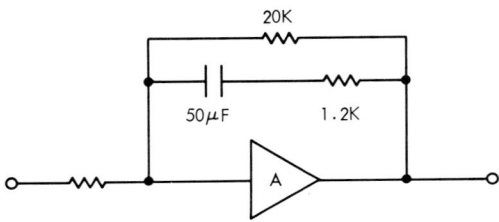

Fig. 9-18. Loop filter components for design example (from Klapper et al.[8]).

Step 4. Required E/η: In Fig. 9-2, we find that for $P_e = 10^{-3}$, the required E/η for an LD is 10 dB. The PLL, being about 0.5 dB better in performance, requires an E/η of about 9.5 dB. Some safety margin in E/η would normally be used to assure the required reliability.

B. Theoretical Considerations

This section provides some justification for the design procedure outlined earlier and arrived at through experimental tests.[8,22] The two basic questions are: (1) why a modified first-order loop appears to be the best choice, and (2) what are the constraints that lead to the optimum PLL bandwidth?

Why a first-order loop? The phase error in a first-order loop will generally, for the same input signal, be considerably different than in the second-order loop. In the first-order loop, the phase error is the same as the VCO input (ignoring the rf components), i.e., the same as the demodulator output signal. On the other hand, in the second-order loop the phase error tends toward a differentiated version of the output, inasmuch as the loop filter intercedes between the phase error and the loop output. The behavior of the phase error is illustrated graphically in Figs. 9-19, 9-20, and 9-21. Figure 9-19 presents a plot of normalized phase error as a function of normalized time, for an applied frequency step, with the loop damping factor ξ as a parameter. The

9.3. Angular Feedback Demodulators

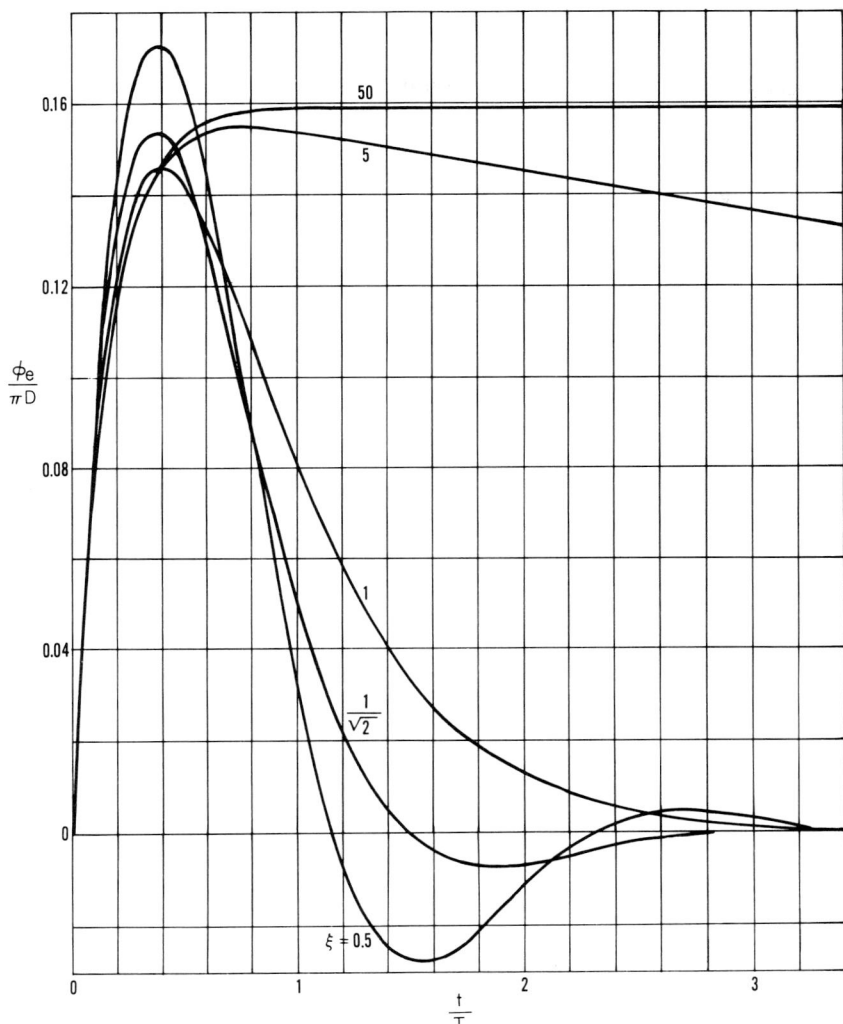

Fig. 9-19. Phase error for frequency step input with the damping factor as a parameter ($B_n T = \pi/2$; $D = 2\Delta fT$; from Klapper *et al.*[8]).

noise bandwidth is kept constant. A first-order loop is obtained for $\xi \to \infty$. The ordinate and abscissa are normalized with respect to $\pi D = 2\Delta fT$ and T, respectively, for convenience. It is observed that for a step in input frequency, the peak phase error is lowest for the critically-damped loop ($\xi = 1$). This is why critical damping is generally recommended for frequency step tracking. In our case, as will be seen, it is of greater importance to note that

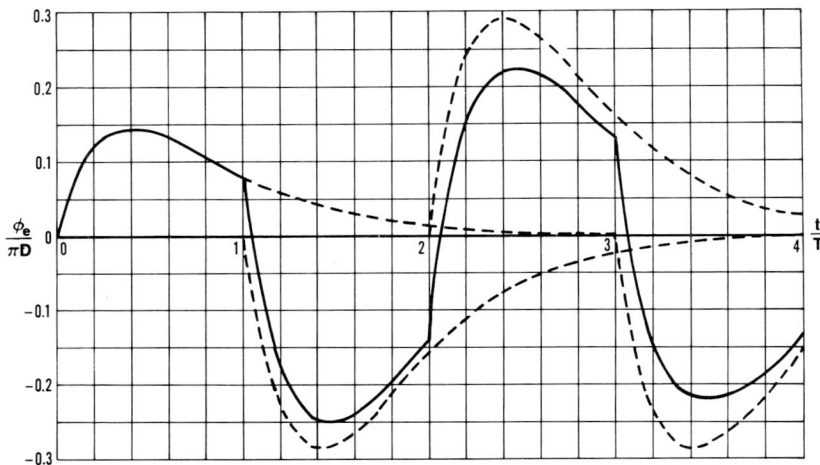

Fig. 9-20. Phase error for square-wave input in second-order unity-damping PLL. Dashed curves are individual step responses; solid curve gives the total normalized error (from Klapper et al.[8]).

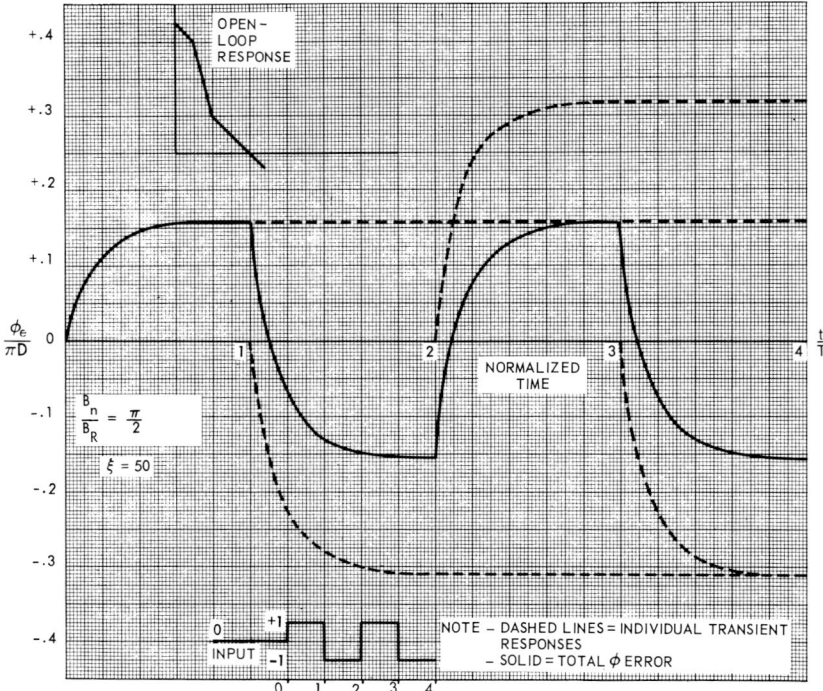

Fig. 9-21. Phase error for square-wave input in modified first-order PLL. Dashed curves are individual transient responses; solid curves are the total phase error (from Klapper et al.[8]).

the phase error of the first-order loop remains at its peak value at the end of the bit period ($t/T = 1$), while that of the critically-damped loop has dropped considerably. Again, for the tracking of a single frequency step, the critically-damped loop is preferable because the phase error diminishes quickly.

It is, however, a different situation in binary FM where a series of steps appears in alternate directions. This is illustrated in Figs. 9-20 and 9-21, where the phase error build-up to steady state is given for a square-wave FM. Figure 9-20 is for the critically-damped second-order loop ($\xi = 1$). Figure 9-21 is for the modified first-order loop ($\xi = 50$), for which the response to a square wave is essentially the same as that of a first-order loop. The noise bandwidth is the same in both loops. It is observed that in steady state, the peak phase error is larger for the second-order loop due to the rapid dropoff of phase error toward the end of a bit. It is recalled that the larger the peak phase error, the more prone the PLL is to lose lock and produce LLI, under interference conditions. Therefore, for the same noise bandwidth, the rate of LLI is generally reduced if the PLL with the smaller peak phase error is chosen, leading to the first-order PLL.

Now, why a "modified" first-order loop? As far as the signal is concerned, the modified first-order loop behaves identically to the first-order loop, provided the time between frequency changes is small compared to the loop filter time constants. The "modification" provides for a much higher DC gain for the same loop bandwidth, reducing the phase error due to VCO instabilities and, thus, improving performance in a practical situation. The higher DC gain also increases the hold-in range of the loop.

How is the experimentally determined optimum loop bandwidth justifiable? The optimum bandwidth formula requires the 0-dB gain frequency in the open-loop response to be approximately $2 \Delta f$, except for deviation indices less than one, where the crossover frequency equals the bit rate. This appears reasonable, inasmuch as the frequency change is always $2 \Delta f$, and for smaller D, the transient effect must be considered. Note that this results in a bandwidth specification that resembles Carson's rule.

9.4. Other PLL Applications

A number of other useful PLL applications will now be described. The treatment is very brief, of an "abstract" type, and its purpose is to call to the reader's attention the much wider application grounds and usefulness of the PLL principle. It is emphasized that the list is representative, but not complete. The reader is at this point believed to be sufficiently well prepared to understand and follow many of the application papers referenced below.

A. PSK Demodulation

The PSK wave may be demodulated as an FM signal with "impulsive" modulation, or as a phase-inscribed signal using correlation techniques.[23,24]

Meyerhoff and Mazer[5] developed a PSK system where the demodulator is the conventional LD. (Clearly, the demodulator may be replaced by a PLL or any of the other angular-feedback types.) They derived an error rate performance indicating that the system may be somewhat better than the conventional binary FM type.

The performance indicated for coherent PSK in Fig. 9-2 is obtainable by correlation techniques, using a perfect phase reference generally unavailable at the receiver. The phase reference is usually extracted by a narrow-band PLL, and the purity of the reference improves as the PLL bandwidth is narrowed. Opposing a narrow-band design are phase and frequency variations introduced by the channel or the system. Thus, a design is a compromise between phase-reference purity and ability to adjust to received signal variations. The PLL offers greater flexibility and adaptibility in this application than does a conventional bandpass filter.

A block diagram for phase-reference extraction from a binary PSK stream is shown in Fig. 9-22. The incoming wave is squared to eliminate the modulation contained in the polarity of the bits, thus creating a strong component at $2f_0$. The PLL performs here two functions: (1) it filters the wave, and (2) it divides the frequency by two, providing the reference at the required f_0 frequency.

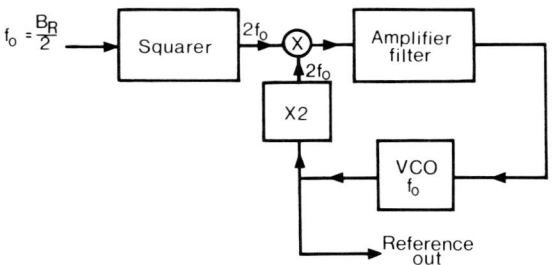

Fig. 9-22. PSK phase-reference extraction.

B. Frequency Synthesizer Applications

Two salient considerations in frequency synthesis are frequency stability and output purity. Both of these are greatly facilitated by a PLL which permits "locking" to a crystal-controlled stable frequency source and, by its relatively narrow bandwidth, cleanses the output signal. PLL's are used both in analog[25] and digital[26,27] types of frequency synthesizers.

C. Regenerative Frequency Dividers and Multipliers

One of the early applications of the PLL principle was the implementation of regenerative frequency dividers and multipliers.[28-31] By placing a frequency divider between the VCO and the phase detector, the output of the VCO is *multiplied* up in frequency by the division factor. Hence, it is possible from a stable low-frequency reference applied to the phase-detector input to generate a coherent harmonic signal (available at the VCO output). A principle application for this circuit is in frequency synthesizers. Design considerations are basically governed by the frequency of the primary reference, the stability and noise content of the VCO, and timing jitter of the frequency-division circuit. The closed-loop bandwidth is restricted to be narrower than the primary reference frequency so that it, and its harmonics, do not propagate around the loop and produce incidental FM on the "multiplied" output. The restriction on closed-loop bandwidth results in minimum relocking time if the division factor is switched to realize a new VCO output frequency. Open-loop VCO jitter is suppressed only to the extent that the jitter spectral components fall within the closed-loop bandwidth of the system.

By placing a frequency multiplier between the VCO and phase detector it is possible to divide coherently an input frequency reference by the multiplier factor. Again, the principal application of this technique is in frequency synthesizers. Assuming the multiplier factor is N, the VCO may lock in a stable manner in any one of N phases; hence, a fundamental phase ambiguity exists in the system. By designing a narrowband closed-loop system, a relatively jitter-free subharmonic coherent signal can be derived from a noisy reference signal. This has an important application in PSK systems, where it is common practice to multiply up the signal in frequency to create a reference component that is then divided down by a narrowband regenerative divider to form a stable reference for PSK demodulation.

D. Carrier Reference Extractor

The PLL principle applied to carrier reference extraction[32] is basically a method to filter out a spectral component from a noisy background. For this case, the design objective is a narrowband closed-loop response with large tracking range. The white-noise input analysis and design, discussed in the earlier chapters, relates directly to this application, and is used to calculate output reference (VCO output signal) jitter and spurious content. The extracted reference is an integral part of coherent AGC systems, suppressed-carrier AM or SSB detection systems, and systems to detect PSK signals. The SNR enhancement provided by the extractor is proportional to the reciprocal of loop noise bandwidth. Hence, to achieve a specific SNR gain, pull-in range and pull-in time are sacrificed. Modifications to the basic PLL can, however, improve pull-in range and time by externally sweeping the VCO over the desired frequency band until frequency lock is achieved and/or by increasing the loop bandwidth during acquisition. The loop bandwidth is narrowed once frequency lock is achieved.

The statistical characteristics (probability of lock) of the "swept" loop have been examined by Frazier and Page,[33] providing empirical design equations to achieve a specified performance. Improved acquisition due to loop bandwidth widening has also been examined by Hiroshige.[34]

E. Tracking Bandpass or Notch Filters

The closed-loop frequency response characteristic of a PLL permits the realization of high-Q bandpass filters with center frequency tracking characteristics.[35] The filter is amenable to microminiaturization—a distinct advantage in present-day communications equipment. As the filter performs a demodulation process, interfering components are suppressed in accordance with loop filter design that may be tailored to exacting narrow-band requirements. In addition, the loop capture and tracking ranges can be controlled to minimize adjacent-channel interference. This type of filter exhibits superior center frequency stability, a problem usually associated with high-Q bandpass filters.

A narrow-band notch filter can be formed simply by phase-shifting the VCO output by 90°, weighting, and subtracting from the input to null the input component that captures the loop. Again, baseband loop filter design directly controls the IF characteristics, permitting high-Q, stable performance not readily attainable by passive filters.[36]

F. Phase and Frequency Modulation

By locking the PLL to a stable reference signal and varying the DC control voltage in the loop, it is possible to obtain coherent phase or frequency modulation of the reference[37] (available at the VCO output). This technique permits the generation of highly stable frequency or phase modulated signals, and by utilizing phase detectors of 2π linear range, peak phase deviations of π rad are possible. Greater deviation ranges can be had by incorporating a frequency divider between the VCO and the phase detector. Design considerations are linearity of the loop components and loop filter design to achieve desirable modulation response characteristics. An example is an RCA-developed phase-locked Klystron modulator which handles 120 FDM channels with a worst-channel NPR of 55 dB at an output center frequency of 4.0 GHz.[38]

G. Phased Array Antenna Control

The phased array antenna[39] is a group of independent antennas, usually spaced more than several wavelengths apart. It is desired to add the signals from the individual antennas to provide a resultant signal greater than the individual received signals. Assuming that the noise components from each antenna, or those introduced prior to signal addition, are independent, and the signals are of equal level, then the SNR gain is $N^{1/2}$ (N is the number of independent antenna signal sources) if the signals are added coherently. This system is equivalent to an antenna of aperture N times that of the individual antennas, thus resulting in more realistic-sized antennas to achieve large signal gains. The application of the PLL principle is then to phase the individual signals so they add coherently. Several techniques are available to achieve this result; however, the most fruitful technique at present is to phase-lock each of the antenna channels to the resultant that is formed in the summing process. This minimizes the tracking range (and, hence the noise bandwidth) required to phase the individual signals. Design considerations are basically pull-in time, threshold, and maximum tracking rate.

H. Coherent Ranging Systems

The coherent ranging system[40,41] is based on generating a code with a very low periodicity. The code is transmitted to the vehicle (whose range is desired), converted to another carrier in the vehicle that is phase coherent but not of the same frequency as the base transmitted carrier, and retransmitted back to the base.

The phase of the original base code is phase-locked to the incoming code (transmitted back from the vehicle) with the delay (or phase shift) being related to the two-way distance between vehicle and base. The design considerations are acquisition time and tracking rate (how fast the vehicle may travel or how fast the phase of the two codes may change in the code phase synchronizer). Phase-locked circuits are also used to generate the coherent carrier in the vehicle transmitter.

I. PCM Timing

A PCM signal is a string of binary digits occurring at a fixed rate. For detection and processing of the signal it is necessary to have a "clock" that is synchronized with the bit rate, and it is convenient to obtain this clock from the incoming data stream. Phase-lock techniques are widely used to recover the clock directly from the data stream.[42]

Timing information is inherent in the bit transitions. Transitions can have either positive or negative direction, with both polarities equally affecting timing recovery. A series of unidirectional pulses is generated from the incoming data stream to mark data transition times. Since there is a discrete component of the bit frequency in the pulse train, a PLL can be locked to it.

J. Oscillator Stabilization

Crystal oscillators used as frequency standards have their best long-term stability if they are operated at extremely low rf power levels. However, best short-term phase stability is obtained at an intermediate power level, where the rf signal is much greater than the circuit noise.

Advantageous results are obtained if two separate oscillators are used: A very low-level one for good long-term stability; and a second oscillator, phase-locked to the first, operated at a higher level for good short-term stability. The bandwidth of the loop would be as narrow as possible, consistent with maintaining a reliable lock. The output is taken from the locked oscillator.

A similar application exists, e.g., at microwave frequencies, where direct crystal control at the final output frequency is not feasible. There, an oscillator at UHF or microwaves is made to lock to a harmonic of a crystal-controlled source by means of a PLL.[43,44]

Frequency synchronization in television played a major role in the development of the PLL, and probably represents the major commercial use of the circuit.[32,45] The PLL principle is also used in precise frequency measurement.[46]

References

1. W. R. Bennett and S. O. Rice, Spectral density and autocorrelation functions associated with binary frequency-shift keying. *Bell Syst. Tech. J.* **42**, No. 5, 2355–2386 (1963).
2. V. A. Kotelnikov, "Optimum Noise Immunity" (translated by R. A. Silverman). McGraw-Hill, New York, 1959.

3. G. L. Turin, Error probabilities for binary symmetric ideal reception through communication systems. *IEEE Trans.* **IT-10**, No. 1, 1603 (1958).
3a. S. Stein and J. J. Jones, "Modern Communication Principles with Application to Digital Signaling." McGraw-Hill, New York, 1967.
4. L. R. Brown, Experimental determination of signal-to-noise relationships in PCM/FM and PCM/PM transmission. N62-13483. Electro-Mech. Res., Inc., for NASA, October 21, 1961.
5. A. A. Meyerhoff and W. M. Mazer, Optimum binary FM reception using discriminator detection and IF shaping. *RCA Rev.* **22**, No. 4, 698–728 (1961).
6. P. D. Shaft, Error rate of PCM-FM using discriminator detection. *IEEE Trans.* **SET-9**, 131–137 (1963).
7. W. R. Bennett and J. Salz, Binary data transmission by FM over a real channel. *Bell Syst. Tech. J.* **42**, 2387–2426 (1963).
8. J. Klapper, G. Aaronson, A. Acampora, J. Frankle, and P. McLaughlin, Error rates with angular feedback demodulators. Conf. Rec. *IEEE Telemetering Conf. Houston, Texas, 1968*, pp. 23–28. J. Klapper and A. Acampora, Improved communication Techniques. AFAL-TR-66-202, AD484503. RCA Commun. Syst. Div., June 1966.
9. J. Salz, Performance of multilevel narrow-band FM digital communication systems, *IEEE Trans.* **COM-13**, No. 4, 420–442 (1965); J. Salz and V. G. Koll, An experimental digital multilevel FM modem. *IEEE Trans.* **COM-14**, No. 3, 259–265 (1966).
10. J. Klapper, The effect of the integrator-dump circuit on PCM/FM error rates. *IEEE Trans.* **COM-14**, No. 3, 349 (1966).
11. J. G. Lawton, Comparison of binary data transmission systems. *Proc. Nat. Conv. Military Electron.*, pp. 54–61 (1958).
12. E. F. Smith, Attainable error probabilities in demodulation of random binary PCM/FM waveforms. *IEEE Trans.* **SET-8**, No. 4, 290 (1962).
13. J. Klapper, Demodulator threshold performance and error rates in angle-modulated digital signals. *RCA Rev.* **27**, No. 2, 226–244, (1966); *in* "Selected Papers on Frequency Modulation" (J. Klapper, ed.). Dover, New York, 1970.
14. J. Klapper and J. Creutz, Minimization of cycle slipping rate in first-order PLL with frequency offset. Record *Asilomar Conf. Circuits Syst., 3rd, Pacific Grove, California, December* 10–12, 1969; J. Creutz, M.S. Thesis, Newark College of Eng., Newark, New Jersey, June 1969.
15. A. B. Carlson, "Communication Systems," McGraw-Hill, New York, 1968.
16. J. E. Mazo and J. Salz, Theory of error rates in digital FM. *Bell Syst. Tech. J.* **45**, No. 9, 1511–1535 (1966).
17. S. O. Rice, Noise in FM receivers. *In* "Selected Papers on Frequency Modulation" (J. Klapper, ed.). Dover, New York, 1970.
18. I. Ringdahl and D. L. Schilling, On the distribution of the spikes seen at the output of an FM discriminator below threshold. Correspondence, *IEEE Proc.* **52**, 1756–1757 (1964).
19. A. Papoulis, "Probability, Random Variables and Stochastic Processes." McGraw-Hill, New York, 1965.
20. H. Nyquist, Certain topics in telegraph transmission theory. *AIEE Trans.* **47**, 617–644 (1928).
21. A. J. Viterbi, Phase-locked loop dynamics in the presence of noise by Fokker-Planck techniques. *Proc. IEEE* **51**, No. 12, 1737–1753 (1963).
22. J. Klapper, A. Newton, and P. McLaughlin, FM demodulator impulse cancellation techniques. Final Rep., Contract No. F33-615-67-C-1317. RCA for WPAFB, Avionics Lab., 1967.
23. J. P. Costas, Synchronous communications. *Proc. IRE* **44**, No. 12, 1713–1718 (1956).

24. W. C. Lindsey, Phase-shift-keyed signal detection with noisy reference signals. *IEEE Trans.* **AES-2**, No. 4, 393–401 (1966); W. Hannon and T. Olson, An automatic-frequency-controlled phase-shift-keyed demodulator. *RCA Rev.* **22**, No. 4, 729–752 (1961).
25. J. Klapper and B. Rabinovici, A frequency synthesizer study for the proposed AN/GRC-103, AD289562. RCA Rep. under contract with USASRDL. RCA, July 1962.
26. E. D. Menkes and I. Harmon, A digital frequency synthesizer for a UHF communications transceiver. *Proc. NATCOM* pp. 41–46 (1965).
27. A. F. Evers, A versatile digital frequency synthesizer for use in mobile radio communication sets. *Electron. Eng.* **38**, 296–303 (1966).
28. R. L. Miller, Fractional frequency generators utilizing regenerative modulation. *Proc. IRE* **27**, 446–457 (1939).
29. J. K. Clapp and F. D. Lewis, A unique standard-frequency multiplier. *IRE Nat. Conv. Rec.* Pt. 5 (1957).
30. C. J. Byrne, Properties and design of the phase-controlled oscillator with a sawtooth comparator. *Bell Syst. Tech. J.* **41**, No. 2, 559–602 (1962).
31. H. P. Stratemeyer, A low-noise phase locked oscillator multiplier. *Symp. Definition and Measurement of Short-Term Frequency Stability, Part 3, December 1964*, pp. 121–136. Goddard Space Flight Center.
32. D. Richman, Color-carrier reference phase synchronization accuracy in NTSC color television. *Proc. IRE* **42**, No. 1, 106–133 (1954); The DC quadricorrelator: A two-mode synchronization system. *Proc. IRE* **42**, 288–299 (1954).
33. J. P. Frazier and J. Page, Phase-lock loop frequency acquisition study. *IRE Trans.* **SET-8**, 210–227 (1962).
34. K. Hiroshige, A simple technique for improving the pull-in capability of phase-lock loops. *IEEE Trans.* **SET-11**, No. 1, 40–46 (1965).
35. G. S. Moschytz, Miniaturized RC filters using phase-locked loop. *Bell Syst. Tech. J.* **44**, No. 5, 823–870 (1965).
36. W. B. Warren, Jr., Tracking notch filter for the rejection of CW interference. *Proc. Tri-State Conf. Electromagn. Compatibility, 9th, Chicago, Illinois, October 1963*, pp. 326–339.
37. G. R. Vaughan, E. F. Osborne, and G. S. Entwistole, Locked oscillator phase modulator. NASA Rep. N65-29140, Appendix D. NASA, August 25 1964.
38. F. Lefrak and L. Ozolins, *Microwave phase-lock FM exciter*. RCA Internal Mem., EM-64-419-3, RCA, 1964.
39. J. H. Schader, A phase-lock receiver for the arraying of independently directed antennas. *IEEE Trans.* **AP-12**, No. 2, 155–161 (1964); S. T. Cost, A description of the cohering receiver to be used in the OSY satellite communications facility. RADC-TR-66-666, AD647274, January 1967.
40. J. C. Springett, Telemetry and command techniques for planetary spacecraft. Jet Propul. Lab. Rep. No. TR-32-495. Jet Propul. Lab., California, Inst. Of Technology, January 15 1965.
41. F. M. Gardner and S. S. Kent, Theory of phaselock techniques as applied to aerospace transponders. NASA Contract No. NAS8-11509, N67-13247. Resdel Corp., 1967.
42. F. M. Gardner, "Phaselock Techniques," Chapter 8. Wiley, New York, 1966.
43. M. Peter and M. W. P. Strandberg, Phase stabilization of microwave oscillators. *Proc. IRE* **43**, 869–873 (1955).
44. A. Benjaminson, Phase-locking microwave oscillators to improve stability and frequency modulation. *Microwave J.* **6**, 88–92 (1963).

45. K. R. Wendt and G. L. Fredendall, Automatic frequency and phase control of synchronization in television receivers. *Proc. IRE* **31**, 7–15 (1943).
46. J. V. Murphy, Frequency measurement using the phase-controlled oscillator. *Proc. IEEE* **55**, No. 7, 1144–1153 (1967).

BOOKS
(References 47 and 48 are recommended for material on digital FM discriminator detection.)
47. W. R. Bennett and J. R. Davey, "Data Transmission." McGraw-Hill, New York, 1965.
48. R. W. Lucky, J. Salz, and E. J. Weldon, Jr., "Principles of Data Communications." McGraw-Hill, New York, 1968.

CHAPTER
10

Testing and Evaluation Procedures

This chapter presents procedures for aligning and evaluating PLL and FMFB systems.† These procedures are divided into three major groups, namely: component tests, loop tests, and system tests. Component tests cover the important parameters, and their measurement, associated with the PLL or FMFB loop functional elements; they determine the suitability of each element for a particular loop design. Loop tests provide the means of evaluating overall performance of the loop with respect to the design specifications. Within this group, open- and closed-loop tests are discussed in terms of directly verifying the loop design. System tests consider the PLL or FMFB as a part of a communication system; evaluation is in terms of system performance indexes, such as SNR and CNR. The system test procedures are actually applicable to any FM demodulator, whether it be an LD, a PLL, an FMFB, or of any other design. These overall system tests are particularly valuable as a basis for consistent demodulator testing and performance evaluation.

10.1. Component Tests

A. VCO

The significant characteristics of the VCO affecting the performance of FMFB and PLL systems are:

(1) Presence of spurious components in the output, or of irregular variation of output frequency or amplitude with input voltage, including hysteresis.

† The reader is referred to Sections 6.4 and 7.4 for component and system specifications.

Shortcomings in this area may cause incorrect locking of AFC systems as well as poor threshold and distortion performance of demodulators.

(2) Deviation linearity; i.e., linearity of output frequency as a function of input voltage. Poor linearity over the bandwidth occupied by the VCO output signal results in a distorted baseband output signal and, in extreme cases, may also affect threshold performance.

(3) Modulation sensitivity, defined as the slope of the curve of VCO output frequency versus input voltage. This sensitivity, in hertz per volt, is a factor contributing to loop gain. With a given loop gain requirement, higher VCO sensitivity permits a gain-bandwidth tradeoff in other loop components, such as the baseband amplifier, to favor increased bandwidth, hence lowered loop delay and improved threshold.

(4) Delay. In the interest of best threshold performance, delay in the VCO, as in other loop elements, should be minimized. VCO delay occurs in two areas: In circuits affecting modulation bandwidth (i.e., the baseband bandwidth over which VCO modulation sensitivity is constant), and in circuits which pass the carrier signal generated by the VCO (e.g., VCO output filtering). Generally, for a VCO design exhibiting low delay with respect to threshold performance requirements, the modulation sensitivity is constant through frequencies well beyond the actual baseband.

(5) Incidental amplitude modulation. AM should be low over the deviation range experienced by the VCO in the loop, so that the phase detector sensitivity in the PLL, or mixer conversion loss in the FMFB, is sufficiently constant not to affect loop gain (in the PLL, especially) or limiter performance (in the FMFB). Usually, requirements on AM in the VCO are not stringent. This is because the VCO drive to the phase detector or mixer is usually chosen as a high-level signal which performs a "switching" rather than a multiplying function; therefore, variation in the drive level has little effect.

Testing a VCO for adequate performance with respect to these characteristics can be performed as follows:

A. SPURIOUS COMPONENTS, DEVIATION LINEARITY, MODULATION SENSITIVITY, AND AM

Means should be provided for manual variation of output frequency via control of the VCO's DC input voltage. Static plots of output frequency versus input voltage and output frequency versus output power should then be made, with the data plotted over a range of about ten times the expected VCO output signal bandwidth, and the voltage scale chosen for approximately unity slope at the operating carrier frequency. The input voltage should be both increased and decreased, while the output waveform is monitored on an oscilloscope. Moderate harmonic distortion seen on the oscilloscope is not detrimental (it simply adds to the harmonics generated in the phase detector or mixer). However, envelope ripple, or other departure from constant envelope of the output waveform, indicates a spurious component.

As a first approach to judging VCO linearity, the static plot of output frequency versus input voltage should "look linear" over the VCO output signal bandwidth. Modulation sensitivity at the operating carrier frequency is the slope of the foregoing curve at that frequency.

With respect to AM, the first requirement is that there be no abrupt changes in output power as a function of output frequency over the entire range of input voltage the loop can apply to the VCO. Such changes are indicative of spurious modes of oscillation. Second, variation of output power over the actual VCO deviation range to be utilized must be evaluated in terms of phase detector or mixer characteristics.

B. FINAL TESTS OF DEVIATION LINEARITY AND MODULATION SENSITIVITY

Evaluation of VCO linearity can be achieved in an open-loop test. This is done by biasing the VCO to the center frequency used in the loop, and applying to the VCO input a baseband signal similar to that which will be used in the actual system, at an amplitude such that VCO deviation is that value which will be encountered in the loop. The VCO output is fed to a linear discriminator, using a frequency converter if necessary. A distortion test is then performed, similar to that by which overall demodulator performance will be evaluated (e.g., harmonic or intermodulation distortion for sinusoidal modulation, and NPR for noise modulation).

Modulation sensitivity can be measured accurately by applying sinusoidal modulation, again in an open-loop test, but feeding the VCO output to a spectrum analyzer (or other device capable of displaying or selecting and metering individual frequency components). The modulation amplitude is then increased until a convenient Bessel function zero is observed. If the first carrier null is chosen (the usual practice), the modulation sensitivity μ is given by

$$\mu = 2.40 f_m / V_p \qquad (10\text{-}1)$$

where f_m is the modulating frequency applied, and V_p is the peak voltage of the modulating signal. For high accuracy this measurement should be repeated at various values of f_m over the baseband frequency range to be used.

C. DELAY

The wider the baseband bandwidth for which a demodulator is designed, the more a given amount of delay degrades threshold performance. To measure delay in the nanosecond range, which is significant for basebands of about 1 MHz or more, is particularly difficult. While techniques exist for measuring even small delays of a network whose input and output are of the same type (e.g., a baseband or IF amplifier), a direct method of measuring the delay between the baseband input of a VCO and the frequency deviation at its output is not apparent. One approach is to feed the VCO output to a wide-band LD whose delay, if not negligible, is calculable. Delay from baseband input to baseband output of the combination can then be measured readily.

For wideband systems, cascading the VCO with an LD may not give useful results, because the delay of the LD may swamp the small delay being examined in the VCO. Considering separately the two areas of VCO circuitry that contribute to delay may be more profitable. Modulation bandwidth can be measured by performing the modulation sensitivity test already described, up to modulating frequencies high enough that the response falls off significantly, say 3 dB.

The effect of output filtering can be determined by static frequency-response measurements. If the VCO output network is simple, one can rely on a value of delay based on network parameters. Delay for a single-tuned filter, for example, is given by

$$\tau_d = 1/(\pi B) \tag{10-2}$$

where τ_d is the delay in microseconds, and B is the 3-dB bandwidth in megahertz.

B. Loop Filter

The PLL and the FMFB generally incorporate a loop filter of the form shown in Figs. 10-1 and 10-2, respectively. Either of these filters can be tested by measuring the voltage/frequency response, then comparing this to the

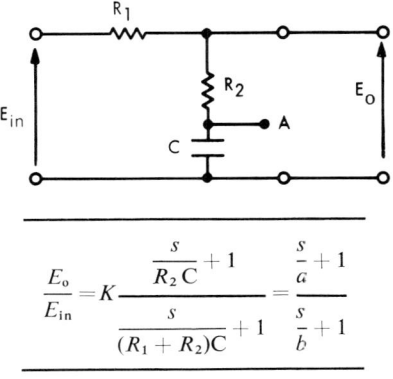

Fig. 10-1. PLL loop filter.

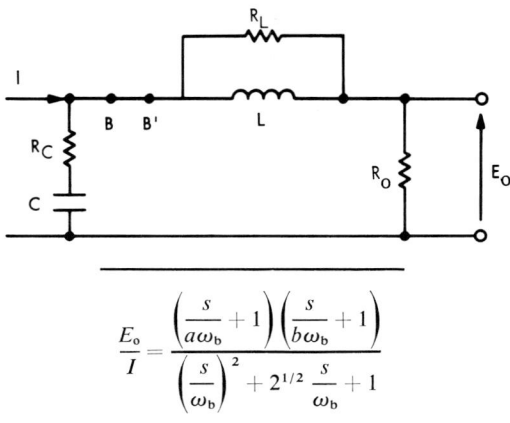

Fig. 10-2. FMFB loop filter (from Enloe[1b]).

theoretical response for these networks, which can be generated, for example, from the asymptotic responses shown in Fig. 10-3.

An alternative is to isolate the corner frequencies, i.e., the poles and zeros of the networks. This is done as follows:

(1) For the PLL filter, frequency-response measurements at point A yield a low-pass characteristic whose 3-dB point occurs at b rad/sec. Shorting resistor R_1 and again measuring the response at A gives a low-pass characteristic whose 3-dB point is at a rad/sec.

(2) For the FMFB filter, if points B-B′ are opened and the voltage/frequency response is measured at point B, a characteristic is obtained whose response is 3 dB higher at $b\omega_b$ rad/sec than at a frequency a decade (or more) greater. Similarly, by restoring connection B-B′, opening R_C, and shorting R_0, the voltage response measured at point B yields a high-pass characteristic whose 3-dB point is at $a\omega_b$ rad/sec. Finally, by shorting R_C, opening R_L, and restoring R_0, the voltage response measured at the filter output is the well-known second-order Butterworth type, with cutoff at ω_b rad/sec.

C. Phase Detector

The phase detector in the PLL loop can be tested by applying two cw signals, one to simulate the input carrier, and the other to simulate the VCO reference. If the normal operating levels are used, and a small frequency offset exists between the cw sources, then the following characteristics of the phase detector can be obtained by observing the beat note at the detector ouput: (1) symmetry and DC balance; (2) sensitivity (by noting the peak swing A of the beat note, the sensitivity is defined as A V/rad); (3) influence of signal

10.1. Component Tests

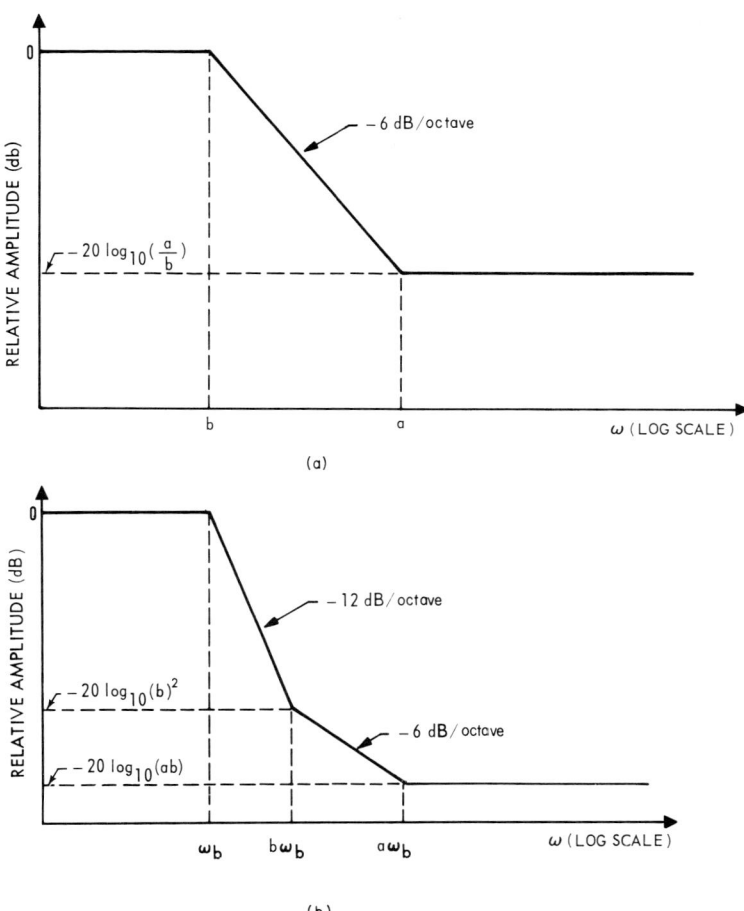

Fig. 10-3. Asymptotic responses for the loop filters. (a) PLL filter response. (b) FMFB loop filter response.

drive level on the above characteristics and, hence, the device dynamic range; and (4) frequency reponse of the detector, which is obtained by observing the beat note amplitude as a function of the frequency offset.

D. Input Mixer

The most important characteristic of the mixer is amplitude linearity with respect to signal and noise applied to the port used as the FMFB loop input terminal. Nonlinearity ahead of the selective IF filter degrades the demodulator threshold performance. The linear dynamic range of the mixer is related to the power of the signal applied to the VCO or "local oscillator" port. A test should be performed to measure IF output power versus input carrier power, with the normal VCO drive applied (the VCO can be unmodulated, or a signal generator can be used). To account for noise peaks, the nominal operating carrier input power should be at least 10 dB below the 1-dB compression point.

Another important mixer characteristic is bandwidth. Input and IF output bandwidths should be great enough for flat response over the respective signal bandwidths. Input bandwidth can be measured by varying the input frequency and tracking the frequency with a signal generator substituted for the VCO, for constant IF, and measuring the IF output power. Output bandwidth with a given load impedance can be measured by varying the input frequency, with constant frequency at the VCO port, and measuring the IF output power versus frequency. Each mixer input should be at normal power for these tests.

E. IF FILTER

Specified bandwidth, response characteristic and symmetry, freedom from spurious responses, and center frequency stability are requisite characteristics of the IF filter.

To determine the bandwidth and other response features, the IF filter should be swept with a generator which allows a bandwidth many times greater than the desired 3-dB bandwidth to be observed on an oscilloscope. Precise measurement of the 3-dB bandwidth is best made by a point-by-point response test. If the IF filter is embedded in the IF amplifier, an overall check for spurious responses should be made, and sufficient stages before and after the IF filter should be included in the bandwidth measurement to avoid errors due to instrumentation loading. If disconnecting the limiter is inconvenient, the bandwidth measurement must be made at an IF input signal power well below limiting level. Identical bandwidth readings at input powers differing by at least 3 dB should assure an accurate result.

The respective stabilities of the IF center frequency, discriminator tuning, and AFC† reference voltage are interdependent, as each of these parameters affects signal centering in the IF passband. Lack of IF signal centering results in rapid degradation of threshold performance. IF filter stability as a function of time and temperature may be measured directly by comparing relative input and output signal phase; this provides a more sensitive measurement than does measurement of peak amplitude. It may be sufficient, however, to check the tuning stability by simply measuring demodulator output noise below threshold, since this provides a sensitive measure of changes in threshold CNR.

† A separate AFC loop is often used within the FMFB to stabilize the VCO with respect to the input signal center frequency. For illustration, see Fig. 10-7.

F. IF Amplifier

The IF amplifier must have wide bandwidth consistent with the low-delay requirement for the loop. The dynamic range must be sufficient to avoid unintentional limiting.

Bandwidth can be measured with sufficient precision by a swept-frequency test. If incorporated into the IF amplifier, the IF filter must be disabled, and the input signal must be below the limiting level.

Nonlinearity in IF amplifier stages ahead of the IF filter is similar in its effect to nonlinearity in the mixer, in that it can degrade threshold performance. Unintentional limiting either before or after the IF filter may be nonsymmetrical, in which case it gives rise to AM-to-PM conversion. The principal effect of this is to convert additive input noise to noise which angle-modulates the carrier; hence, the noise improvement factor for SNR in the thermal region is degraded. Rather than test specifically for AM-to-PM conversion, it is best to check for overdrive in each amplifier stage, and leave the final judgment on performance to a comparison between the theoretical value of noise improvement factor to that achieved with the demodulator. Overdrive in a transistor stage is detectable as a shift in the DC base-to-emitter voltage. As a guide, no stage in the IF amplifier should show a shift of more than 0.01 V at an input power 10 dB above the normal operating value.

G. Limiter

The effectivenesss of limiting can be evaluated by measuring output power versus input power. In this test the limiter output should be filtered to eliminate harmonics. The initial value of input power should be 10 dB or more above the point at which limiting begins, with the final value chosen to minimize threshold CNR and distortion in the completed demodulator.

Limiting should be symmetrical to prevent AM-to-PM conversion as mentioned in connection with the IF amplifier. Visual examination of the limited IF waveform should suffice to ascertain symmetry, as well as to reveal nonideal limiting characteristics. For example, the output of a shunt-diode limiter may have rounded peaks increasing in amplitude with input power, due to excessive diode forward resistance; while the output of a series-diode limiter may have sharp peaks at the beginning of the flat-top portions of the waveform, due to excessive reverse recovery time (relative to the period of the IF signal).

An indication of the ability of the limiter to clip signals effectively over a sufficient bandwidth can be obtained by cascading the limiter with the IF amplifier and filter (in their normal configuration) and observing their combined swept-frequency response. At low input power, the response of the IF filter is seen. As the input power is increased, the response broadens, with the maximum amplitude reaching a constant value. The frequency range over which the amplitude is constant increases above and below the center frequency. At high input levels, if the amplitude is constant within approximately 1 dB over a frequency range several times the 3-dB bandwidth of the IF filter, then as a general guide, the limiting is effective over a sufficient bandwidth.

H. Discriminator

Pertinent discriminator characteristics to be checked are center frequency, peak-to-peak bandwidth, linearity, sensitivity (in volts per hertz), and IF leakage at the output.

The bandwidth can be measured by using a swept-frequency response, showing the "S" curve.

Linearity is not easily judged by observing the "S" curve, because the eye is not sufficiently sensitive to variations in slope. If the output is differentiated (e.g., by an operational differentiator accurate at the sweep frequency, usually 60 Hz) variation in slope is translated into variation in amplitude, which is more easily discernible.

An alternative method of measuring nonlinearity is to combine a large-amplitude low-frequency tone and a small-amplitude high-frequency tone, and apply the combination to the input of a linear frequency modulator. The deviation should be set so that the low-frequency tone causes excursion over the entire central portion of interest in the discriminator response. A filter is used at the discriminator output to reject the low frequency, and the observed variation in amplitude of the high-frequency tone is directly proportional to the variation in discriminator slope.

Discriminator sensitivity, which is a factor of the loop gain constant, is the slope of the "S" curve at the frequency at which the output voltage is zero. This value can be obtained from a calibrated oscilloscope display, or more accurately by a point-by-point measurement.

IF leakage at the discriminator output is a particular problem in FMFB demodulators. In a conventional (open-loop) LD, strong filtering may be employed to eliminate the IF in the baseband output. In the FMFB, however, this would introduce excessive delay; yet, the IF must be attenuated so as not to overload the baseband amplifier or VCO, causing distortion or producing spurious components in the VCO output. Three techniques for suppressing the IF are available: (1) trapping, appropriate for narrow-band signals; (2) canceling the IF by introducing a properly attenuated and phased sample of the discriminator input signal; and (3) using push–pull detection in the discriminator. A trap is sufficient to remove the second harmonic. To test for IF leakage, an oscilloscope is connected to the discriminator output, through an amplifier if necessary, and the pertinent circuits are adjusted for minimum indication. The oscilloscope is preferable to a meter because it permits fundamental and harmonic components to be distinguished. The foregoing tests should be performed at the normal input power to the discriminator.

Drift of the frequency at which the zero output of the discriminator occurs is an important consideration. The consequence of this, and a test for it, are discussed in connection with the IF filter tests.

Performance of the FMFB IF amplifier, filter, limiter, and discriminator combination can be checked as follows: Following the initial setup of these loop elements, tests are made in an open-loop manner, with an input deviation equal to the compressed value expected in actual loop operation. Next, tests are performed on a standard LD-type demodulator, preceded by an identical IF filter, under the same conditions of deviation and range of CNR. The two demodulators should compare favorably with respect to threshold CNR, noise improvement factor, and distortion. Note that best FMFB performance *cannot* be realized without a high-quality limiter and discriminator in the loop.

Distortion levels measured in this open-loop test may be higher than those specified for the overall FMFB demodulator, because in the closed-loop mode the levels will be lower by approximately the feedback factor.

10.2. Loop Tests

A. PLL

Open- and closed-loop amplitude and phase responses provide very useful information in checking for proper PLL performance. The test setup for both open- and closed-loop responses is basically the same, and relies on supplying a carrier to the PLL, with the carrier lightly frequency modulated by a fixed amplitude tone as shown in Fig. 10-4. The amplitude and phase responses are obtained by comparing the PLL demodulated output with the modulating tone, as a function of modulating tone frequency.

The measurement assumes that the frequency modulator provides fairly flat frequency modulation for tone frequencies ranging from low to well in excess of the top baseband frequencies for which the loop has been designed. Also, the loop response measurements are rendered inaccurate by band-limiting elements, such as predetection filters which precede the PLL; hence, removal of these elements from the test setup is necessary.

If the above qualifications present no problem, or are adequately accounted for, then the test setup shown in Fig. 10-4, is indeed, a simple laboratory instrumentation. However, if removal of the predetection filter is not possible, or if the frequency range of interest exceeds the capabilities of the ordinary FM signal generator, then an alternative setup, shown in Fig. 10-5, can be used.

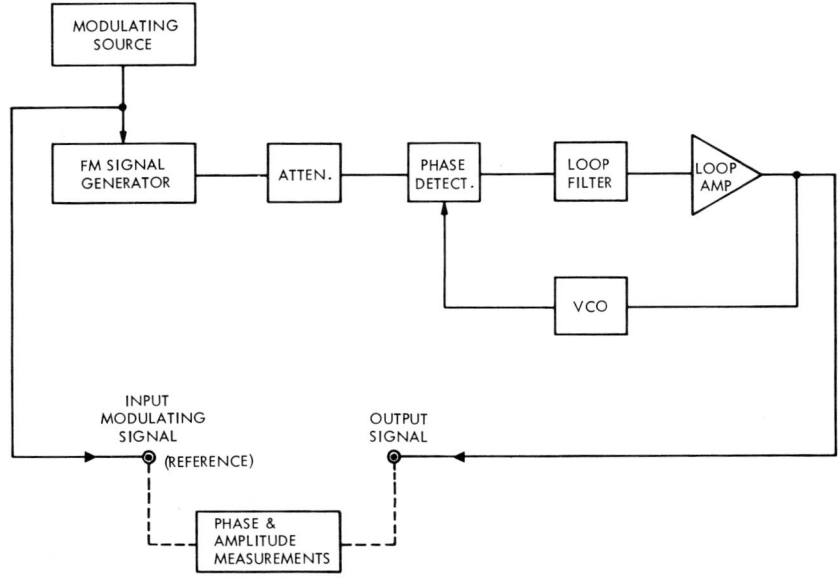

Fig. 10-4. PLL loop response test setup.

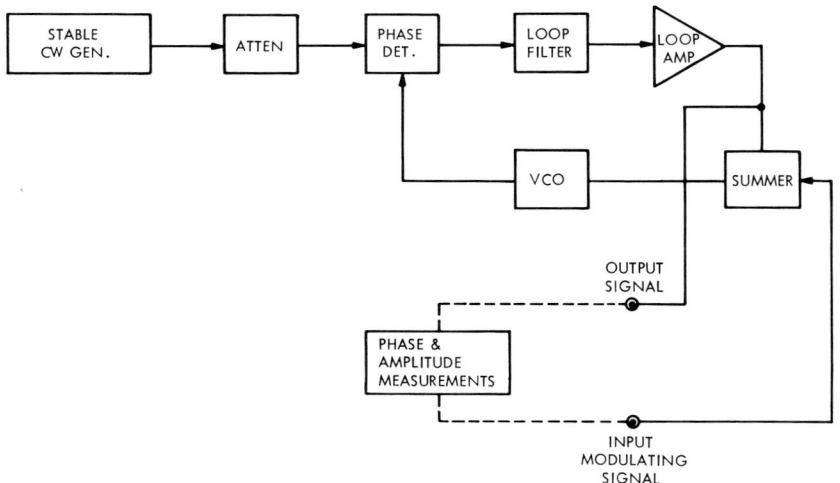

Fig. 10-5. Alternate PLL loop response measurement setup.

This setup uses a stable cw signal generator while the modulation is introduced into the loop itself. The input carrier signal is unmodulated, and used only as the system reference. Precautions which must be observed, however, are that the summing device used to introduce the modulation into the loop must have a flat frequency response, and that any variation in loop gain caused by the introduction of the summer be accounted for.

A. OPEN-LOOP TESTS

Open-loop tests, in the true sense, are not practical with the PLL. Unlocked or opened loops result in unwanted beat notes due to the frequency offset which exists between the input and VCO signals. On the other hand, open-loop amplitude and phase measurements can provide useful information concerning the existence of spurious poles or abnormal delays in the loop. The ordinary PLL open-loop asymptotic response, shown in Fig. 10-6, is composed of the integration effect of the VCO, and the corner frequencies provided by the loop filter. The open-loop gain at 1 rad/sec is the designed gain constant K provided b is much greater than 1 rad/sec.

The open-loop test is performed in an artificial manner by reducing the loop gain to a point where there is sufficient DC gain to lock the loop, but insufficient loop gain for the modulating frequencies, so that the loop is effectively opened and does not "track" the modulation. Also, a low modulation index must be used. The gain reduction requirement is accomplished simply by reducing the input signal by an amount X dB from the nominal operating level, where the X dB is related to actual design as shown in Fig. 10-6. The low index is necessary to provide an input signal with a relatively strong carrier component (for loop locking). The criterion is that phase modulation of the carrier must be less than or equal to $\pi/4$ rad. The output level and phase, as functions of frequency, provide a quasi-open-loop amplitude and phase response. This is so because, for extremely low-frequency modulation, the loop will have sufficient gain to track. Hence, the output for these frequencies will be relatively flat. However, for frequencies considerably higher than the new 0-dB gain frequency, the response will be the actual open-loop response. This is illustrated in Fig. 10-6.

The object of this test is to uncover any radical change between measured and theoretical responses, such a change indicating unknown and spurious poles and zeros in the loop. Furthermore, the phase margin in the region of a rad/sec (which ordinarily should not be less than $\pi/4$ rad) can be determined from the phase response.

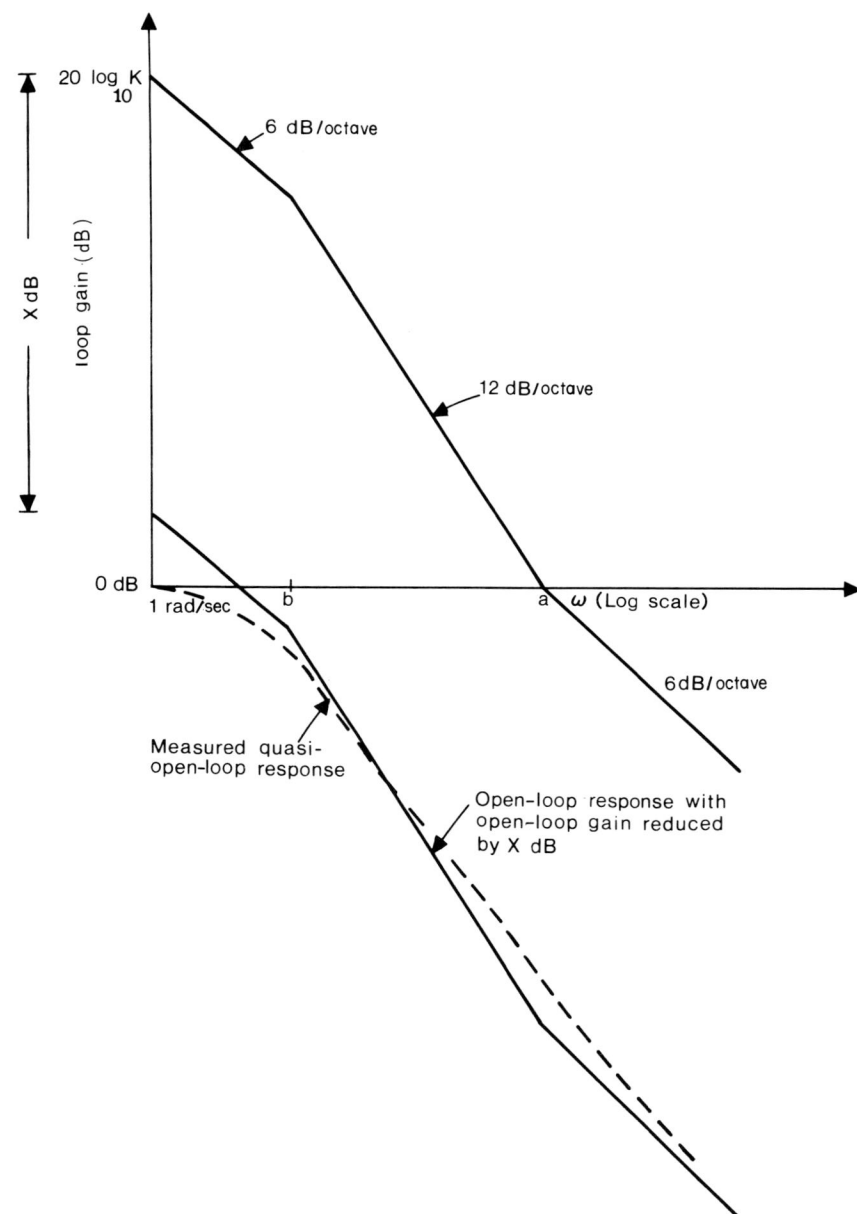

Fig. 10-6. PLL open-loop response.

B. CLOSED-LOOP TESTS

After restoring the loop gain to the design value and, again using a low-deviation input modulation, measurement of the loop output amplitude and phase response provides the closed-loop characteristics. A suggested method for insuring that the deviation is low enough is to set the modulation frequency at about a rad/sec, where a is the loop filter corner frequency; by slowly increasing the modulation deviation, a point will be reached where the output signal becomes perturbed by a loss-of-lock phenomenon; decreasing the deviation by a factor of 2 from this point, usually results in a satisfactory deviation for the tests.

The closed-loop amplitude response (see Fig. 5-5) obtained by varying the modulating frequency, should have the following characteristics for $a/\omega_n = 1$: (1) a peak of 3.3 dB at 0.855 ω_n rad/sec; (2) a smooth 6-dB per octave rolloff at the high end; (3) a noise bandwidth (obtained, for example, by numerical methods) of approximately the designed B_n (i.e., $\omega_n/2$ Hz).

C. TRANSIENT RESPONSE TESTS FOR DIGITAL SIGNALS

For digital FM demodulation, transient response tests are very helpful indicators of the state of the loop. The following procedure is suggested:

(1) Apply a square-wave modulated FM carrier of a frequency equal to one half the bit rate for which the demodulator is designed. The predetection filter is not inserted in this test, to prevent "shaping" of the signal by this filter. The output of the demodulator is taken before any post-detection filtering.

(2) Observe the shape of the square wave at the demodulator output. It should have no ringing and a reasonable rise time, on the order of 0.35 divided by the closed-loop 3-dB bandwidth in hertz.

B. FMFB

A block diagram of an FMFB demodulator with an internal AFC loop is shown in Fig. 10-7.

A. INITIAL SETUP TESTS

It should first be established that the individual loop elements are performing satisfactorily according to the criteria given in Section 10.1, as well as that the open-loop threshold, noise improvement factor, and distortion tests recommended for the IF filter and amplifier, limiter, and discriminator combination are satisfactory. Then, when the loop is first closed, before

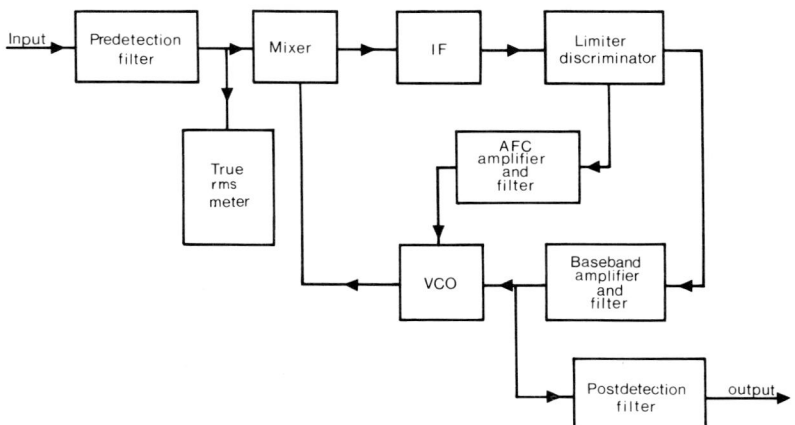

Fig. 10-7. Block diagram of FMFB demodulator with AFC loop.

running the actual performance tests to be described later (e.g., SNR vs CNR), the following checks should be made to ascertain proper adjustment of the demodulator. Some of the tests serve a debugging function, and some indicate how to optimize for best threshold and distortion performance.

B. DC AND IF CHECK

To ascertain whether the AFC system operates properly, the DC input voltage to the VCO is measured with the AFC loop closed. It should be close to the correct value for the desired VCO output frequency, established in the VCO component test. If convenient, this measurement should be made with the baseband feedback path open; if not convenient, the loop response should be temporarily narrowed (to insure stability in the event that the baseband filter or feedback factor F is initially out of adjustment) by introducing a single real pole well below the top baseband frequency in the baseband filter. This test should be made both without an input to the demodulator and with an unmodulated noise-free carrier at the correct frequency and power. When the carrier is applied, adjustment of the AFC reference control may be necessary.

Next, with the unmodulated carrier applied at the input, measure the amplitude and frequency of the IF signal after the IF filter. If the frequency is correct, tune the IF filter for maximum drive to the limiter.

C. IF SYMMETRY AND FEEDBACK FACTOR CHECK

With the AFC and baseband paths around the loop closed, but with the baseband filter narrow-banded as described above, modulate the input carrier with a low-frequency tone (no greater than one third the cutoff frequency of the temporary baseband filter). Observing the demodulator baseband output on an oscilloscope, increase the deviation until the output is clipped. The clipping should be fairly symmetrical, with the peaks of the sinusoid turning inward at high deviation. This determines that, when modulated, the IF is approximately centered in the discriminator response. Next, increase and decrease the deviation; the clipping should respectively increase and decrease smoothly. If a distorted sinusoid of original amplitude is observed when restoring a lower deviation, AFC instability is evident. The reason for the considerable attention to AFC is that in the FMFB loop, as opposed to a conventional LD in which only an AFC voltage is fed back, the differing requirements for the two loops necessitate more careful design of the AFC system for overall stability. The clipping in the present test should occur at a deviation well above the specified value for the demodulator, provided the feedback factor is near the correct value. If difficulty is experienced, repeat the test with the baseband path open, in which case clipping should start at a lower deviation.

With the peak deviation of the tone set approximately F dB below the specified peak deviation for the demodulator, measure the feedback factor as the ratio of baseband output amplitude with the loop open (by opening the baseband path) to the amplitude with the loop closed, and adjust the loop gain for the design value of feedback factor.

Caution is needed in choosing a method of opening the loop, to obtain a correct value of feedback factor, and for later tests involving frequency response measurement. The loop should be opened at a point between the baseband filter and the VCO. If this point is after the point at which the demodulator output is normally taken, the same output terminal can conveniently be used for the present feedback factor measurement, with the test signal applied to the normal loop input terminal under both open- and closed-loop conditions. It must be established that opening the loop does not change the gain of an amplifier stage (e.g., by changing its load or source impedance); or, if such a change does occur, that it is taken into account in the feedback factor calculation. An alternative method of operating the demodulator in the open-loop mode will be presented in connection with frequency response measurements.

D. THRESHOLD AND DISTORTION OPTIMIZATION

The design optimizations presented in Chapter 7 are, by necessity, approximate. In cases where maximum performance improvement is required, or for different modulation conditions, the experimental optimization procedure described below will be found helpful.

The first step is to reestablish the design values of the baseband filter parameters (i.e., remove the temporary filter used above). Apply an unmodulated carrier to the closed loop. Without added input noise, ascertain that the loop is stable, according to the procedures described earlier. An oscilloscope at the loop output will usually reveal the presence of oscillation. Reduce the feedback factor or narrow the baseband filter to eliminate such oscillation.

Apply the intended modulation and add input noise to the carrier signal. Reduce the CNR toward the value at which threshold is expected. The output signal amplitude should be nearly constant even below threshold in an FMFB demodulator. If it decreases as CNR is reduced, or if the output SNR (or NPR in a noise-loaded baseband test) decreases sharply above where threshold is expected, adjust the IF filter tuning to maximize SNR at the value of CNR at which the rapid SNR decrease is first noted. Set the CNR at a value about 2 dB below threshold, and readjust the IF filter tuning for maximum SNR. This is a very sensitive adjustment for lowest threshold CNR. Minimum distortion, measured at high CNR, should occur at essentially the same IF filter tuning position as that yielding lowest threshold CNR. If difficulty is experienced with these adjustments, run through the sequence with an unmodulated carrier (looking for minimum baseband output noise), followed by adjustment with modulation.

Now the demodulator is ready for an experimental optimization of the feedback factor, IF filter bandwidth, baseband filter parameters, and input power (carrier-plus-noise, keeping CNR constant) for best threshold. Once threshold has been made to occur at the lowest possible CNR, distortion in the demodulator should be measured; if poorer than specified, it can often be improved at the expense of threshold by increasing the IF filter bandwidth with subsequent reoptimization of the feedback factor and baseband filter. If the IF filter is implemented in such a way as to make the IF amplifier gain dependent upon IF filter bandwidth, the input power should be readjusted as well. This threshold-distortion tradeoff is possible in the event that nonlinearity in the VCO, discriminator, and baseband amplifier is so low that the distortion due to nonlinearity in the IF filter phase response predominates. If nonlinearity in the discriminator or baseband amplifier predominates, demodulator distortion can be reduced by increasing the feedback factor, with resultant degradation of threshold.

It should be noted that adjustments of the feedback factor, the IF, and baseband filters usually interact, and that an adjustment in one element can often be complemented by adjustment of another, with little change in closed-loop response and hence, threshold performance. The recommended scheme of adjustment, which can be altered according to the needs of specific circuit implementations, is as follows:

Begin with the design values of all loop parameters. Adjust the baseband filter [for this purpose employing variable elements is convenient; making C, R_C, and R_L in the Enloe configuration (Fig. 10-2) variable should suffice] for maximum SNR at a CNR 1–3-dB below threshold, with normal modulation applied. Adjust the input power (carrier-plus-noise, keeping CNR constant) for maximum SNR, and recheck the baseband filter adjustment. Record the SNR.

Choose another value of feedback factor, differing from the first by approximately 2 dB and repeat the above steps. Continue to do this until a feedback factor is established which maximizes SNR at a given value of CNR. If SNR increases significantly in the process, reflecting lower threshold CNR, move to a lower CNR to remain in the high-slope region of the SNR vs CNR characteristic below threshold.

Now, change the bandwidth of the IF filter by about 30%, and repeat the entire procedure of optimizing the baseband filter, feedback factor, and input power. Continue with different IF filter bandwidths until an optimum combination of all loop parameters is achieved. Note that in the presence of modulation it is not expected that threshold CNR will continue to improve as the IF filter bandwidth is reduced; beyond a certain point, threshold CNR will increase again.

Throughout the optimization procedure the IF filter tuning should be adjusted periodically for maximum SNR below threshold, to compensate for possible drift.

E. Closed-Loop Response Check

A simple test can be performed to indicate whether the loop response is as expected, following the adjustment just described. It can be performed earlier, if there is reason to suspect that threshold performance is far from that expected.

With no added input noise, sinusoidal modulation is applied to the closed-loop demodulator, at low enough deviation such that no nonlinear effects take place at any frequency, including the range above the top of the actual baseband. Vary the frequency and check that the response peaks smoothly at a high frequency, and falls off rapidly above this. The frequency of the peak is usually on the order of 2 to 3 times the top of the baseband frequency. It has been found that a loop adjusted experimentally for lowest threshold CNR in the presence of modulation tends to have a greater peak amplitude in its closed-loop response than that of the conventional Enloe design.

F. Further Tests for Optimization and Analysis

The tests described here are intended to aid optimization of the demodulator in the event that difficulty is experienced in the procedures given earlier, as well as to supply documentation of demodulator characteristics other than those dealing with SNR, threshold, and distortion, which are considered later. The characteristics of present concern are open- and closed-loop amplitude and phase responses as functions of frequency, and hold-in and pull-in ranges which are associated with AFC characteristics.

Closed-loop frequency response tests can be performed in a conventional manner, remembering that the data obtained reflect the combined performance of the demodulator and a test modulator. If preemphasis and deemphasis are used, or if the loop in the demodulator is followed by a filter outside the loop, then the measured response will, of course, differ from that predicted for the closed loop on the basis of the loop elements alone.

The same remarks apply to open-loop response measurements, but consideration must be given to the method of opening the loop. The simplest is the method described earlier (see Section 10.2,B,C); however, this omits the VCO transfer function. If moderate data accuracy is sufficient, the estimated VCO phase shift can be added to the phase response data obtained by this test method, the baseband amplitude response of the VCO being assumed to be flat over all frequencies of interest in the response. Similarly, the separately measured VCO sensitivity must be used to calculate the open-loop gain constant. Care must be exercised to keep the deviation low enough to insure linear operation.

A more elegant method of measuring the open-loop response is as follows: Apply an unmodulated carrier to the demodulator input, and open the loop as described earlier. Apply a baseband tone at the break, feeding the VCO, and measure the output appearing at the "output end" of the break. Provide a means to apply AFC, and ensure that a sufficiently low amplitude tone is used. The entire open-loop response will now be obtained directly, including the gain constant, if care is taken to maintain the correct baseband amplifier gain in spite of the interface of instrumentation with the stages near the break.

The hold-in and pull-in ranges are usually defined in terms of the frequency ranges over which the AFC system maintains and establishes, respectively, tracking by the VCO of slow variations of input frequency. Hold-in range is measured by first establishing lock, and then varying the input frequency in each direction from center frequency until the VCO carrier frequency no longer tracks. Pull-in range is measured by varying the input frequency until the loop does not track, and then returning toward center frequency until tracking is established. This is carried out above and below center frequency. Note the similarity with the identically named quantities for the PLL.

In an FMFB receiver, an additional characteristic associated with AFC loop gain is significant. Since threshold and distortion performance are critically dependent upon IF filter tuning, changes in the IF carrier frequency accompanying a change in loop input frequency gives rise to increased threshold CNR and increased distortion. Thus, the tolerance on input frequency may be more meaningfully associated with an allowable degradation of threshold or distortion performance than with the actual pull-in or hold-in range as defined above.

10.3. System Tests of FM Demodulators

A. Significant Performance Characteristics

The characteristics by which the performance of a transmission system are usually evaluated are:
(1) output SNR and threshold CNR (probability of error in the case of digital signals),
(2) distortion due to baseband-amplitude and carrier-signal phase nonlinearity, and
(3) transient or reliability characteristics, such as acquisition time and mean time between failures (MTBF).

This section is concerned with measuring performance related to the first two stated characteristics. Measurement of output SNR as input CNR is varied will at the same time reveal the threshold level. Depending upon the type of modulation, suitable experimental procedures yield distortion data either from separate tests or from SNR vs CNR data.

To determine performance with respect to SNR, threshold, and distortion, a demodulator is tested in conjunction with a modulator. Threshold is a function of the demodulator only, and SNR in the linear range above threshold is primarily determined by the demodulator's AM rejection. Distortion, however, is equally likely to be generated by the modulator and the demodulator. Therefore, to ascertain the distortion produced by the demodulator, a modulator known to be sufficiently linear for the application must be used, and vice versa. It is possible for complementary nonlinearities in the modulator and demodulator to cancel, yielding low overall distortion; this may be satisfactory for some applications. Ultimately, the distortion produced by either unit alone may be related to the static amplitude nonlinearity of baseband circuits and phase nonlinearity of carrier-signal circuits.

B. Tests Appropriate for Analog Signals

FM systems transmit analog signals of various types, such as voice, FDM, and video. Alternative approaches to testing the modulator-demodulator portion of these systems utilize (1) instrumentation and modulating signals which simplify test procedures, but which may produce experimental data that are difficult to relate to intended performance; (2) more complex instrumentation, and modulating signals chosen to represent more closely the baseband signals encountered in practice, with the result that the data more readily predict actual system performance; or (3) subjective evaluation of output signal quality, with modulating signals representative of those actually encountered in practice.

The most common modulating signal of the first category is the sinusoidal tone, used to test for SNR, threshold CNR, and harmonic or intermodulation distortion in systems designed for all types of basebands. For types of modulating signals for which specialized objective testing techniques are prohibitively difficult to implement, or where standardized tests have not been adopted, tone modulation offers a means for obtaining quantitative data.

A. Voice Signals

For voice signals, articulation testing, a subjective approach in which the fraction of words heard correctly from standard lists (as a function of CNR, for example), is a preferred measure of system effectiveness, even more so for the feedback type of demodulators than for the conventional LD. However, because of the involved and complex nature of these tests, it is more common to evaluate a demodulator in terms of test-tone modulation. The frequency of the test tone is usually chosen as 1 kHz, and the peak frequency deviation is that specified for the system.

B. FDM Signals

Tests of transmission systems for FDM, including FM modulator–demodulator systems, are standardized. The baseband signal is represented by a band of Gaussian noise, which has similar statistical properties to FDM, especially when the number of multiplexed channels is large. While usually applied to FDM voice, the technique is also valid for an array of tones or data channels. SNR is measured in terms of NPR, defined as the ratio of output noise power in a given channel when that channel is loaded (along with the other channels), to the output noise power in that channel when the loading is removed from it by means of a bandstop filter ahead of the transmission system under test. Given a relation between the mean-squared frequency deviation due to the noise loading and the average signal power in a typical channel based on observed traffic conditions (this relation is standardized in a CCIR recommendation),[2] NPR is uniquely related to SNR in the channel in question. If NPR is measured in high, low, and middle channels in the baseband, over a wide range of CNR, a full set of data is available regarding threshold CNR and distortion, as well as SNR in the linear region above threshold. The harmonic distortion of baseband noise spectral components and intermodulation distortion among these components (intermodulation greatly outweighs harmonic distortion in FDM) cause noise products to fall in the channel made idle by inserting the bandstop filter. This effect is seen as a leveling of the NPR vs. CNR curves at high CNR. Under actual traffic conditions, the intermodulation products falling into a given channel are uncorrelated with the signal in that channel, and appear as noise.

Below threshold, impulsive noise appears in the baseband output of the demodulator. In systems in which the entire baseband output is observed at once, such as single-channel voice or data or video, these impulses may be disruptive to a degree out of proportion to the noise power contained in them. For FDM, however, any given channel is observed through a narrowband filter, whose effect is to integrate the threshold impulses so that their amplitude probability distribution approaches the Gaussian. The character of the output noise in each channel is, therefore, indistinguishable over the entire range of CNR, and operation below threshold is permissible, provided that the NPR (and corresponding SNR) is great enough.

C. VIDEO SIGNALS

Demodulated video signals are most meaningfully evaluated subjectively, since the eye can best judge and classify the relative disturbances introduced by the noise. This is especially valuable in the threshold region where the statistics of the noise vary rapidly. However, entirely different types of demodulators (such as feedback and nonfeedback) may produce threshold noise of different character, and a visual judgement of relative quality becomes a function of the observer's preference and, therefore, debatable. Because of this, as well as for quantitative measurements, a tone may be used to represent the video signal, and the performance evaluated in terms of SNR. Most often, however, video transmission system performance is measured in terms of the ratio of peak video signal amplitude to rms noise.

Distortion of a video signal is also measured in the time domain in terms of the nonlinearity in the waveform. Special waveforms are used, such as a staircase with initially equal amplitude steps, to check amplitude linearity.

C. Tests Appropriate for Digital Signals

In digital FM, the demodulator output quality cannot be measured in terms of the power ratios of signal and noise, as is conventional for analog signals. Here the rare events, which may not necessarily contribute measurably to the rms value, are generally of prime significance. Thus, the demodulator output quality is best judged directly in terms of the probability of mistaking one symbol for another, i.e., bit error rate (BER).

If the system is fixed, then a proper measurement is BER vs received CNR. However, to facilitate a comparison with other systems (theoretically derived or measured) on a normalized basis, the data is often converted to BER vs the energy ratio E/η, where E is the rf energy per bit and η is the noise power density (a flat noise power spectrum is assumed). The energy ratio is akin to "CNR in twice the baseband" often used with analog signals. The conversion from measured CNR to E/η is given by

$$E/\eta = (CNR)B_n T, \qquad (10\text{-}3)$$

where B_n is the noise bandwidth of the predetection filter in hertz and T is the duration of a bit in seconds (see also Section 9.2).

D. Analog-Signal Test Procedures

A. Basic System

The basic system under consideration is shown in Fig. 10-8. Depending upon the test method, the baseband signal source may be a tone, video, noise generator, or a human speaker (for subjective voice tests). The FM modulator

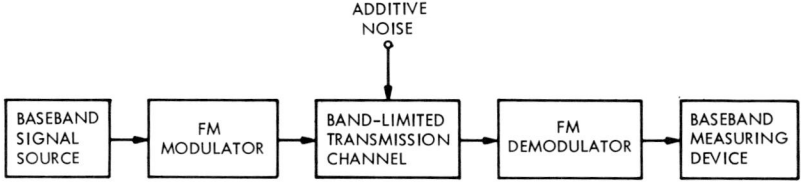

Fig. 10-8. Basic test configuration.

includes any premodulation processing such as preemphasis. The transmission channel is bandlimited by a filter following the point at which Gaussian noise is added. If the effect of the filter on the carrier signal is not of concern for a particular test, the filter may be placed in the noise path prior to addition. As demodulator performance is generally affected by the noise bandwidth, the filter bandwidth should be commensurate with that which will be employed ahead of the demodulator in an actual receiver system.

The baseband measuring device is either a true rms voltmeter for SNR measurement, or the receiver portion of a noise-loading test set for NPR tests. Deemphasis, if required, is included in the FM demodulator. It is convenient to provide a common setup in the case where both SNR and distortion tests are to be conducted. Distortion is then measured by adjusting the controls for essentially infinite CNR.

B. Modulator Sensitivity Calibration

Tone modulation is utilized to calibrate the sensitivity of the modulator, regardless of what type of modulating signal is applied for SNR or other tests. A sine generator with low distortion is connected to the modulator input, with an accurate voltmeter to read the input voltage. A modulating frequency in the baseband frequency range is chosen, and if the modulation is suspected of not having constant response over this range, the test is repeated at several tone frequencies. If preemphasis is to be utilized in SNR or other tests, calibrate the modulator sensitivity at a frequency high in the baseband, because the high-frequency baseband energy stresses the demodulator performance, and the precise value of deviation in this region is critical in some cases. The sensitivity of the modulator is measured as for the VCO, defined by Eq. (10-1).

C. INSTRUMENTATION COMMON TO VARIOUS TEST METHODS

In Fig. 10-9, those portions of the test configuration in Fig. 10-8 that are universal among tests using different modulating signals and baseband measuring equipment, are shown in greater detail. Note that only the noise is passed through a bandpass filter. This relaxes the requirement for a well-equalized predetection filter passband. The carrier-setting attenuator is adjusted for correct input power to the demodulator.

The following precautions must be observed:

(1) The noise must have a Gaussian amplitude distribution, with positive-negative symmetry. While it is not likely to be a problem when amplifier self-noise is used, it requires particular checking when the source is a noise diode. As a minimum, an oscilloscope could be used to check the positive-negative symmetry.

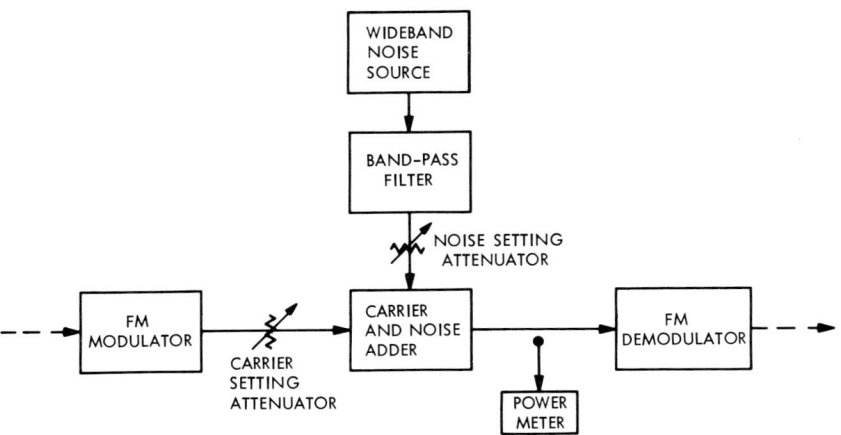

Fig. 10-9. Instrumentation common to various measurements.

(2) The input spectrum noise should be flat over the 20-dB bandwidth of the bandpass filter which follows it. That the swept response of the filter is flat should not be taken as assurance that the noise applied to the demodulator is flat. It is best to measure the spectrum directly at the demodulator input by means of a tunable voltmeter. Lacking this, a flat spectrum analyzer with linear or squaring vertical deflection (rather than log) connected at this point generally suffices as a means of assuring that the noise is flat.

(3) The dynamic range of the amplifier in the wide-band noise source, as well as any other amplifier added to the noise or carrier-plus-noise channel in Fig. 10-9, must be sufficient to avoid compression of noise peaks. If compression occurs ahead of the carrier-and-noise adder, threshold CNR will appear optimistic (too low). If compression occurs after the carrier-and-noise adder, threshold CNR will appear pessimistic if a feedback-type demodulator is being tested. A guide for an amplifier which has a fairly sharp overload characteristic is that the noise output, as measured by a thermal-type power meter, be 15 dB below the 1-dB compression point in the amplifier output-versus-input characteristic.

(4) Check for spurious components at the input to the demodulator, using an oscilloscope (at various sweep speeds to reveal hum as well as parasitic oscillation) and a spectrum analyzer. One source of spurious components is pickup of strong FM or TV signals in the high-gain amplifier used in the noise source.

The band-pass filter requires accurate calculation of its noise bandwidth. A point-by-point plot of the response is required, in situ; that is, the effects of source and load terminations actually provided in the setup should be present at the time of measurement. It is preferable to make the final bandwidth measurement by means of a tunable voltmeter, measuring the noise spectrum at the demodulator input, as mentioned above. A 3-pole Butterworth design for the band-pass filter gives adequate selectivity for most applications, but the flat portion of the response must extend over the entire frequency range occupied by significant sideband components in the carrier signal (even though the carrier signal does not pass through the filter; the concern here is representative operation of the demodulator).

The carrier-and-noise adder is most readily implemented as a resistive device, to avoid nonlinear amplifier effects. A network of miniature film resistors connected in a delta configuration with almost no lead length results in a low VSWR to at least 100 MHz.

The power meter is used for calibrating CNR. This calibration is performed at 0-dB CNR, to eliminate relative scale readings as a possible source of inaccuracy. Introduce a minimum of 30-dB attenuation, and adjust the noise-setting attenuator (0.1-dB steps should be provided) to obtain a reference reading on the power meter (portion of scale where divisions are 0.2 dB or less apart) within 0.05 dB. Repeat the carrier power reading and the noise power reading at least once to insure against meter drift and attenuator interaction. Check the power meter reading with both carrier and noise present. It should be 3 dB higher than the reference. If not within meter tolerance, check the dynamic range of the amplifiers preceding the power meter.

D. SPECIFIC TECHNIQUES FOR SNR VS. CNR AND DISTORTION MEASUREMENTS

Those aspects of instrumentation common to all tests under consideration for FM demodulators have already been described. Specific aspects of the test equipment and procedures applicable to various classes of tests will now be outlined. The classes of tests most universally applicable or the most standardized are as follows:

(1) SNR vs. CNR tests using a sinusoidal modulating tone, from which threshold CNR data may be obtained; also distortion tests with tone modulation.

(2) NPR vs. CNR tests with noise loading of the baseband, from which may be obtained worst-channel performance at specified channel SNR values, as affected by threshold, above threshold, and intermodulation noises. "Worst channel" is defined as that channel for which the greatest CNR is required to achieve the specified channel SNR, and is generally a function of the SNR.

(i) *Tone modulation.* For tests utilizing tone modulation, the FM modulator in Fig. 10-9 is preceded by a sine generator (or two of them, with means for isolating their outputs for intermodulation distortion measurements) and a voltmeter to set the frequency deviation. A preemphasis network can be installed, if one is normally part of the system; however, the tone amplitude can, instead, be set to the preemphasized-value, depending upon the frequency of the tone. The tone frequency may be at the top of the baseband frequency range, but if applied at the full deviation rating of the system, it may stress the demodulator more than the normal baseband signal, resulting in pessimistic threshold and distortion performance. This is especially true of a feedback-type demodulator, such as the FMFB or PLL. Often, the tone used in testing a single-channel voice system is placed where most of the voice spectral energy lies, at 800 or 1000 Hz; rather than at the band edge, which may be 4000 Hz. Normally, the peak deviation of the tone is set equal to the expected peak deviation of the voice signal.

A deemphasis network should be provided at the output of the FM demodulator, corresponding to the preemphasis utilized. The baseband signal is passed through a sharp-cutoff low-pass filter defining the baseband bandwidth, and fed to a true rms (not only rms-indicating) voltmeter. If it is desired to confirm experimentally the theoretical value of noise improvement factor relating SNR to CNR above threshold, the low-pass filter noise bandwidth is determined as follows:

(1) Measure the response, point by point.
(2) Weight the response linearly with frequency; that is, in a region where the response is flat, the response at a frequency f is 6 dB less than at $2f$. This accounts for the parabolic power spectrum of the output noise above threshold.
(3) Square and normalize the weighted response and plot as relative power density versus frequency (numerically, not in decibels), with relative power density equal unity at the frequency at which the modulating tone is located. Calculate (graphically) the noise bandwidth (area) of the latter response. This bandwidth is the quantity b_n in the theoretical noise-improvement-factor† (NIF) equation,

$$\text{NIF} = (B_n/2b_n)(\Delta f/f_T)^2 \qquad (10\text{-}4)$$

where B_n is the noise bandwidth of the predetection filter (the bandpass filter in the noise-channel, Fig. 10-9), and Δf and f_T are the peak frequency deviation and test-tone frequency, respectively, in hertz.

To measure SNR as a function of CNR (i.e., to measure the NIF), there are two principal methods:

(1) The first method requires the least equipment, but yields the threshold value of CNR "in absence of modulation"; that is, the effect of modulation in increasing the incidence of threshold impulses, which is particularly significant in the case of feedback-type demodulators, is ignored. A conventional LD tested by this technique yields more useful data because its threshold is less sensitive to modulation.

In this technique, signal-plus-noise is measured at each value of CNR by the true rms meter following the low-pass filter at the demodulator output, followed by a measurement of noise on the same meter when modulation is removed from the input of the modulator. A refinement on the method is to substitute a wave analyzer (selective voltmeter) for the output signal measure-

---

† The NIF is the ratio of the SNR to CNR measured in the bandwidths of the actual filters in the receiver.

10.3. System Tests of FM Demodulators

ment to measure signal instead of signal plus noise. The true rms meter must still be used to measure the output noise.

(2) The second method allows measurement of output noise with modulation present. It requires interposing a narrow-bandstop filter ahead of the true rms meter to reject the tone. To calculate the NIF accurately, as described above, the combined response of the bandstop and lowpass filters must be used in arriving at the noise bandwidth b_n, as the bandstop filter is also present in the measured NIF data. Relative power equal to unity is taken from the tone-frequency response in absence of the bandstop filter. A suitable bandstop filter is the type supplied with a noise-loading test set, as described below, which has a high degree of rejection and a narrow bandwidth.

Distortion measurements are made with the wide-band noise source of Fig. 10-9 removed, or the noise-setting attenuator at high attenuation. For harmonic distortion measurement, one modulating tone is applied, and at the demodulator output either the total harmonic distortion is registered on a fundamental-nulling-type distortion meter, or individual components are measured with a wave analyzer. For intermodulation measurement, two modulating tones at f_1 and f_2 are applied, usually such that the third-order products $(2f_2 - f_1$ and $2f_1 - f_2)$ fall within the baseband. The distortion products can be measured with a wave analyzer.

(ii) *Noise Loading.* For FDM systems, special noise-loading test sets are available to generate a baseband signal resembling an array of multiplexed channels and to measure NPR. The test set consists of a noise generator with high-pass and low-pass filters to define the baseband, a bank of bandstop filters inserted one at a time to simulate an idle channel, and a noise receiver with switch-selectable tuning to frequencies corresponding to the bandstop filters. The noise generator, followed by the bandstop filter bank, precedes the transmission system to be tested, in this case the FM modulator with preemphasis if required. The noise receiver follows the transmission system. The noise receiver may be placed directly after the demodulator with no deemphasis network, even if preemphasis is used, because only a ratio is measured of loaded-to-idle noise power in a given channel, and deemphasis effects noise and "signal" of a narrow-band channel equally. Preemphasis, if the system calls for it, should be used because the distribution of the noise-loading energy in the baseband spectrum affects the NIF, threshold CNR, and distortion.

The principles and practices of NPR measurement, evaluation of the results in terms of per-channel SNR, and a description of available equipment are given in the publications of the instrument manufacturers (see for example reference 3), and the reader is referred to them for detailed instructions. Present comments will be limited to some of the more specialized areas, or problems which are less apparent.

As stated earlier in this chapter, NPR, and hence SNR can be found by the same measurement technique at all CNR values, irrespective of whether the CNR value is below threshold, in the "thermal" region (the linear region above threshold), or the high CNR region where intermodulation noise predominates. From the curves of NPR vs. CNR, worst-channel CNR at specified channel SNR values can be read as in the case of tone tests, where the entire baseband output is measured at once. In the low channels, when large frequency deviation is employed, the "above threshold" noise power is very low and obscured by the intermodulation noise. It is, therefore, not possible to measure directly the conventionally defined threshold. However, threshold improvement is still important for these channels since the below threshold noise power is correspondingly reduced.

In making a particular NPR measurement, after obtaining a reference reading for the loaded channel on the noise receiver, the pertinent bandstop filter is inserted. The baseband power fed to the modulator is thereby reduced, due to the nonzero bandwidth of the filter. This reduction is generally not negligible, and when a high channel is being measured and preemphasis is used, the reduction is as great as 1–3 dB. It is accepted practice to increase the overall deviation (by means of an attenuator ahead of the modulator) to that without the bandstop filter. The justification for this is as follows: Fair evaluation of demodulator performance requires that full modulation be applied, because threshold and distortion are affected by the amount of deviation. The amount of idle noise (due to intermodulation or threshold) in a channel depends only secondarily upon the exact distribution of the baseband energy, however, and except for very low channel capacities (12 channels or less for typical filter characteristics), replacing baseband energy in the region of the channel being measured with energy elsewhere in the baseband is a good approximation.

When obtaining a reference reading on the noise receiver, CNR should be set at that value which will be used for the corresponding idle-channel measurement. This is opposite to the usual procedure of setting the reference at high CNR for measurement at all CNR values. The method advised here is based on the fact that the user of a real transmission system is presented with a given value of CNR over which he has no operational control. He wants to know the output SNR at that CNR, rather than at an inaccessible ratio of signal-without-rf noise to output noise. The required SNR is directly related to NPR measured per the present instructions. Numerically, a difference arises between the two methods of setting the loaded-channel reference because output signal power decreases below threshold (note that with FDM, the below-threshold region is an acceptable region of operation).

Along with each value of NPR and throughout the range of CNR, it is desirable to measure the corresponding *baseband intrinsic noise ratio*

(BINR) defined as the ratio of loaded-channel noise to noise in the channel with *all* modulation removed. Its measurement reveals the following:

(1) By comparison with the NPR measurements at high CNR, it shows to what extent the leveling of the NPR vs. CNR curve is due to intermodulation noise, and to what extent it is due to residual additive noise in the system, including the effect of oscillator jitter.

(2) By comparison with the NPR at and below threshold, it shows the extent to which threshold CNR is degraded by the presence of modulation, giving a useful indication on demodulator performance.

(3) By comparison with the theoretical NIF, it checks the accuracy of deviation, preemphasis, and the predetection bandwidth calibrations. In the highest channel, with a properly functioning FM demodulator, AM rejection should be sufficient for BINR to correspond to theory within the tolerance of these calibrations, plus an additional correction for the high- and low-pass filter response in the noise generator portion of the noise-loading test set. The low-pass filter, especially, must be taken into account. CCIR recommendations on test-set filter response permit a skirt such that with 4-dB preemphasis (which accentuates the extra baseband power in the skirt response above the nominal top of the baseband), approximately 0.5 dB extra noise is presented to the modulator, and is registered on the meter indicating the frequency deviation. Deviation in a given channel is, therefore, less than the correct value by this amount. It is not appropriate to compensate by raising the overall deviation so that the per-channel value is correct, because the overall deviation will then stress the modulator and demodulator more than in normal operation. For a critical evaluation of SNR in each channel, it is necessary to measure the spectrum of the modulating noise with a wave analyzer after it has passed through the high- and lowpass filters and the preemphasis network (if any is used). The noise bandwidth of this spectrum should be calculated (graphically) with relative power equal to unity at each pertinent channel frequency. This bandwidth should be compared with that which would result using ideal filters and preemphasis, and the difference (in decibels) applied to the NPR values as an upward correction.

E. TEST EQUIPMENT FOR ANALOG-SIGNAL TESTS

The manufacturers and model number of equipment listed in Table VI are given for illustrative purposes only, and many other units of comparable performance will function as well. The intent is to clarify the discussion of test procedures given for analog signals by identifying, via commercially available instruments, the desirable technical features. The order is that in which the instruments are mentioned in the text, and only those in critical portions of the test setup are included.

TABLE VI

Typical Test Equipment for Analog Signals

Instrument	Manufacturer and model	Main features
FM modulator	RCA Victor Co., Ltd., MM1200	Standard for basebands to 5 MHz, deviations to 15 MHz; 70 MHz output
Wide-band amplifiers (for wide-band noise source, etc.)	Avantek, C-Cor, Conductron	Gains of 20 to 60 dB, bandwidth to 200 MHz
Noise diode (for wide-band noise source)	Solitron "Sounvistor"	Wide-band units available
Tunable voltmeter (for measuring noise spectrum at IF)	Stoddart Field Intensity Meter, NM series or Hewlett Packard Wave Analyzer, 3590A, 312A	Various models for a wide range of frequency. Calibrated gain and bandwidth. To 620 kHz and 18 MHz, respectively
Spectrum analyzer (for checking flatness of wide-band noise)	Hewlett Packard 8553L/8552A	
Power meter	Hewlett Packard 432A General Microwave 454A	10 MHz and up
True rms voltmeter	Ballantine 320, 320A, 323; Hewlett Packard 3400A	Bandwidths of 500 kHz to 20 MHz
Wave analyzer (for baseband spectrum measurement)	Hewlett Packard 3590A, 312A	To 50 kHz, 620 kHz, and 18 MHz, respectively
Noise-loading test set	Marconi 2090	12–2700 channels

E. Digital-Signal Test Procedures

A. The Test Setup

The important performance measurement for binary FM is BER as a function of CNR or E/η. A block diagram of a test setup for this measurement is given in Fig. 10-10. Basically, it is comprised of a simulated transmitter, receiver, channel, timing circuits, error detection circuits, and the associated instruments. There exists commercial equipment which lumps a number of blocks into a single unit, simplifying the setup. Clearly, a number of variations of the setup will suggest themselves.

In the transmitter the data source is a word generator, with the symbol stream selected in a random or quasi-random manner. A random stream may be obtained by triggering the word generator by a random noise source. A quasi-random stream is a repetition of a long word, with the word appropriately selected. Care should be taken that the ratio of marks to spaces be close to unity, or close to that expected in the message for which the system is designed. The transitions in the bit stream are the same as those expected in the system. An attenuator adjusts the modulation amplitude thereby controlling the deviation index.

The channel, as indicated, corrupts the FM wave with additive white Gaussian noise. In a more general way, the channel may corrupt the signal

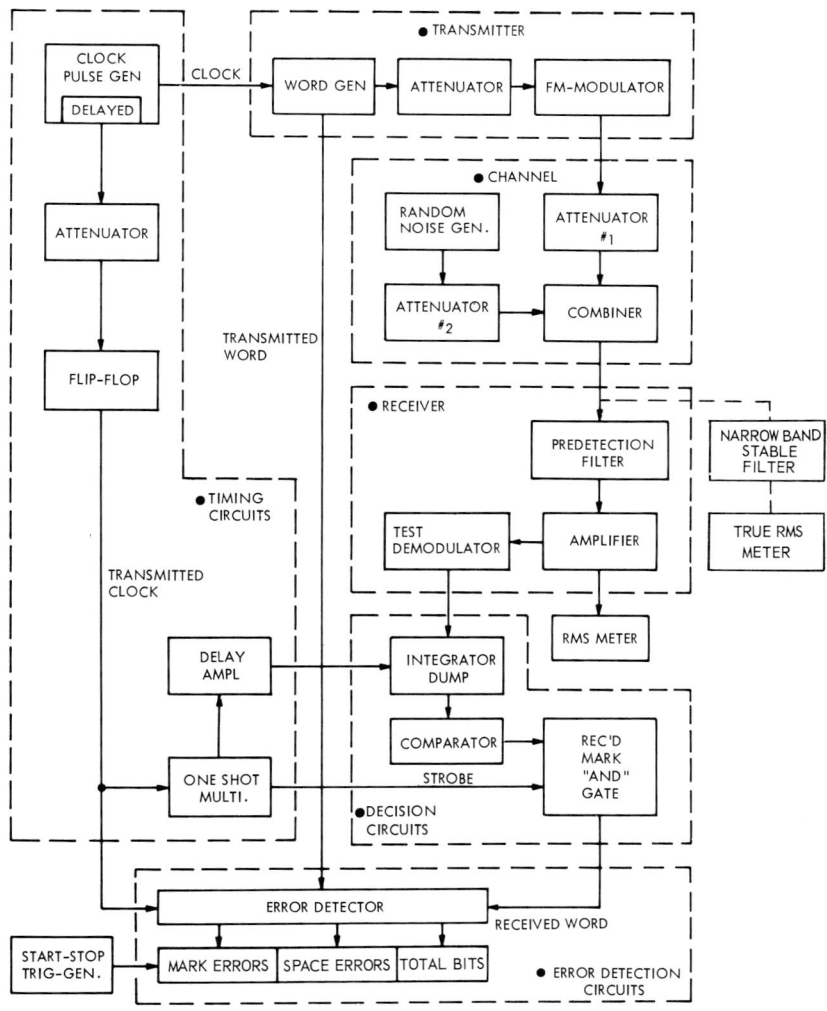

Fig. 10-10. Block diagram of test setup for BER measurements.

with both additive and multiplicative noise (fading), and the distribution of the noise may be non-Gaussian.

The receiver is comprised of a predetection filter and the demodulator under test. Further elements are the amplifiers and attenuators to adjust the levels to their appropriate value, and those providing some gross filtering.

The decision on whether a mark or a space was received is made by the decision circuits. In the indicated setup, the decision processor is of the integrate-dump type, i.e., the output of the demodulator is integrated over a bit period and at the end of the bit period a decision is made on the basis of the polarity of the integrator output. The contents of the integrator are then "dumped" to prepare for the next integration interval. The decision circuits also comprise a comparator for the purpose of regenerating the data stream. The decision processor may also be a conventional lowpass filter, and the decision is then performed by sampling the output of the filter at a time when the peak of the signal is expected.

The error detection circuits receive both the regenerated data stream from the receiver decision circuits and a delayed version of the modulation stream, with the delay exactly compensating for the signal delay through the transmitter, channel, receiver, and decision circuits. Errors are detected by comparing the two data streams. It is useful to monitor separately the mark and space errors, as any appreciable imbalance indicates some bias in the circuits.

Lastly, the timing circuits provide the timing for the whole system, including word generation, sampling time, and dumping time. An additional critical duty of these circuits is to provide appropriate compensating delays. In an actual system, the timing for the decision circuits would have to be generated from the received data stream; therefore, the setup will often obtain the timing in that manner, to simulate more readily the actual receiver. It is generally considered that for system error rates of 10^{-4} or lower, relatively error-free timing is obtained in the receiver, and therefore direct timing as it is done in this setup is quite appropriate. For coded speech, where higher-error rates are permissible, it may be advisable to include the actual timing extraction circuit in the setup, to include its effect. Commercial word analyzers are available that comprise the decision circuits, generate a duplicate of the transmitted quasi-random word, and automatically adjust themselves to the timing of the demodulated signal.

B. TEST PROCEDURE

The measurement consists of BER as a function of CNR, which can be subsequently converted to BER as a function of E/η. After the test setup has been found to work correctly, the test procedure is as follows:

(1) Adjust the demodulator signal input level for the design value, with the random noise generator highly attenuated (attenuator #2).

10.3. System Tests of FM Demodulators

(2) With the signal input highly attenuated (attenuator #1), adjust the noise level to the desired number of decibels below the design signal level. For better accuracy, adjust the noise level to the same level as the design signal level, and then reduce the noise to the desired level using a carefully calibrated attenuator.

(3) Bring back the signal level to the desired value and note the BER. Allow a sufficient number of errors (at least 30) to accummulate, in order to obtain a representative average rate of errors. Further accuracy is obtained by repeating the measurement for the same CNR and averaging the results. A substantial difference between a plurality of measurements indicates a drift in the system which should be eliminated. When the error rates are very low (caused by low bit rates, high CNR, or both), it may be difficult to stabilize the system for a sufficient length of time to accumulate a sufficient number of errors and make meaningful measurements.

(4) Further test points are obtained by varying attenuator #2 to obtain the desired CNR value.

(5) To convert the data to BER vs E/η, it is required to know the noise bandwidth of the predetection channel and the bit rate, as given by Eq. (10-3). The E/η is referred to the input of the predetection filter. In this manner, any losses in the predetection filter are charged as a penalty to the demodulator.

(6) In some cases it is convenient to precede the true rms meter (Fig. 10-10) by a narrowband, stable filter of appropriate center frequency, and the combination is connected before the predetection filter. The E/η is then measured through that filter, with the carrier power measured with no modulation. This method is convenient when a series of tests are to be made, that normally require repeated recalibration of the noise bandwidth of the predetection filter due to suspected drift or due to a willful change of the filter parameters.

C. Precautions and Adjustments

Following is a checklist of factors that often contribute to improper tests of performance:

(1) Has the demodulator been checked as described in Sections 10.1 and 10.2?

(2) Does the receiver have sufficient dynamic range to prevent clipping of noise peaks? (Noise peaks may be considered to reach 15 dB above the rms value).

(3) Have the meters and attenuators been appropriately terminated for accuracy?

(4) Is the input level to the demodulator the design value?

(5) Are the strobe pulse and the dump pulse correctly timed with respect to the end of the bit period?

(6) Are the center frequency of the VCO in the demodulator, the frequency of the input signal, and the frequency of the predetection filter correct? Center frequency offsets will generally be detected by an unequal number of mark and space errors.

References

1. L. H. Enloe, The synthesis of frequency feedback demodulators. *Proc. Nat. Electron. Conf.* **18**, p. 477 (1962).
2. Radio-relay systems for telephony using frequency-division multiplex. *Recommendations, Doc. Plenary Assembly*, 9th, **1**, CCIR Rec. No. 287, pp. 256–258.
3. W. Oliver, White noise loading of multi-channel communications systems. Marconi Instruments, Englewood, New Jersey, September 1964.

APPENDIX A

Derivation of Angle-Modulation Improvement Equations

The sensitivity of the frequency or phase detector does not enter into the results, since both the signal and the noise are affected by it equally. Therefore, we shall assume below that the sensitivity is unity. For further details on the signal models, see Section 6.3.

A. Single-Channel Speech

The signal of the speech channel is modeled as follows:

(1) It is random with Gaussian statistics.
(2) It has a power spectral density that is nonzero between frequencies f_a and f_b only, and within this band, the power spectral density (PSD) varies inversely with f^2. The frequencies f_a and f_b are generally taken as 1 kHz and 4 kHz, respectively.
(3) Alternatively, the system performance is calculated and tested with a test-tone, usually of 1 kHz.

A. SPEECH MODEL: FM

The signal power P_s is

$$P_s = (\Delta\omega_{\text{rms}})^2 \qquad (A\text{-}1)$$

while the noise power is

$$\overline{n_o^2(t)} = 2\eta f_b^3 (2\pi)^2 / 3A^2 \qquad (3\text{-}18')$$

Therefore,

$$\text{SNR}_{\text{SP}} = \frac{P_s}{\overline{n_o^2(t)}} = 6\left(\frac{\Delta\omega_{\text{rms}}}{\omega_b}\right)^2 \left(\frac{A^2/2}{2\eta f_b}\right) = 6\sigma^2 (\text{CNR})_{\text{AM}} \quad (3\text{-}19)$$

B. SPEECH MODEL: PM

The output is proportional to phase. The power spectral density is given by

$$W_{\phi_i}(f) = \eta_m/\omega^2 \quad \text{W/Hz}, \quad f_a \le f \le f_b \quad (\text{A-2})$$

Now,

$$P_s = \int_{f_a}^{f_b} W_{\phi_i}(f)\,df = \int_{f_a}^{f_b} \frac{\eta_m}{\omega^2}\,df = \frac{\eta_m}{2\pi}\left(\frac{1}{\omega_a} - \frac{1}{\omega_b}\right) \quad (\text{A-3})$$

and

$$\overline{n_o^2(t)} = (2\eta/A^2) f_b \quad (\text{A-4})$$

Therefore,

$$\text{SNR}_{\text{SP}} = \frac{\eta_m}{2\pi}\left(\frac{1}{\omega_a} - \frac{1}{\omega_b}\right) \frac{2(A^2/2)}{2\eta f_b} = \frac{\eta_m}{\pi}\left(\frac{\omega_b - \omega_a}{\omega_a \omega_b}\right)(\text{CNR})_{\text{AM}}$$

Since the PSD of the frequency deviation is $\omega^2 W_{\phi_i}(f)$, the mean-square-value of the frequency deviation is

$$(\Delta\omega_{\text{rms}})^2 = \int_{f_a}^{f_b} \omega^2 W_{\phi_i}(f)\,df \quad (6\text{-}86)$$

$$= \int_{f_a}^{f_b} \eta_m\,df = \eta_m(f_b - f_a) \quad (\text{A-5})$$

or,

$$\eta_m = (\Delta\omega_{\text{rms}})^2/(f_b - f_a)$$

Hence,

$$\text{SNR}_{\text{SP}} = \frac{1}{\pi}\frac{(\Delta\omega_{\text{rms}})^2}{(f_b - f_a)} \frac{(\omega_b - \omega_a)}{\omega_b \omega_a} \frac{\omega_b}{\omega_b}(\text{CNR})_{\text{AM}} = 2\sigma^2\left(\frac{\omega_b}{\omega_a}\right)(\text{CNR})_{\text{AM}} \quad (3\text{-}20)$$

C. TEST-TONE: PM

The signal power is proportional to the phase deviation, i.e., to the modulation index, viz.,

$$P_s = m^2/2, \quad (\text{A-6})$$

where $m = \Delta f_P/f_T$. The noise power is given by Eq. (A-4). Therefore,

$$(\text{SNR})_{\text{TT}} = \frac{m^2/2}{2\eta f_b/A^2} = m^2 \frac{A^2/2}{2\eta f_b} = m^2 (\text{CNR})_{\text{AM}} \qquad (A\text{-}7)$$

In terms of $m_P = \Delta f_P/f_b$, we have

$$(\text{SNR})_{\text{TT}} = m_P^2 (\omega_b/\omega_T)^2 (\text{CNR})_{\text{AM}} \qquad (3\text{-}21)$$

B. FDM Signals

The FDM signal is modeled as follows:

(1) It is random with Gaussian statistics.
(2) It has a power spectral density which is flat within the frequency band f_a to f_b, and zero elsewhere. The performance is generally given in terms of NPR, which for the case of high-Q channels assumed here (i.e., the channel bandwidth is much smaller than its center frequency) reduces to the ratio of the power spectral densities of signal and noise.
(3) Alternatively, the system performance is calculated and tested with a test-tone at the channel frequency ω_{CH}, and given in terms of $(\text{SNR})_{\text{TT}}$.

A. FDM Model: FM

The PSD of signal is η_m; the PSD of noise is $(2\eta/A^2)\omega_{\text{CH}}^2$. Therefore,

$$\text{NPR} = \frac{\eta_m}{2\eta\omega_{\text{CH}}^2/A^2} = \frac{2\eta_m f_b}{\omega_{\text{CH}}^2} \frac{A^2/2}{2\eta f_b} = \frac{2\eta_m f_b}{\omega_{\text{CH}}^2} (\text{CNR})_{\text{AM}}$$

However,

$$\eta_m = (\Delta\omega_{\text{rms}})^2/(f_b - f_a)$$

Hence,

$$\text{NPR} = \frac{2f_b}{f_b - f_a} \frac{(\text{CNR}_{\text{AM}})}{\omega_{\text{CH}}^2} (\Delta\omega_{\text{rms}})^2 \left(\frac{\omega_b}{\omega_b}\right)^2$$

$$= 2 \frac{\sigma^2}{(1 - \omega_a/\omega_b)} \left(\frac{\omega_b}{\omega_{\text{CH}}}\right)^2 (\text{CNR})_{\text{AM}} \qquad (3\text{-}22)$$

B. FDM Model: PM

The PSD of signal is η_m; the PSD of noise is $2\eta/A^2$. Therefore,

$$\text{NPR} = \frac{\eta_m}{2\eta/A^2} = \eta_m 2f_b \frac{A^2/2}{2\eta f_b} = \eta_m 2f_b (\text{CNR})_{\text{AM}}$$

Now,

$$(\Delta\omega_{rms})^2 = \int_{f_a}^{f_b} \omega^2 W_{\phi_i}(f)\,df \qquad (6\text{-}86)$$

$$= \int_{f_a}^{f_b} (2\pi)^2 f^2 \eta_m\,df = \frac{(2\pi)^2 \eta_m}{3}(f_b^3 - f_a^3)$$

or,

$$\eta_m = 3(\Delta\omega_{rms})^2/[(2\pi)^2(f_b^3 - f_a^3)]$$

Hence,

$$\text{NPR} = 2f_b(\text{CNR})_{AM}\,\frac{3(\Delta\omega_{rms})^2}{(2\pi)^2(f_b^3 - f_a^3)}\,\frac{\omega_b^2}{\omega_b^2}$$

$$= \frac{6\sigma^2}{[1 - (\omega_a/\omega_b)^3]}(\text{CNR})_{AM} \qquad (3\text{-}23)$$

C. Test-Tone: FM

The signal power is given by

$$P_s = (\Delta\omega_p)^2/2$$

The PSD of the noise is $(2\eta/A^2)\omega_{CH}^2$.
For a channel of bandwidth B (Hz), the noise power is

$$\overline{n_o^2(t)} = (2\eta/A^2)\omega_{CH}^2\,B \qquad (A\text{-}8)$$

Therefore,

$$\text{SNR}_{TT} = \frac{(\Delta\omega_p)^2/2}{2\eta\omega_{CH}^2\,B}\,A^2 \qquad (A\text{-}9)$$

$$= \frac{(\Delta\omega_p)^2}{\omega_b^2}\,\frac{\omega_b^2}{B\omega_{CH}^2}\,\frac{(A^2/2)}{2\eta}\,\frac{f_b}{f_b}$$

$$= m_p^2\left(\frac{\omega_b}{\omega_{CH}}\right)^2\left(\frac{f_b}{B}\right)(\text{CNR})_{AM} \qquad (3\text{-}24)$$

D. Test-Tone: PM

The signal power is given by:

$$P_s = m^2/2,$$

while the noise power is

$$\overline{n_o^2(t)} = (2\eta/A^2)B$$

Therefore,

$$\text{SNR}_{\text{TT}} = \frac{m^2/2}{2\eta B} A^2 = \frac{(\Delta\omega_p)^2/2}{\omega_{\text{CH}}^2 \, 2\eta B} A^2$$

which is identical to Eq. (A-9). The result is that Eq. (3-24) holds for both FM and PM signals.

APPENDIX

B

Derivation of the Discriminator Baseband Response

Zinn's Eq. (38)† giving the response of a discriminator to an impulse in frequency (or step in phase), is

$$V_o(t) = K\{[1 + 2m \sin(\Delta t - \theta/2) + m^2]^{1/2} - [1 - 2m \sin(\Delta t + \theta/2) + m^2]^{1/2}\} \quad \text{(B-1)}$$

where K is a gain constant (given explicitly by Zinn), $m = 2e^{-\alpha t} \sin(\theta/2)$, $\alpha = 1/2RC$ is half the angular 3-dB bandwidth of each tuned circuit, θ is the height of the phase step (weight of the frequency impulse), Δ is the bandwidth in radians per second from the center frequency to either of the peaks in the discriminator.

The low deviation in the FMFB IF allows for that application the assumption that the phase step is small; hence, $\sin(\theta/2)$ is small and $m^2 \ll 1$. Equation (B-1) becomes

$$V_o(t) = K\{[1 + 4e^{-\alpha t} \sin(\theta/2) \sin(\Delta t - \theta/2)]^{1/2} -$$
$$[1 - 4e^{-\alpha t} \sin(\theta/2) \sin(\Delta t + \theta/2)]^{1/2}\} \quad \text{(B-2)}$$

Further, for $x \ll 1$,

$$(1 + x)^{1/2} \cong 1 + (x/2)$$

† M. K. Zinn, Transient response of an FM receiver, *Bell. Syst. Tech. J.* **27**, 714–731 (1948).

Discriminator Baseband Response

Identifying x with $4e^{-\alpha t} \sin(\theta/2) \sin(\Delta t - \theta/2)$ leads to

$$V_o(t) = 2Ke^{-\alpha t} \sin(\theta/2) [\sin(\Delta t - \theta/2) + \sin(\Delta t + \theta/2)]$$
$$= 4Ke^{-\alpha t} \sin(\theta/2) \sin(\Delta t) \cos(\theta/2) \qquad \text{(B-3)}$$

For θ small, $\sin(\theta/2) \cong \theta/2$ and $\cos(\theta/2) \cong 1$. Hence,

$$V_o(t) = K\theta e^{-\alpha t} \sin(\Delta t) \qquad \text{(B-4)}$$

The Laplace transform is

$$\mathscr{L}[V_o(t)] = \frac{K\theta\Delta}{(s+\alpha)^2 + \Delta^2} = \frac{K\theta\Delta}{s^2 + 2\alpha s + \alpha^2 + \Delta^2} \qquad \text{(B-5)}$$

resulting in the transfer function

$$H_d(s) = \frac{K\Delta}{s^2 + 2\alpha s + \alpha^2 + \Delta^2} \qquad \text{(3-34)}$$

APPENDIX

C

The Ideal Demodulator

To measure the efficiency of the various demodulation techniques, an ideal demodulator is formulated to serve as a reference. This demodulator is one that can compress (or decode) transmitted signals to the base bandwidth with minimum loss of information. The performance boundary for such an ideal demodulator will now be derived.†

The information channel capacity at the demodulator input is

$$C_i = B_i \log_2[1 + S_i/(\eta_i B_i)] \tag{C-1}$$

where C is channel capacity in bits per second, S is average signal power in watts, η is noise power spectral density in watts per hertz, B is noise bandwidth in hertz, and the subscript i denotes an input quantity. At the output, for a given output SNR, the information rate must exceed

$$R \geq f_b \log_2(S_o/\eta_o f_b) \tag{C-2}$$

where the subscript o refers to output quantities, and f_b is the base bandwidth.

While the channel capacity is the boundary of errorless transmission, it is a very sharp bound, with errors increasing very rapidly if it is violated. Therefore, the channel capacity must equal or exceed the minimum rate given in Eq. (C-2), and we have

$$B_i \log_2[1 + S_i/(\eta_i B_i)] \geq f_b \log_2[S_o/(\eta_o f_b)] \tag{C-3}$$

† T. J. Goblick, Theoretical limitations on the transmission of data from analog sources, *IEEE Trans.* **IT-11**, No. 4, 558–567 (1965).

The Ideal Demodulator

resulting in

$$S_o/(\eta_o f_b) \leq [1 + (S_i/(\eta_i B_i))]^{B_i/f_b} \tag{C-4}$$

In order to interpret these results in terms of FM, we relate the rms modulation parameter (for noiselike signals) σ to the bandwidth B_i by

$$B_i = 2f_b(1 + 10^{1/2}\sigma) \tag{C-5}$$

resulting in the ideal performance curve plotted in Fig. 3-8.

The transmission mode in the ideal demodulator analysis is quite general; There is no implication that FM could satisfy the "coding and decoding" requirements to achieve the ideal performance boundary. The results indicate, however, that FM coupled to the more esoteric demodulators discussed in this text proves to be a relatively efficient transmission system in the presence of Gaussian noise.

APPENDIX

D

Varactor VCO Distortion

The output frequency of the varactor-controlled VCO is $f(v_i)$,

$$f(v_i) = K[C_o + C(v_i)]^{-1/2} \tag{D-1}$$

where v_i is the input voltage, C_o is the stray and fixed capacitance of the circuit, K is a proportionality constant, and

$$C(v_i) = C_{\min}\left[\frac{V_o + v_i}{V_o + V_B}\right]^{-\gamma} \tag{D-2}$$

where V_o, V_B, and γ are constants of the varactor diode used in the VCO.

If the driving voltage v_i is a bias plus a function of time

$$v_i = E_b + v(t) \tag{D-3}$$

then we may write the capacitance as a function of the time-varying signal

$$C(v) = C_{\min}\left[\frac{V_o + E_b + v(t)}{V_o + V_B}\right]^{-\gamma} = \frac{C_{\min}(V_o + E_b)^{-\gamma}}{(V_o + V_B)^{-\gamma}}\left[1 + \frac{v(t)}{V_o + E_b}\right]^{-\gamma} \tag{D-4}$$

Let $C_{\min}[(V_o + E_b)/(V_o + V_B)]^{-\gamma} = C_b$ (varactor quiescent capacitance). Assuming $|v(t)/(V_o + E_b)| < 1$, to prevent diode conduction, the following expansion is carried out†

† B. O. Peirce, "A Short Table of Integrals," Entry No. 748. Ginn, New York, 1929.

$$C(v) = C_b\left[1 - \gamma\frac{v(t)}{V_o + E_b} + \frac{\gamma(\gamma+1)}{2}\left(\frac{v(t)}{V_o + E_b}\right)^2\right.$$
$$\left. - \frac{\gamma(\gamma+1)(\gamma+2)}{6}\left(\frac{v(t)}{V_o + E_b}\right)^3 + \cdots\right] \quad \text{(D-5)}$$

We may now write the VCO output frequency as a function of time $f_r(t)$

$$f(v) = f_r(t) = K[(C_o + C_b) - C_b C_1(t)]^{-1/2} \quad \text{(D-6)}$$

where

$$C_1(t) = \gamma\frac{v(t)}{V_o + E_b} - \frac{\gamma(\gamma+1)}{2}\left(\frac{v(t)}{V_o + E_b}\right)^2 + \frac{\gamma(\gamma+1)(\gamma+2)}{6}\left(\frac{v(t)}{V_o + E_b}\right)^3 - \cdots \quad \text{(D-7)}$$

Assuming that $|C_1(t)| \ll (C_o + C_b)/C_b$, which implies that the variable part is much smaller than the average capacitance and is usually required for good linearity, then†

$$f_r(t) = K(C_o + C_b)^{-1/2}\left[1 - \frac{C_b}{(C_o + C_b)}C_1(t)\right]^{-1/2}$$
$$\cong \frac{K}{(C_o + C_b)^{1/2}}\left[1 + \frac{C_b}{2(C_o + C_b)}C_1(t)\right] \quad \text{(D-8)}$$

or

$$f_r(t) = f_o\left\{1 + \frac{C_b}{2(C_o + C_b)}\left[\gamma\left(\frac{v(t)}{V_o + E_b}\right) - \frac{\gamma(\gamma+1)}{2}\left(\frac{v(t)}{V_o + E_b}\right)^2\right.\right.$$
$$\left.\left. + \frac{\gamma(\gamma+1)(\gamma+2)}{6}\left(\frac{v(t)}{V_o + E_b}\right)^3 - \cdots\right]\right\} \quad \text{(D-9)}$$

where $f_o = K/(C_o + C_b)^{1/2}$ is the undeviated or center frequency. Defining

$$\mu = 2\pi\gamma C_b f_o/[2(C_o + C_b)(V_o + E_b)] \quad \text{(D-10)}$$

then

$$f_r(t) = \left\{f_o + \frac{\mu}{2\pi}\left[v(t) - \frac{\gamma+1}{2}\frac{(v(t))^2}{V_o + E_b} + \frac{(\gamma+1)(\gamma+2)(v(t))^3}{6(V_o + E_b)^2} - \cdots\right]\right\} \quad \text{(D-11)}$$

where μ is sensitivity of VCO in radians per second per volt.

Equation (D-11) may be written in the form of a power series

$$2\pi f_r(t) = \omega_r(t) = \omega_o + \Delta\omega(t) = \omega_o + a_1 v(t) + a_2 v^2(t) + a_3 v^3(t) + \cdots \quad \text{(D-12)}$$

† Ibid.

where

$$a_1 = \mu$$
$$a_2 = -\mu(\gamma + 1)/[2(V_0 + E_b)] \tag{D-13}$$
$$a_3 = \mu(\gamma + 1)(\gamma + 2)/6(V_0 + E_b)^2 \tag{D-14}$$

etc. These coefficients may be utilized in the text in Eqs. (6-33) and (7-71).

The above equations would indicate that there is no distortion for $\gamma = -1$. This incorrect result stems from approximation (D-8). However, the equations do indicate the directions for γ, V_0, and E_b for minimum distortion.

APPENDIX

E

Baseband-Equivalent Response of a Single-Tuned Circuit to a Small-Index Off-Tuned FM Carrier

For a single-tuned circuit, the lowpass equivalent is an RC network with the response given by Eq. (4-61)

$$H_{L1}(j\omega, 0) = 1 \Big/ \left(\frac{j\omega}{b\omega_b} + 1\right)$$

$$= \exp\left(-j\tan^{-1}\frac{\omega}{b\omega_b}\right) \Big/ \left[\left(\frac{\omega}{b\omega_b}\right)^2 + 1\right]^{1/2}$$

This is now applied to Eqs. (4-17), (4-18), and (4-19) to determine the baseband equivalent response which includes the effect of an off-tuned FM carrier. Now,

$$|H_{L1}(\Delta\omega_e)| = 1 \Big/ \left[\left(\frac{\Delta\omega_e}{b\omega_b}\right)^2 + 1\right]^{1/2} \equiv \frac{1}{\gamma} \qquad \text{(E-1)}$$

$$H_x(j\omega) = \exp\left(j\tan^{-1}\frac{\Delta\omega_e}{b\omega_b}\right) \Big/ \left[\frac{j(\omega + \Delta\omega_e)}{b\omega_b} + 1\right] \qquad \text{(E-2)}$$

$$H_x^*(-j\omega) = \exp\left(-j\tan^{-1}\frac{\Delta\omega_e}{b\omega_b}\right) \Big/ \left[\frac{-j(-\omega + \Delta\omega_e)}{b\omega_b} + 1\right] \qquad \text{(E-3)}$$

Since the exponent is a unit vector, it can be written

$$\exp\left(j\tan^{-1}\frac{\Delta\omega_e}{b\omega_b}\right) = \left(1 + \frac{j\Delta\omega_e}{b\omega_b}\right)\bigg/\left[1 + \left(\frac{\Delta\omega_e}{b\omega_b}\right)^2\right]^{1/2}$$

$$= \left(1 + j\frac{\Delta\omega_e}{b\omega_b}\right)\bigg/\gamma \qquad (E\text{-}4)$$

Let

$$j\omega/b\omega_b = Z \quad \text{and} \quad j\Delta\omega_e/b\omega_b = Z_e \qquad (E\text{-}5)$$

Then from Eqs. (E-2), (E-4), and (E-5)

$$H_x(j\omega) = \frac{1}{\gamma}\frac{(1+Z_e)}{(1+Z+Z_e)} \qquad (E\text{-}6)$$

$$H_x^*(-j\omega) = \frac{(1-Z_e)}{\gamma(1+Z-Z_e)} \qquad (E\text{-}7)$$

and

$$H_x(j\omega) + H_x^*(-j\omega) = \frac{2(Z+1-Z_e^2)}{\gamma(Z^2+2Z+1-Z_e^2)} \qquad (E\text{-}8)$$

But

$$1 - Z_e^2 = \gamma^2 \qquad (E\text{-}9)$$

Therefore,

$$H_x(j\omega) + H_x^*(-j\omega) = \frac{2\left(\dfrac{Z}{\gamma^2}+1\right)}{\gamma\left[\left(\dfrac{Z}{\gamma}\right)^2 + \dfrac{2Z}{\gamma^2}+1\right]} \qquad (E\text{-}10)$$

The desired result is, from Eq. (4-17) and Eqs. (E-1) and (E-10), given by

$$H_{b1}(j\omega, \Delta\omega_e) = \frac{1 + \dfrac{Z}{\gamma^2}}{1 + \dfrac{2Z}{\gamma^2} + \dfrac{Z^2}{\gamma^2}}$$

Letting $j\omega = s$, resulting in $Z = s/(b\omega_b)$, we obtain Eq. (4-63).

APPENDIX

F

Calculation of Minimum Noise Bandwidth and Loop Filter Zero Constant for the FMFB

We wish to calculate the minimum noise bandwidth of the following second-order transfer function $H(s)$ in terms of the FMFB parameters

$$H(s) = \frac{K_0[s/(a_0\,\omega_n) + 1]}{(s/\omega_n)^2 + 2\xi_0(s/\omega_n) + 1} \qquad \text{(F-1)}$$

with the noise bandwidth being

$$B_n = \frac{1}{K_0^2} \int_0^\infty |H(j\omega)|^2\, df \qquad \text{(F-2)}$$

The solution to Eq. (F-2) is tabulated in Eq. (2-56), giving

$$B_n = (\omega_n/8\xi_0)[(1/a_0)^2 + 1] \qquad \text{(F-3)}$$

In terms of FMFB theory, the parameters ω_n, a_0, and ξ_0 are related to the loop parameters K, a, and ξ [refer to Eqs. (4-50)–(4-56)] by

$$\omega_n = \omega_b(1+K)^{1/2}, \quad K_0 = \frac{K}{1+K}, \quad a_0 = \frac{a}{(1+K)^{1/2}}, \quad \xi_0 = \frac{\xi + K/(2a)}{(1+K)^{1/2}},$$

$$F = 1 + K \qquad \text{(F-4)}$$

Substitution of the loop parameters into Eq. (F-3) results in

$$B_n = \frac{\omega_b F}{4} \frac{[a^2 + F]}{[2\xi a^2 + (F-1)a]} \qquad \text{(F-5)}$$

which we wish to minimize for fixed ω_b, F, and ξ by varying a. Taking the partial derivative $\partial B_n/\partial a$ produces

$$\frac{\partial B_n}{\partial a} = \frac{\omega_b F}{4} \frac{[2\xi a^2 + (F-1)a][2a] - [a^2 + F][4\xi a + (F-1)]}{[2\xi a^2 + (F-1)a]^2} \quad \text{(F-6)}$$

which, for a minimum, requires the numerator to equal zero, resulting in

$$(F-1)a^2 - 4F\xi a - F(F-1) = 0$$

with solution

$$a = 2\xi F/(F-1) \pm [4\xi^2(F/(F-1))^2 + F]^{1/2} \quad \text{(F-7)}$$

The solution with the minus sign results in a negative a that has no physical meaning to the problem. The solution with the plus sign is taken, yielding

$$a = 2\xi F/(F-1) + [4\xi^2(F/(F-1))^2 + F]^{1/2} \quad \text{(F-8)}$$

This, applied to Eq. (F-5) yields the minimum noise bandwidth. Following Enloe,† we let $\xi = 2^{1/2}/2$, which modifies Eq. (F-8) and ξ_0 to be

$$a_0 = \frac{a}{F^{1/2}} = 2^{1/2}\left(\frac{F^{1/2}}{(F-1)}\right) + \left[2\frac{F}{(F-1)^2} + 1\right]^{1/2} = 2^{1/2}\left(\frac{F^{1/2}}{F-1}\right)$$

$$\times \left\{1 + \left[1 + \frac{(F-1)^2}{2F}\right]^{1/2}\right\} \quad \text{(F-9)}$$

and

$$\xi_0 = \frac{1}{2}\left[\frac{2^{1/2} + (F-1)/(a_0 F^{1/2})}{F^{1/2}}\right] \quad \text{(F-10)}$$

This is the result stated in the text, Eqs. (7-11) and (7-13).

† L. H. Enloe, The synthesis of frequency feedback demodulators. *Proc. Nat. Electron. Conf.* **18**, 477 (1962).

Bibliography

Some books are to be tasted, others to be swallowed, and some to be chewed and digested...

Francis Bacon in the essay *Of Studies*

Note: Some of the methods and results presented in the papers and reports listed in this bibliography have not gained acceptance, or have in practice been found to apply only to a restricted range of parameters. They have been included here for a variety of reasons, such as a new point of view, useful data, historical development, discussion of a problem, etc.

1. PLL—Books

Gardner, F. M., "Phaselock Techniques." Wiley, New York, 1966.
Viterbi, A. J., "Principles of Coherent Communication." McGraw-Hill, New York, 1966.
Van Trees, H. L., "Detection, Estimation, and Modulation Theory." Wiley, New York, Part I—1968, Part II—1971.

2. PLL—Papers

1932

De Bellescize, H., La reception synchrone. *Onde Elec.* **11**, 230–240 (1932).

1939

Sakaroff, S., Frequency-controlled oscillators. *Communications* **19**, 7–9, No. 50 (1939).

1940

Tucker, D., Carrier frequency synchronization. *Post Office Elec. Eng.* **33**, 75–81 (1940).

1943

Tucker, D., The synchronization of oscillators. *Electron. Eng.* **16**, 26–30 (1943).

Wendt, K. R., and Fredendall, G. L., Automatic frequency and phase control of synchronization in television receivers. *Proc. IRE* **31**, 7–15 (1943).

1944

Beers, G. L., A frequency-dividing locked-in oscillator FM receiver. *Proc. IRE* **32**, 730–737 (1944).

1949

Labin, E., Theory of synchronization by control of phase. *Philips Res. Rep.* **24**, 291–315 (1949).

Schlesinger, K., Locked oscillator for television synchronization. *Electronics* **22**, 112–117 (1949).

1951

Costas, J. P., synchronous detection of amplitude-modulated signals. *Proc. Nat. Electron. Conf.* **7**, 121–129 (1951).

George, T. S., Analysis of synchronizing systems for dot-interlaced color television. *Proc. IRE* **39**, 124–131 (1951).

1952

Golay, M. J. E., Automatic frequency control. *Proc. IRE* (Correspondence) **40**, 996 (1952).

1953

Gruen, W. J., Theory of AFC synchronization. *Proc. IRE* **41**, 1043–1048 (1953). [Correction: *Proc. IRE* **41**, 1171 (1953).]

Preston, G. N., and Tellier, J. C., The lock-in performance of an AFC circuit. *Proc. IRE* **41**, 249–251 (1953).

Richman, D., APC color sync for NTSC color television. *IRE Nat. Conv. Rec.* Part 4, Broadcasting and Television, 13–17 (1953).

1954

Camfield, G. J., A frequency generating system for UHF communication equipment. *Proc. IEE(Brit.)* **101**, Part 3, 85–90 (1954).

Jelonik, Z., Celinski, O., and Syski, R., Pulling effect in synchronized systems. *Proc. IEE(Brit.)* **101**, Part 4, 108–17 (1954).

Ordung, P. F., Gibson, J. E., and Shinn, B. J., Closed loop automatic phase control. *Trans. Amer. Inst. Elec. Eng.* **73**, Part 1, 375–381 (1954).

Richman, D., The DC quadricorrelator: A two-mode synchronization system. *Proc. IRE* **42**, 288–299 (1954).

Richman, D., Color-carrier reference phase synchronization accuracy in NTSC color television. *Proc. IRE* **42**, 106–133 (1954).

1955

Jaffe, R., and Rechtin, E., Design and performance of phase-lock circuits capable of near optimum performance over a wide range of input signal and noise levels. *IRE Trans.* **IT-1,** 66 (1955).

Strandberg, P. M., Phase stabilization of microwave oscillators. *Proc. IRE* **43,** 869–873 (1955).

1956

Costas, J. P., Synchronous communications. *Proc. IRE* **44,** 1713–1718 (1956).

Kapranov, M. V., Pull-in range in phase-controlled AFC systems. *Radiotekhnika (Moscow)* **11,** 37–52 (1956).

Kapranov, M. V., Noise filtering and phase locked automatic frequency control. *Radio Eng. Electron. USSR* **1,** (1956).

1957

Bershtein, I. L., and Sibiryakov, V. L., Phase stabilization of microwave oscillators. *Radio Eng. Electron. USSR* **2,** No. 7, 184–185 (1957).

Clapp, J. K., and Lewis, F. D., A unique standard-frequency multiplier. *IRE Nat. Conv. Rec.* Part 5, 131–136 (1957).

Leek, R., Phase-lock AFC loops. *Electron. Radio Eng.* **34,** pp. 141–146 (April 1957); pp. 177–183 (May 1957).

Margolis, S. R., The response of a phase locked loop to a sinusoid plus noise. Jet Propul. Lab., External Publ. No. 348; *IRE Trans.* **IT-3,** 136–142 (1957).

Woodyard, J. R., Application of the autosynchronized oscillator to frequency demodulation. *Proc. IRE* **25,** 612–619 (1957).

1958

Artym, A. D., The applications of a phase automatic frequency controller. *Radio Eng. USSR* **13,** No. 8, 48–59 (1958).

Bakaev, Yu. N., Study of television flywheel synchronization system. *Radio Eng. Electron. USSR* **3,** No. 2, 315–329 (1958).

Bershtein, I. L., On the theory of automatic phase control of frequency. *Radio Eng. Electron. USSR* **3,** No. 2, 410–414 (1958).

Bershtein, I. L., and Sibiryakov, V. L., Automatic phase control of microwave oscillator frequency. *Radio Eng. Electron. USSR* **3,** No. 2, 415–418 (1958).

Crafts, C. A., Phase multilock communication. *Conf. Proc. Nat. Conv. Mil. Electron.* 2nd, *Washington D.C., June 1958*, pp 262–265.

Gilchriest, C. E., Application of the phase-locked loop to telemetry as a discriminator or tracking filter. *IRE Trans.* **TRC-4,** 20–35 (1958).

McRae, D. D., Phase-locked demodulation in telemetry receivers. *Proc. Nat. Symp. Telemetering, Miami Beach, Florida, September 1958.*

Zanadvorov, P. N., On the synchronization of oscillators by periodic pulse trains. *Radio Eng. Electron. USSR* **3,** No. 2, 281–296 (1958).

1959

Costas, J. P., Some notes on space communications. *Proc. IRE* **47,** 1383–1385 (1959).

Loutit, J. A., and Story, R. F., Flexible phase lock frequency control by analysis procedures. *Proc. Annu. Symp. Frequency Control, 13th, Asbury Park, New Jersey, May 12–14, 1959,* 371–383.

McAleer, H. T., A new look at the phase-locked oscillator. *Proc. IRE* **47,** 1137–1143 (1959). [Correction: *Proc. IRE* **48,** 1771 (1960).]
Preston, G. W., Basic theory of locked oscillators in tracking FM signals. *IRE Trans.* **SET-5,** 30–32 (1959).
Runyan, R. A., Factors affecting choice of loop filters in phase-locked loop discriminators. *Proc. Nat. Symp. Space Electron. Telemetry, San Francisco, California, September 1959.*
Sampson, W. F., and Ruegg, F. A., Phase-lock in space communications. *Proc. Nat. Symp. Space Electron. Telemetry, San Francisco, California, September 1959.*
Sanders, R. W., Digilock telemetry system. Presented at *Nat. Symp. Space Electron. Telemetry, San Francisco, California, September 1959.*
Stratonovich, R. L., Synchronization of an oscillator in the presence of interference. *Radio Eng. Electron. USSR* **3,** No. 4, 54–68 (1958).
Thomson, D. N., Performance of the Self Cohered Detector. *Annu. Radar Symp. 5th, Willow Run Lab., Univ. of Michigan, January 1959*, 85–91.
Tikhonov, V. I., The effect of noise on phase locked oscillator operation. *Automat. Remote Contr. USSR* **20,** No. 9, 1160–1168 (1959).
Viterbi, A. J., System design criteria for space television. *J. Brit. Inst. Radio Eng.* **19,** No. 9, 561–570 (1959).
Weaver, C. S., A new approach to the linear design and analysis of phase-locked loops. *IRE Trans.* **SET-5,** No. 4, 166 (1959).

1960

Brockman, M. H., Buchanan, H., Choate, R., and Malling, L. R., Extra-terrestial radio tracking and communication. *Proc. IRE* **48,** 643–654 (1960).
Choate, R. L., Analysis of a phase-modulation communications system. *IRE Trans.* **CS-8,** 221–227 (1960).
deBey, L. G., Tracking in space by DOPLOC. *IRE Trans.* **MIL-4,** 332–335 (1960).
Goldstein, A. J., and Byrne, C. J., The phase-controlled loop with a sawtooth comparator. *Northeast Electron. Res. Eng. Meeting, Boston, Massachusetts, Conv. Rec., November 1960.*
Martin, B. D., Threshold improvement in an FM subcarrier system. *IRE Trans.* **SET-6,** 25–33 (1960). [Comment by J. J. Spilker appears in *IRE Trans.* **SET-7,** 55 (1961).]
Merrick, W., Rechtin, E., Stevens, R., and Victor, W., Deep space communications. *IRE Trans.* **MIL-4,** No. 2–3, 158–162 (1960).
Nag, B. R., Locking range of an oscillator for different nonlinearities. *Trans. Amer. Inst. Elec. Eng. Part I* **79,** 134–136 (1960).
Pullen, K. A., A theory of frequency tracking for narrowband communications. *Nat. Commun. Symp. 6th, Utica, New York, October 1960*, 83–89.
Rey, T. J., Automatic phase control: Theory and design. *Proc. IRE* **48,** 1760–1771 (1960). [Correction: *Proc. IRE* **49,** 590 (1961).]
Samoilenko, I. I., The reliability of the synchronization of an auto generator in the presence of modulated oscillations. *Radio Eng. USSR* **15,** No. 7, 61–68 (1960).
Terent, C. V. P., and Shakhgilldyan, V. V., Obtaining highly stable variable frequency by the use of automatic phase adjustment. *Telecommunications USSR (London)* 1194–1202 (1960).
Thirup, G., The application of phase-locking techniques to the design of apparatus for measuring complex transfer functions. *J. Brit. Inst. Radio Eng.* **20,** 387–396 (1960).
Tikhonov, V., Phase lock automatic frequency control operation in the presence of noise. *Automat. Remote Control. USSR* **21,** No. 3, 200–214 (1960).

Viterbi, A. J., Acquisition and tracking behavior of phase-locked loops. *Proc. Symp. Active Networks and Feedback Systems, Polytech. Inst. of Brooklyn, New York, April 1960*, **10**, 583–619.

Weaver, C. S., Increasing the dynamic tracking range of a phase lock loop. *Proc. IRE* **98**, 952 (1960).

Westlake, P. R., Digital phase control techniques. *IRE Trans.* **CS-8**, No. 4, 237–246 (1960).

1961

Breese, M., Colbert, R., Rubin, W., and Sferrazza, P., Phase-locked loops for electronically scanned antenna arrays. *IRE Trans.* **SET-7**, 95–100 (1961).

Breese, M., et al., Phase locked loops for electronically scanned antenna arrays. *Globecom V Conv. Rec.* 85–89 (1961).

Carnt, P. S., and Ribchester, E., The synthesis of high purity oscillations suitable for single sideband receivers. *J. Brit. Inst. Radio Eng.* **21**, No. 3, 237–240 (1961).

Finden, H. J., The problem of frequency synthesis. *J. Brit. Inst. Radio Eng.* **21**, No. 1, 95–103 (1961).

Goldstein, A. J., and Byrne, C. J., Pull-in frequency of the phase-controlled oscillator. *Proc. IRE* letters **49**, 1209 (1961).

Hannan, W., and Olson, T., An automatic-frequency-controlled phase-shift-keyed demodulator. *RCA Rev.* **22**, No. 4, 729–752 (1961).

Henderson, R. E., Measuring the doppler frequency shift on satellite transmissions. *Brit. Commun. Electron.* **8**, No. 7, 506–512 (1961).

Rue, A. K., and Lux, P. A., Transient analysis of a phase-locked loop discriminator. *IRE Trans.* **SET-7**, 105–111 (1961).

Schanne, J., and Hannan, W., Use of a phase-locked oscillator in PSK demodulators. *RCA Eng.* **6**, 27–28 (1961).

Weaver, C. S., Thresholds and tracking ranges in phase-locked loops. *IRE Trans.* **SET-7**, No. 3, 60–70 (1961).

1962

Booton, R. C., Jr., Demodulation of wideband frequency modulation utilizing phase-lock technique. *Proc. Nat. Telemetering Conf. Washington, D.C., May 1962*.

Byrne, C. J., Properties and design of the phase-controlled oscillator with a sawtooth comparator. *Bell Syst. Tech. J.* **41**, 559–602 (1962).

Cahn, C. R., Piecewise linear analysis of phase-lock loops. *IRE Trans.* **SET-8**, No. 1, 8–13 (1962).

Chang. S. S. L., and Harris, B., An optimum self-synchronized Communication system. *Trans. Amer. Inst. Elec. Eng. Part 1.* **81**, 110–116 (1962).

Debey, A. L., and Richard, V. W., The doploc dark satellite tracking system. *Proc. Army Sci. Conf., West Point, New York, June 1962*, **1**, *September 1962*, 199–210.

Develet, J. A., Coherent FDM/FM telephone communication. *Proc. IRE* **50**, No. 9, 1957–1966 (1962).

Frazier, J. P., and Page, J., Phase-lock loop frequency acquisition study. *IRE Trans.* **SET-8**, 210–227 (1962).

Goldstein, A. J., Analysis of the phase-controlled loop with a sawtooth comparator. *Bell Syst. Tech. J.* **41**, 603–633 (1962).

Runyan, R., A., Technique in the application of phase lock demodulators to data processing. *Proc. Nat. Telemetering Conf. May 1962*, **1**, p. 10.

1963

Benjaminson, A., Phase-locking microwave oscillators to improve stability and frequency modulation. *Microwave J.* **6**, 88–92 (1963).
Develet, J. A., An analytic approximation of phase-lock receiver threshold. *IEEE Trans.* **SET-9**, 9–12 (1963).
Develet, J. A., The influence of time delay on second-order phase-lock loop acquisition range. *Proc. Int. Telemetering Conf.*, *London, England, September 1963*, **1**, 432–437.
Develet, J. A., A threshold criterion for phase-lock demodulation. *Proc. IEEE* **51**, No. 2, 349–356 (1963). [Correction: *Proc. IEEE* **51**, 580 (1963).]
Frank, J., Newton, A., and Pugliese, L., Solid state, phase-locked, tunable UHF power oscillator. *Proc. Annu. East Coast Conf. Aerospace and Navigational Electron.*, *10th, Baltimore, Maryland, October 1963*, 3.3.2 to 3.3.2–8.
Holzmann, E. G., Pulsed phase-lock loops. *Joint Automat. Control Conf.*, *4th, Univ. of Minnesota, Minneapolis, Minnesota, June 1963*, 398–403.
Kapranov, M. V., Method of calculating the locking range of a phase sensitive automatic frequency control system. *Telecommun. Radio Eng. USSR* **17**, Part I 13–22 (1963).
Laughlin, C. R., The diversity-locked loop—A coherent combiner. *IEEE Trans.* **SET-9**, 84–92 (1963).
Nelson, W. L., PLL design for coherent angle-error detection in telstar satellite tracking system. *Bell Syst. Tech. J.* **42**, 1941–1975 (1963).
Schilling D. L., The response of an automatic phase control system to FM signals and noise. *Proc. IEEE* **51**, 1306–1315 (1963).
Uyeda, H., Accuracy of frequency comparison. *J. Radio Res. Lab.* **10**, 335–345 (1963).
Van Trees, H. L., A lower bound on stability in phase-locked loops. *Inform. Contr.* **6**, 195–212 (1963).
Van Trees, H. L., Functional techniques for the analysis of the nonlinear behavior of phase-locked loops. *WESCON, San Francisco, California, August 1963*.
Viterbi, A. J., Phase-locked loop dynamics in the presence of noise by Fokker-Planck techniques. *Proc. IEEE* **51**, No. 12, 1737–1753 (1963).
Warren, W. B., Tracking notch filter for the rejection of CW interference. *Proc. Tri-Service Conf. Electromagn. Compatibility, Chicago, Illinois, 9th, October 1963*, 326–339.
Young, J. R., Reamer, E. D., and Craft, R., Detection and measurement of cycle skipping and phase offsets in frequency multiplying phase-lock filters. *IEEE Nat. Space Electron. Symp.*, *1963*, p. 14.

1964

Altman, F. J., Power control for Comsat multiple access. *Rec. Int. Space Electron. Symp., Las Vegas, Nevada, October 1964*, 6-E-1 to 6-E-5.
Baker, T. S., Synchronization of phase-locked loops. *NEREM Rec.* 44–45, Boston, Mass., Nov. (1964).
Balodis, M., Laboratory comparison of tanlock and phase lock receivers. *Proc. Nat. Telemetering Conf. Los Angeles, California, June 1964*, p. 11.
Barry, J. G., Dalrymple, G. F., Fielding, J. C., Goldstein, B. S., and Higgins, W. F., A proposed spacecraft to earth communications link-the unified carrier approach. *IEEE Trans.* **CE-83**, 593–603 (1964).
Frankle, J., Lefrak, F., Mehlman, S., and Newton, A., Phase locked FM demodulator for 600 channel FDM. *Conv. Mil. Electron. 8th, Washington, D.C., September 1964*.
Frenkel, G., Oscillator stability and the second order phase-locked loop. *IEEE Trans.* **SET-10**, 65–69 (1964).

Golay, M. J. E., Normalized equations of the regenerative oscillator-noise, phase-locking, and pulling. *Proc. IEEE* **52**, 1311–1330 (1964).

Gupta, S. C., Transient analysis of a phase-locked loop optimized for a frequency ramp input. *IEEE Trans.* **SET-10**, 79–84 (1964).

Hartl, P., The phase locked loop principle and its application to communication receivers for space flight. *Raumfahrtforschung* **8**, 55–64 (1964).

Heinemann, H. N., Threshold extension applied to single channel FM receivers. *Conf. Proc. Int. Conv. Mil. Electron. 8th, Washington, D.C., September 1964.*

Holtsman, J. M., and Rue, A. K., Regions of asymptotic stability for phase-lock loops. *IEEE Trans.* **SET-10** 45–46 (1964).

Sanneman, R. W., and Rowbotham, J. R., Unlock characteristics of the optimum type II phase-locked loop. *IEEE Trans.* **ANE-11**, 15–24 (1964).

Sassler, M, A phase-locked demodulator for multichannel telephone traffic from satellites. *Proc. Nat. Electron. Conf.* **20**, 481–485 (1964).

Sassler, M., and Surenian, R., Communication receiver for satellite ground station. *Elec. Commun.* **39**, No. 1, 89–97 (1964).

Schrader, J. H., A phase-lock receiver for the arraying of independently directed antennas. *IEEE Trans.* **AP-12**, 155–160 (1964).

Shakhgil'dyan, V. V., and Liakhovkin, A. A., Filtering of a monochromatic signal by a phase-locked oscillator. *Elektrosvyaz* **18**, 11–18 (1964).

Svoboda, D. E., A phase-locked receiving array for high-frequency communications use. *IEEE Trans.* **AP-12**, 207–215 (1964).

Van Trees, H. L., Functional techniques for the analysis of the nonlinear behavior of phase-locked loops. *Proc. IEEE* **52**, 894–911 (1964).

Viterbi, A. J., and Cahn, C. R., Optimum coherent phase and frequency demodulation of a class of modulating spectra. *IEEE Trans.* **SET-10**, 95–102 (1964).

Williams, W. J., Selection of phase sensitive detectors for space radar. *IEEE Trans.* **ANE-11**, 230–234 (1964).

1965

Alexander, P. H., and Kalra, S. N., Unlock characteristics of a phase-locked loop. *Proc. IEEE* (letters) p. 1138 (1965).

Bakaev, Iu. N., Synchronizing properties of a phase type AFC of the third order. *Radio Eng. Electron. Phys. USSR* **10**, No, 6, 926–929 (1965).

Bakaev, Iu. N., Synchronizing properties of a third-order phase-locked oscillator system. *Radiotekh. Elektron.* **10**, 1083–1087 (1965).

Dye, R. A., Phase-lock loop swept-frequency synchronization analysis. *Rec. Int. Symp. Space Electron., Miami Beach, Florida November 1965.*

Enloe, L. H., and Rodda, J. L., Laser phase-locked loop. *Proc. IEEE* **53**, 165–166 (1965).

Graefe, P. W. U., and Loh, N. K., Stability criteria of phase-locked loops. *Proc. Annu. Allerton Conf. Circuit and System Theory, 3rd, October 1965.*

Gupta, S. C., and Solem, R. J., Optimum filters for second- and third-order phase-locked loops by an error-function criterion. *IEEE Trans.* **SET-11**, 54–62 (1965).

Hiroshige, K., A simple technique for improving the pull-in capability of phase-lock loops. *IEEE Trans.* **SET-11**, 40–46 (1965).

Jankovich, J. L., Phase-locked interferometer. *Int. Astronaut. Congr. Proc. 16th, Athens, Greece, September 1965*, p. 33.

Kine, A. J., Jr., and Moore, W. C., Concepts and computational techniques used in the design of phase-lock circuits. *Proc. Nat. Telemetering Conf. Houston, Texas, April 1965*, p. 39–45.

Long, L. L., and Rutledge, R. B., A digital computer simulation for comparative phase-locked loop analysis. *Proc. Nat. Electron. Conf.* **21**, 319–324 (1965).

Lyness, H. L., Oscillator stability in phase-locked loops. *Electro-Technol. New York*, **75**, 34–36 (1965).

Magill, D. T., Noise theory of tracking cross-correlators. *IEEE Int. Conv. Rec.* **13**, Part 2, 158–167 (1965).

Meer, S. A., A class of Wiener filters useful in PLL applications. *Proc. IEEE* **53**, No. 12, 2121 (1965).

Moschytz, G. S., Miniaturized RC filters using phase-locked loop. *Bell Syst. Tech. J.* **44**, 823–870 (1965).

Nesvizhskii, Iu. B., Pulse-phase locked automatic frequency control. *Telecommun. Radio Eng. Part 2* **20**, 95–102. (1965).

Nikitin, N. P., Probability of signal acquisition by a phase-locked oscillator system operating in the frequency search mode. *Radiotekhnika (Kiev)* **8**, 696–703 (1965).

Rey, T. J., Further on the phase-locked loop in the presence of noise. *Proc. IEEE* **53**, 494–495 (1965).

Schilling, D. L., Abrams, B. S., Oberst, J. F., and Berkoff, M., Phase locked loop threshold. *Proc. IEEE* (letters) **53**, 1673 (1965).

Schilling, D. L., Billig, J., and Kermisch, D., Error rates in FSK using the phase locked loop demodulator. *IEEE Annu. Commun. Conv. 1st, Boulder, Colorado, June 1965*.

Shakhgil'dyan, V. V., Determination of capture band of a phase lock AFC system when the reference signal is phase modulated. *Radio Eng. Electron. Phys.* **11**, No. 10 (1965).

Yen, C.-S. Phase-locked sampling instruments. *IEEE Trans.* **IM-14**, 64–68 (1965).

1966

Arnold, J., and Leclerq, M., Measurements of phase and amplitude with the aid of phase-locked filters. *l'Echo Rech.* April, 70–75 (1966).

Charles, F. J., and Lindsey, W. C., Some analytical and Experimental PLL results for low SNR. *Proc. IEEE* **54**, No. 9, 1152–1166 (1966).

Chen, C., An analysis of performance of phase-locked loop susceptible to interference and noise. Digest of Technical Papers, *IEEE Int. Commun. Conf., Philadelphia, Pennsylvania, June 1966*, 212–213.

Ellis, M. E., and Sage, G. F., Optimum control loops design for synchronous FM tone demodulators. *Proc. Nat. Telemetering Conf., Boston, Massachusetts, May 1966*, 238–242.

Gagliardi, R. M., Error probabilities in PCM/FM with phase lock loop discriminators. *IEEE Trans.* **AES-2**, 608–611 (1966).

Hartl, P., Das Prinzip des "Phase-locked loop" und seine Anwendung in Nachrichten-Empfaengern fuer die raumfahrt. *Raumfahrtforschung* **6**, 55–64 (1966).

Jelonek, F. J., and Khanu, A. H., Synchronized oscillatory systems with nonuniform gain in the feedback loop. *Proc. Brit. Inst. Elec. Eng.* **113**, No. 11, 1769–1774 (1966).

Kline, A. J., and Moore, W. C., Phase locked loop in space communications. *Instrum. Contr. Syst.* **39**, No. 9, 131–137 (1966).

Ku, Y. H., and Su, J. C. C., Comparison of variances evaluated by Kolmogorov and Volterra techniques for phase locked loop subjected to white gaussian input. *Proc. IEEE* **54**, 900–901 (1966).

Lindsey, W. C., Phase-shift-keyed signal detection with noisy reference signals. *IEEE Trans.* **AES-2**, No. 4, 393–401 (1966).

Lindsey, W. C., and Charles, F. J., A model distribution for the phase error in second order PLL's. *IEEE Trans.* **COM-14**, 662–664 (October 1966).

Long, L. L., Rutledge, R. B., and Wallace, N. D., A simulation study of phase-locked loop dynamics in the presence of noise. *Rec. Region Six Annu. Conf. Tucson, Arizona, April 1966*, 705–716.

Meer, S. A., Analysis of phase-locked loop acquisition—A quasi stationary approach. *IEEE Int. Conv. Rec.* **14**, part 7, 85–106 (1966).

Ongano, D., and Rocca, F., Microwave, nearly-optimum phase-lock demodulator. *Alta Freq.* **35**, No. 8, 845–855 (1966).

Rey, T. J., Stability of an APC system for frequency division. *Proc. IEEE* (letters) **54** 73–74 (January 1966).

Ridgway, R., A method of calculating phase-lock threshold. *Proc. IEEE* (letters) **54**, 2024–2025 (1966).

Sanneman, R. W., and Gupta, S. C., Optimum strategies for minimum-time frequency transitions in phase-locked loops. *IEEE Trans.* **AES-2**, 570–581 (1966).

Shakhgil'dyan, V. V., and Ignatov, Yu. F., Effect of noise on the accuracy of operation of a phase-locked automatic frequency control system. *Telecommun. Radio Eng. USSR* Part 1, **20**, No. 3, 32–38 (1966).

Shaktarin, B. I., Filtering ability of a phase-locked AFC system. *Telecommun. Radio Eng.* Part 1, *USSR* **20**, 20–25 (1966).

Shakhtarin, B. I., and Shchepkin, Yu. N., Experimental study of the action of fluctuation noise on a phase-locked AFC system. *Telecommun. Radio Eng. USSR* Part 1, **20**, 15–19 (1966).

Smith, B. M., The phase-lock loop with filter: Frequency of skipping cycles. *Proc. IEEE* (letters) **54**, 296 (1966).

Smith, B. M., A semi-empirical approach to the PLL threshold. *IEEE Trans.* **AES-2**, No. 4, 463–468 (1966).

Smith, B. M., Phase-locked loop threshold. *Proc. IEEE* (letters) **54**, 810–811 (1966). [Comments: *Proc. IEEE* **55**, 82–83 (1967).]

Splitt, F. G., Design and analysis of a linear PLL of wide dynamic range. *IEEE Trans.* **COM-14**, 432–440 (1966).

Thomas, C. M., Optimization of phase-lock demodulator for single-channel voice. *Proc. Int. Commun. Conf. Philadelphia, Pennsylvania, June 1966*; also *Microwave J.* **10**, No. 7, 43–47 (1967).

Vaughan, G. R., and Osborne, E., Phase locked phase modulator. Digest of technical papers, *IEEE Int. Commun. Conf., June 1966*, 206–207.

1967

Anderson, D. R., and Luh, Y. Y. S., An analysis of high order phase-locked loop behavior in the presence of white noise. *Int. Commun. Conf. Minneapolis, Minnesota*, Digest of technical papers, *June 1967*, p. 139.

Biswas, B. N., and Datta, G., Tunable compound phase-locked demodulator. *Proc. IEEE* **55**, 2044–2045 (1967).

Britt, C. L., and Palmer, D. F., Effects of CW interference on narrowband second order PLL. *IEEE Trans.* **AES-3**, No. 1, 123–135 (1967).

Glenn, A. B., Mars voyager-lander direct link communications system. *RCA Eng.* **12**, 70–74 (1967).

Guers, K., Modulation and mode locking of the continuous ruby laser. *IEEE J. Quantum Electron.* **3**, 175–180 (1967).

Ignatov, Iu. F., and Shakhgil,dyan, V. V., Synchronization failure in a phase lock automatic frequency control system. *Elektrosvyaz* **21**, 17–22 (1967).

Judd, L. F., Sample data analysis of digital phase-locked loops. *SWIEEECO Rec., Annu. Southwestern Conf. and Exhibition, April 1967*, 22-4-1–22-4-6.
Lewis, P. H., and Weingarten, W. E., A comparison of second, third, and fourth order phase locked loops. *IEEE Trans.* **AES-3**, No. 4, pp. 720 ff (1967).
Lindsey, W. C., and Tausworthe, R. C., A survey of phase-locked loop theory. *IEEE Int. Commun. Conf., Minneapolis, Minnesota, June 1967*.
McKay, G. A., An extended phase detector for phase-locked receivers. *Region 3ᵉ Meeting, Jackson, Mississippi, April 1967*.
Meer, S. A., A generalized analysis for the acquisition time and pull-in range of phase-locked loops. *Conf. Frequency Generation and Contr. Radio Syst., London, England, May 1967*, 144–148.
Murphy, J. V., Frequency measurement using the phase-controlled oscillator. *Proc. IEEE* **55**, No. 7, 1144–1153 (1967).
Osatake, T., and Fujii, A., A study on FM reception by tracking filters. *Electron. Commun. Jap.* **50**, No. 6, 100–108 (1967).
Ridgway, R. I., and Carter, J. E., More comments on the phase-locked loop threshold. *Proc. IEEE (letters)* **55**, No. 8, 1531–1533 (1967).
Rocca, F., Some properties of optimum unconditionally stable phase-lock demodulators. *Alta Freq.* **36**, 424–442 (1967).
Schilling, D. L., and Smirlock, M., Intermodulation distortion of a phase locked loop demodulator. *IEEE Trans.* **COM-15**, No. 2, 222–228 (1967).
Schilling, D. L., and Smith, B. M., Comments on phase-locked loop threshold. *Proc. IEEE* (letters) **55**, 82–83 (1967).
Stiffler, J. J., On the selection of signals for phase-locked loops. Digest of papers, *Int. Conf. Commun. Minneapolis, Minnesota, June 1967*.
Tausworthe, R. C., A method for calculating phase-locked loop performance near threshold. *IEEE Trans.* **COM-15**, No. 4, 502–506 (1967).
Tausworthe, R. C., Cycle slipping in phase-locked loops. *IEEE Trans.* **COM-15**, 3, 417–421 (1967).
Uhran, J. J., and Lindenlaub, J. C., Effects of a class of phase comparators on the threshold and lock range of phase lock loop systems. *Proc. Int. Conf. Commun. 3rd, Minneapolis, Minnesota, June 1967*.

1968

Anderson, T. O., and Lindsey, W. C., Digital-data transition tracking loops. *Proc. Int. Telemetering Conf., Los Angeles, California, October 1968*, 259–271.
Bambini A., and Burlamacchi, P., Phase locking of a multimode gas laser by means of low-frequency cavity-length modulation. *J. Appl. Phys.* **39**, 4864–4865 (1968).
Carden, F. F., Hintz, T. B., and Kelly, L. R., The FDM demodulating characteristics of non-linear phase-locked loops. *IEEE Nat. Telemetering Conf. Houston, Texas, April 1968*.
Cleland, L. L., and Leon, B. J., Phase-locked loops for large signal tracking. *Proc. Midwest Symp. Circuit Theory 11th, University of Notre Dame, Indiana, May 1968*, 2–9.
Didday, R. L., and Lindsey, W. C., Subcarrier tracking methods and communication system design. *IEEE Trans.* **COM-16**, No. 4, 541–550 (1968).
Gupta, S. C., Bayless, J. W., and Hummels, D. R., Threshold investigation of phase-locked discriminators. *IEEE Trans.* **AES-4**, No. 6, 855–863 (1968).
Gupta, S. C., On optimum digital phase-locked loops. *IEEE Trans.* **COM-16**, No. 2, 340–344 (1968).

Hess, D. T., Cycle slipping in a first-order phase-locked loop. *IEEE Trans.* **COM-16**, No. 2, 255–260 (1968).

Klapper, J., Aaronson, G., Acampora, A., Frankle, J., and McLaughlin, P., Error rates with angular feedback demodulators. *IEEE Nat. Telemetering Conf., Houston. Texas, April 1968.*

Ludwig, D., A general solution for the shortest acquisition time in Type-II phase-lock loops. *IEEE Trans.* (letters) **AES-4**, 639–640 (1968).

Natali, F. D., and Walbesser, W. J., Interference rejection in a phase-locked loop with decision feedback. *EASCON Rec., Washington, D.C., September 1968*, 187–192.

Osatake, T., Fujii, A., and Akutagawa, T., A study on a parametric amplifier as a tracking filter. *Electron. Commun. Jap.* **51B**, No. 6, 63–69 (1968).

Osborne, P., and Schilling, D. L., Threshold performance of phase-locked loop demodulators. *Int. Commun. Conf., Philadelphia, Pennsylvania, June 1968.*

Osborne, P. W., and Schilling, D. L., Expected number of spikes of phase locked loop demodulators. *Proc. Int. Telemetering Conf., Los Angeles, California, October 1968*, 248–258.

Pasternak, G., and Whalin, R. L., Analysis and synthesis of a digital phase-locked loop for FM demodulation. *Bell Syst. Tech. J.* **47**, No. 10, 2207–2239 (1968).

Schmueckle, W., Optimum demodulation of disturbed frequency-modulated signals. *Nachrichtentech. Z.* **21**, 464–470 (1968).

Shaft, P. D., and Dorf. R. C., Minimization of communication-signal acquisition time in tracking loops. *IEEE Trans.* **COM-16**, No. 3, 495–499 (1968).

Stiffler, J. J., On the selection of signals for phase-locked loops. *IEEE Trans.* **COM-16**, No. 2, 239–244 (1968).

Uhran, J. J., Jr., Cycle-slipping effects on the output signal of a phase-locked demodulator. *Proc. IEEE* **56**, No. 1, (letters) 80–81 (1968).

Woodbury, J. R., Phase-locked loop pull-in range. *IEEE Trans.* **COM-16**, 184–186 (1968). [Correction: 495 (1968).]

1969

Carassa, F., and Rocca, F., Advances in phase-lock demodulation. *Int. Commun. Conf., Boulder, Colorado, June 1969*, 12.9–12.14.

Cessna, J. R., Steady state and transient analysis of a digital bit-synchronization phase-locked loop. *Int. Commun. Conf., June 1969.*

Chiang, C.-C., The lock-in range of an automatic phase control system with nonlinear reactance tube characteristics. *Electron. in Mainland China* Part II, 99–108 (1969).

Clarke, J. M., Otero, R. J., and Wanbaugh, W. C., Pulse interference effects in a phase-lock loop. *Rec. Electromagn. Compatibility Symp. 11th, Asbury Park, New Jersey, June 1969*, pp. 207–214.

Evtyanov, S. I., and Snedkova, V. K., Dependence of the hold-in range of a phase lock AFC system on the characteristics of the phase detector with proportional plus integral filter. *Telecommun. Radio Eng.* Part 2, **24**, No. 4 (1969).

Hess, D., Comments on threshold investigation of phase-locked discriminators. *IEEE Trans.* (correspondence) **AES-5**, 877–878 (1969).

Hoffman, E., and Schilling, D. L., Threshold of the FMFB. *Proc. Int. Commun. Conf., Boulder, Colorado, 1969.*

Hussein, F. H., and Rhee, M. Y., The phase error statistics for second order PLL and design of an optimum decision unit for space communications. *Nat. Telemetering Conf., Washington, D.C., April 1969.*

Klapper, J., and Creutz, J., Minimization of cycle slipping rate in a first order PLL with frequency offset. *Asilomar Conf. on Circuits and Systems, 3rd, Pacific Grove, California, December 1969.*

Yang, K.-H., A study of capture bandwidth of phase locked loops with a nonlinear integrating filter. *Electron. Mainland China* Part II 29–42 (1969).

Lindgren, A. G., Pinkos, R. F., and Berube, R. H., Noise dynamics of the phase-locked loop with signal clipping. *IEEE Trans.* **AES-5,** No. 1, 66–76 (1969).

Lindsey, W. C., Phase density distribution of phase-locked loops in cascade. *IEEE Trans.* **COM-17,** No. 4, 503 (1969).

Natali, F. D., and Walbesser, W. J., Phase-locked loop detection of binary PSK signals utilizing decision feedback. *IEEE Trans.* **AES-5,** No. 1, 83–90 (1969).

Protonotarios, E. N., Pull-in performance of a piecewise linear phase-locked loop. *IEEE Trans.* **AES-5,** No. 3, 376–386 (1969).

Rey, T. J., Bozzoni, E., and Mengali, U., Comments on "Comparison between the oscillating limiter and the first-order phase-locked loop." *Proc. IEEE* (letters) **57,** No. 4, 726 (1969).

Thomas, C. M., Study charts phase-locked demodulator distortion in TV, multichannel telephony. *Commun. Designer's Digest* 34–38 (1969).

Viskanta, V. Z., Compound phase-locked loop receiver. *Nat. Telemetering Conf., Washington D.C., April 1969.*

Woodbury, J. R., Phase-locked loop pull-in range. *IEEE Trans.* **COM-16,** 184–186 (1968). [Comments: *IEEE Trans.* **COM-17,** 1, 89–90 (1969).]

1970

Carden, F., and Stewart, I. A., Some solutions and stability criteria for the phase lock loop equation. *Mervin J. Kelly Commun. Conf., Univ. of Missouri-Rolla, October 1970.*

Cleland, L. L., Experimental results on phase locked loops with added nonlinearities. *Mervin J. Kelly Commun. Conf., Univ. of Missouri-Rolla, October 1970.*

Dominiak, K. E., and Pickholtz, R. L., Transient behavior of a phase-locked loop in the presence of noise. *IEEE Trans.* **COM-18,** No. 4, 452–456 (1970).

Holmes, J. K., On a solution to the second-order phase-locked loop. *IEEE Trans.* **COM-18,** No. 2, 119–126 (1970).

Holmes, J. K., On a solution to a digital first order phase lock loop. *Mervin J. Kelly Commun. Conf., Univ. of Missouri-Rolla, October 1970.*

Jones, T. J., Phase-locked loop optimization based upon the mean-square minimization of error and error rate. *Mervin J. Kelly Commun. Conf., Univ. of Missouri-Rolla, October 1970.*

Olsen, D. P., and Lindenlaub, J. C., A phase lock loop with an extended linear range phase detector. *Mervin J. Kelly Commun. Conf., of Univ. Missouri-Rolla, October 1970.*

Protonotarios, E. N., The effect of phase jitter on the performance of a first-order phase-locked loop. *IEEE Trans.* **COM-18,** No. 1, 74–76 (1970).

Rhee, M. Y., Lindauer, C. M., and Gohain, P. K., State modeling of phase-locked loops with random modulation and additive noise. *Mervin J. Kelly Commun. Conf., Univ. of Missouri-Rolla, October 1970.*

Schuchman, L., Time to cycle slip in first and second order phase lock loops. *Int. Commun. Conf., San Francisco, California, June 1970,* 34.1–34.9.

Simon, M. K., On the equivalence in performance of several phase-locked loop configurations. *IEEE Trans.* **COM-18,** No. 4, 449–452 (1970).

3. FMFB-Papers

1939

Carson, J. R., Frequency modulation: Theory of the feedback receiving circuit. *Bell Syst. Tech. J.* **18**, No. 3, 395–403 (1939)

Chaffee, J. G., The application of negative feedback to frequency modulation systems. *Proc. IRE* **27**, 317–331 (1939).

1949

Panter, P. F., and Dite, W., Application of negative feedback to frequency modulation systems. *Elec. Commun. (London)* **26**, 173 (1949).

1958

Buxton, A J., and Felix, M. O., The reduction of threshold by use of frequency compression. *Proc. Inst. Elect. Eng. Suppl.* Part B, 117–121 (1958).

Felix, M. O., and Buxton, A. J., The performance of FM scatter systems using frequency compression. *Proc. Nat. Electron. Conf.*, **14**, 1029–1043 (1958).

1959

Kantor, L. Y., Concerning the noiseproof feature of an FM receiver with frequency feedback. *Telecommunications (Russian)*, No. 10, 1098–1105 (1959).

1961

Ruthroff, C. L., FM demodulators with negative feedback. *Bell Syst. Tech. J.* **40**, 1149–1156 (1961).

1962

Baghdady, E. J., The theory of FM demodulation with frequency-compressive feedback. *IRE Trans.* **CS-10**, No. 3, 226–245 (1962).

Enloe, L. H., Decreasing the threshold in FM by frequency feedback. *Proc. IRE* **50**, 18–30 (1962).

Enloe, L. H., The synthesis of frequency feedback demodulators. *Proc. Nat. Electron. Conf.*, **18**, 477 (1962).

Gagliardi, R. M., The design and capabilities of feedback FM receivers. *WESCON Conv. Rec.* Part 7, **6**, 9 p. (1962).

Rothroff, C. L., and Bodtman, W. F., Design and performance of a broadband FM demodulator with frequency compression. *Proc. IRE* **50**, 2436–2445 (1962).

1963

Develet, J. A., Statistical design and performance of high-sensitivity frequency-feedback receivers. *IEEE Trans.* **MIL-7**, No. 4, (1963).

Downing, J. J., Threshold suppression by FMFB. *Proc. IEEE* (letters) **51**, 387–388 (1963).

Furuya, T., Ito, S., Morita, M., and So, M., High sensitive wide band FM demodulator. *Proc. Int. Symp. Space Technol. Sci. 5th, Tokyo, Japan, September 1963*, 803–813.

Giger, A. J., and Chaffee, J. G., The FM demodulator with negative feedback. *Bell Syst. Tech. J.* **42,** No. 4, Part I, 1109 (1963).

Kallus, S., Rabinovici, B., and Newton, A., Fitting a wide-band signal into a narrow-band receiver. *Electronics* **36,** 47–49 (1963).

Spilker, J. J., Analysis of FM discriminator with frequency feedback. *Proc. IEEE* (letters) **51,** 233–234 (1963).

Wright, J. C., and Blair, W. L., FM feedback stabilizes airborne solid-state VHF transponder. *Electronics* **36,** 66–68 (1963).

1964

Baghdady, E. J., and Enloe, L. H., Decreasing the threshold in FM by frequency feedback. *Proc. IEEE* (letters) **52,** 1039–1044 (1964).

Davis, B. R., Factors affecting the threshold of feedback FM detectors. *IEEE Trans.* **SET-10,** No. 3, 90–94 (1964).

1966

Frutiger, P., Noise in FM receivers with negative frequency feedback. *Proc. IEEE* **54,** No. 11, 1506–1520 (1966). [Correction: *Proc. IEEE* **55,** No. 10, 1674, (1967).]

Kobayashi, S., and Saito, S., Optimal design for frequency compression demodulator. *Electron. Commun. Jap.* **49,** No. 2, 59 (1966).

Lefrak, F., Moore, H., and Newton, A., An FDM-FM feedback demodulator for tropo and satellite relay communications. *U.S. Seminar on Satellite Commun. Earth Station Technol. Washington, D.C., 1966.* Published in Seminar Proc.

Lefrak, F., Moore, H., Newton, A., and Ozolins, L., The frequency feedback discriminator for the lunar orbiter station. *Space Electron. Symp., Miami Beach, Florida, 1965; RCA Rev.* **27,** No. 4, 563 (1966).

1967

Wojnar, A., Noise and threshold in FM systems. *Proc. IEEE* **55,** 1639–1640 (1967).

1968

Anema, S. L., Malinowski, M., and Marchese, J. F., An FMFB demodulator for satellite communications. *Proc. Nat. Electron. Conf. Chicago, Illinois, December 1968,* pp. 450–455.

Bayless, J. W., Concerning noise in FM receivers with negative frequency feedback. *Proc. IEEE* (letters) **56,** 341 (1968).

Roberts, J. H., Frequency-feedback receiver as a low-threshold demodulator in FM/FDM satellite systems. *Proc. IEE (Brit.)* **115,** 1607–1618 (1968).

Shimbo, O., and Loo, C., Noise in FM receivers with negative feedback. *Proc. IEEE* (letters) **56,** 1372 (1968).

1969

Hoffman, E., and Schilling, D. L., Higher order frequency demodulators with feedback. *Proc. Nat. Electron. Conf., Chicago, Illinois, December 1969,* **25.**

Steinbrecher, E., An FM feedback demodulator model for determining performance limits. *Proc. Nat. Electron. Conf., Chicago, Illinois, December 1969,* **25.**

1970

Gerber, M. J., A universal threshold extending FMFB demodulator. *Int. Commun. Conf.*, Boulder, Colorado, June 1969, pp. 12.1–12.8; *IEEE Trans.* **COM-18,** No. 4, 276–280 (1970).

Hoffman, E., and Schilling, D. L., Intermodulation distortion in the FMFB. *Mervin J. Kelly Commun. Conf.*, Univ. of Missouri, Rolla, Missouri, October 1970.

4. Multiple Loops, Comparisons, etc.,—Papers

1960

Morita, M., and Ito, S., High sensitivity receiving system for frequency modulated wave. *IRE Int. Conv. Rec.* **8,** Part 5, 227 (1960).

1961

Spilker, J. J., Jr., Threshold comparison of phase-lock frequency lock and maximum likelihood types of FM discriminators. *IRE Wescon Conv. Rec.* Paper 14/2, (1961).

Spilker, J. J., Jr., and Magill, D. T., The delay-lock discriminator—An optimum tracking device. *Proc. IRE* **49,** 1403–1416 (1961).

1962

Gagliardi, R. M., The design and capabilities of feedback FM receivers. *Western Electron. Show and Conv.*, Los Angeles, 1962, p. 10.

Heitzman, R. E., A study of the threshold power requirements of FMFB receivers. *IRE Trans.* **SET-8,** 249–256 (1962).

O'Sullivan, M. R., Tracking systems employing the delay-lock discriminator. *IEEE Trans.* **SET-8,** 1–7 (1962).

Robinson, L. M., Tanlock: A phase lock loop of extended tracking capability. *Proc. Nat. Winter Conf. Mil. Electron.*, Los Angeles, California, February 1962, pp. 396–421.

Slepian, D., The threshold effect in modulation systems that expand bandwidth. *IRE Trans.* **IT-8,** No. 5, 122–127 (1962).

Weis, W. G., and Evans, M., Application of the delay-lock discriminator to the satellite rendezvous problem. *IRE East Coast Conf. Aerosp. Navigational Electron. 9th*, Baltimore, Maryland, October 1962.

1963

Gagliardi, R. M., Transmitter power reduction with frequency tracking FM receivers. *IEEE Trans.* **SET-9,** 18–25 (1963).

Morita, M., Ito, S., Furuya, T., and So, M., High sensitive wide band FM demodulator. *Proc. Int. Symp. Space Technol. Sci., 5th, Tokyo, Japan, September 1963*, pp. 803–813.

Spilker, J. J., Jr., Delay lock tracking of binary signals. *IEEE Trans.* **SET-9,** 1–8 (1963).

Van Trees, H. L., The structure of efficient demodulators for multidimensional phase modulated signals. *IEEE Trans.* **CS-11,** No. 3, 261–271 (1963).

1964

Akima, H., Theoretical studies on signal to noise characteristics of an FM system. *IEEE Trans.* **SET-10**, 90–94 (1964).

Baghdady, E. J., Theoretical comparison of exponent demodulation by phase-lock and frequency-compressive feedback techniques. *IEEE Int. Conv. Rec., New York, New York*, **12**, Part. 6, 402–421 (1964).

Balodis, M., Laboratory comparison of tanlock and phase-lock receivers. *Nat. Telemetering Conf., Los Angeles, California, 1964*.

Bykov, V. L., Threshold in frequency modulation and methods of reducing it. *Elektrosvyaz* **18**, 22–24 (1964); *Telecommun. Radio Eng.* Part I, **18**, 17–26 (1964).

Darlington, S., Demodulation of wideband low-power FM signals. *Bell Syst. Tech. J.* **43**, No. 1, Part 2, 339–374 (1964).

Davis, B. R., Equivalent centre-frequency amplifiers. *Radio Electron. Eng.* **28**, No. 6, 381–388 (1964).

Powell, N. R., Controlled parameter phase-feedback F. M. demodulation. *Rec. Int. Space Electron. Symp., Las Vegas, Nevada., October 1964*.

Schilling, D. L., and Billig, J., On the threshold extension capabilities of the PLL and the FMFB. *Proc. IEEE* (letters) **52**, 621–622 (1964).

1965

Abbate, J. V., and Schilling, D. L., Estimation of random phase- and frequency-modulating signals using a Bayes estimator. *IEEE Trans.* **IT-11**, No. 3, 462–463 (1965).

Benes, V. F., Index reduction of FM waves by feedback and power law nonlinearities. *Bell Syst. Tech. J.* **44**, 589–601 (1965).

Bykov, V. L., Improving the threshold properties of an FM receiver by means of a frequency divider. *Telecommun. Radio Eng.* Part I, **19**, No. 10 (1965).

Castellani, V., Pent, M., and Zeglio, L. E., Model of a low-threshold FM demodulator with bandwidth subdivision. *Conv. Int. Delle Comunicazioni, Inst. Int. Delle Comunicazioni, Genoa, Italy, October 1965* (in Italian).

Dye, R. A., Performance of the delay-lock tracking discriminator with binary signals. *Milecon/9, Conf. Mil. Electron. Washington, D.C. September 1965*.

Schilling, D. L., and Billig, J., A comparison of the threshold performance of the frequency demodulator using feedback and the phase locked loop. *Rec. Int. Symp. Space Electron., Miami Beach, Florida, November 1965*, 3–E1–3-E9.

1966

Acampora, A., and Newton, A., Use of phase subtraction to extend the range of a phase-locked demodulator. *RCA Rev.* **27**, No. 4, 577–599 (1966).

Afanasyev, Yu. A., and Kantor, L. Ya., Control circuit compensation in an FM-receiver with a tracking filter. *Telecommun. Radio Eng.* Part 2, **21**, 87–93 (1966).

Bykov, V. L., The limit to the improvement of PM and FM receiver threshold properties. *Telecommun. Radio Eng.* **20**, Part 1, No. 5, 41–46 (1966).

Carassa, F., Recent advances in frequency modulators and demodulators. *Colloq. Microwave Commun., 3rd, Budapest, Hungary, April 1966, 37–46*.

Frankle, J., Threshold performance of analog FM demodulators. *RCA Rev.* **27**, No. 4, 521–562 (1966).

Gill, W. J., A comparison of binary delay lock tracking loop implementations. *IEEE Trans.* **AES-2**, 415–424 (1966).

Klapper, J., Demodulator threshold performance and error rates in angle-modulated digital signals. *RCA Rev.* **27**, No. 2, 226–244, (1966).
Muehldoff, E. I., FM threshold reduction by optimum filtering. *Proc. IEEE* **54**, 1972–1973 (1966).
Schilling, D. L., and Hoffman, E., Demodulation of digital signals using an FM discriminator. *Nat. Electron. Conf., Chicago, Illinois, October 1966.*
Thomas, C. M., Principles of threshold extension demodulator operation. *U.S. Seminar on Commun. Satellite Earth Station Technol., Washington, D.C., May 1966*, 531–542.

1967

Baghdady, E. J., and Wachsman, R. H., Effects of random fluctuation noise on FM and FDM/FM reception. *Proc. Int. Telemetering Conf., Washington, D.C., October 1967*, pp. 1–25.
Castellani, V., Digital computer simulation of an FM band-dividing demodulator. *Alta Freq.* **36**, 1048–1062 (1967).
Clarke, K. K., and Hess, D. T., Frequency locked loop FM demodulator. *IEEE Trans.* **COM-15**, No. 4, 518–524 (1967).
Gupta, S. C., and Bayless, J. W., Status of FM feedback in communication systems. *Proc. Eastcon, 1967; IEEE Trans. Suppl.* **AES-3**, 11–23 (1967).
McKay, G. A., An extended phase detector for phase-locked receivers. *IEEE Region 3 Meeting, Jackson, Mississippi, April 1967.*

1968

Bayless, J. W., and Gupta, S. C., Threshold extension using phase lock demodulator in a FM feedback loop. *IEEE Nat. Telemetering Conf., Houston, Texas, April 1968.*
Bloustein, J. L., and Onians, F. A., The economics of satellite communications. *Seminar on Commun. Satellite Earth Station Planning and Operation, Int. Telecommun. Union, London, England, May 1968.*
Boor, S. B., and Pelchat, M. G., Frequency feed-forward: An open loop approach for extending the threshold and linearity of FM demodulators. *Proc. Int. Telemetering Conf., Los Angeles, California, October 1968*, 513–529.
Calandrino, L., and Immovile, G., On the performance of amplitude-phase correlation FM demodulators. *Alta Freq.* (Engl. ed.) 125–131 (1968).
Fudge, R. E., Performance of a threshold extension demodulator. *Seminar on Commun. Satellite Earth Station Planning and operation, Int. Telecommun. Union, London, England, May 1968.*
Heinemann, H., Newton, A., and Frankle, J., Multiple-loop frequency-compressive feedback for angle-modulation detection. *RCA Rev.* **29**, No. 2, 252–269 (1968).
Hess, D. T., and Clarke, K. K., Quantized second order FLL scores over discriminator for FM and FSK detection. *Commun. Designer's Digest*, December (1968).
Hess, D. T., Equivalence of FM threshold extension receivers. *Int. Commun. Conf., Philadelphia, Pennsylvania, June 1968; IEEE Trans.* **COM-16**, No. 5, 746–748 (1968).
Langseth, R. E., and Lambert, R. F., Influence of bandwidth on some nonlinear transformations of a gaussian process. *IEEE Trans.* **IT-14**, No. 1, 88–93 (1968).
Lockyer, K. S., Threshold extension of an FM demodulator using a dynamic tracking filter. *Proc. IEEE Brit.* **115**, 1102–1108 (1968).
Roberts, J. H., Dynamic tracking filter as a low-threshold demodulator in FM/FDM satellite systems. *Proc. IEE Brit.* **115**, 1597–1606 (1968).

Uhran, J. J., and Lindenlaub, J. C., Experimental results for phase lock loop systems having a modified n*th* order tanlock phase detector. *IEEE Trans.* **COM-16**, No. 6, 787–795 (1968).

Unkauf, M. G., and Schulman, R. J., Experimental signal/noise-ratio comparison of the second-order phase-locked loop and the second-order frequency-locked loop. *Electron. Lett.* **4**, No. 26, 585–586 (1968).

1969

Bozzoni, E., and Mengali, U., An analysis of the performance of the oscillating limiter driven by FM signals corrupted by noise. *IEEE Trans.* **AES-5**, No. 3, 537–547 (1969).

Gray, L. F., Threshold extension demodulator performance. *Nat. Electron. Conf., Chicago, Illinois, December 1969.*

Hekimian, N. C., and Mack, W., A new FM threshold extension demodulator for FM multiplex. *EASTCON, Washington, D.C., October 1969*, pp. 53–61.

Hess, D. T., Optimization of the frequency locked loop. *Nat. Electron. Conf., Chicago, Illinois, December 1969.*

Layland, J. W., On optimal signals for phase-locked loops. *IEEE Trans.* **COM-17**, No. 5, 526–531 (1969).

Lindsey, W. C., Nonlinear analysis of generalized tracking systems. *Proc. IEEE* **57**, No. 10, 1705–1722 (1969).

Lob, W. H., The distribution of FM-discriminator click widths. *Proc. IEEE* (letters) **57**, 732–733 (1969).

Locke, F., Threshold extension technique using impulse noise elimination. *Nat. Electron. Conf., Chicago, Illinois, December 1969.*

Osborne, P. W., and Schilling, D. L., FM receiver evaluation using computer techniques. *Proc. Symp. Comput. Process. Commun., Polytech. Inst. of Brooklyn, April 1969.*

Pelchat, G. M., Boor, S. B., and Allen, D. B., Distortion in varicap FM oscillators. *IEEE Trans.* **COM-17**, No. 1, 49–53 (1969).

Proni, E., FM demodulator employing an injection locked oscillator. *Alta Freq.* (Engl. ed.) **38**, 95–103 (1969).

Stewart, T. L., and Hommond, W. M., A model distribution for a hybrid phase-locked loop. *Conf. Rec. Asilomar Conf. on Circuits and Syst., 3rd, Pacific Grove, California, December 1969*, 149–152.

Viskanta, V. Z., Compound phase-locked loop receiver. *Nat. Telemetering Conf. Washington, D.C., April 1969.*

Wang, C. C., An exact solution of injection phase-locking. *Proc. Int. Telemetering Conf., Washington, D.C., September 1969*, pp. 94–104.

Yavuz, D., and Hess, D. T., FM noise and clicks. *IEEE Trans*, **COM-17**, No. 6, 648–653 (1969).

1970

Arndt, G. D., and Loch, F. J., A comparative analysis of frequency modulation threshold extension techniques. *Int. Commun. Conf., San Francisco, California, June 1970*, pp. 21.20–21.26.

Quinn, M. J., An FM click eliminator. *Mervin J. Kelly Commun. Conf., Univ. of Missouri-Rolla, October 1970.*

Unkauf, M. G., Binary FM demodulation using the frequency locked loop. *Mervin J. Kelly Commun. Conf., Univ. of Missouri-Rolla, October 1970.*

5. PLL—Reports

1954

Banta, E. D., Lock-in in APC systems. Rep. No. 38. Philco Math. Group. 1954.

Victor, W., Minimum bandwidths of phase-lock loops using crystal-controlled oscillators. Sect. Rep. 8-496. Jet Propul. Lab., Pasadena, California, March 15, 1954.

1955

White, E. L. C., The pull-in range of an A.P.C. loop. Rep. RK/94. EMI Res. Lab., Middlesex, England, November 1955.

1956

Ingham, W. E., The design of an APC synchronizing loop. Rep. No. RW/8. EMI Res. Lab., Hayes, Middlesex, England, April 1956.

1957

Gilchriest, C. E., The application of phase-locked loop discriminators for threshold improvement and error reduction in FM/FM telemetry. External publ. 364. Jet Propul. Lab. Pasadena, California, January 7, 1957.

Nielson, C. L., Principles and applications of phase-lock detection in phase-coherent systems. Tech. Note HTR 57-003. Hallamore Electron. Corp., Anaheim, California, April 12, 1957.

Rechtin, E., Design of phase-lock oscillator circuits. Sect. Rep. 8-566. Jet Propul. Lab., Pasadena, California, February 7, 1957.

1958

Gilchriest, C. E., Design and operations handbook for phase-locked-loop discriminator. Publ. 127. Jet Propul. Lab., Pasadena, California, May 30, 1958.

Rey, T. J., Effects of the filter in oscillator synchronization. Tech. Rep. 181. Lincoln Lab., M.I.T., Lexington, Massachusetts, May 1958.

Sampson, W. F., Comparative noise performance of phase-lock and pulse-counting discriminators. Tech. Note HTR 58-007. Hallamore Electron. Corp., Anaheim, California, February 28, 1958.

Viterbi, A. J., Functional design of telemetering discriminators. Tech. Mem. 8-1. Jet Propul. Lab., Pasadena, California, August 1958.

1959

Martin, B. D., A coherent minimum-power lunar probe telemetry system. External Publ. 610 (Revised). Jet Propul. Lab., Pasadena, California, August 12, 1959.

Stevens, R., and Brockman, M. H., Design and performance of a deep space tracking and telemetry system. External Publ. 629. Jet Propul. Lab., Pasadena, California, May 1959. Presented at the *Region 7 Conf. Electron. Exhibit, 1959* (sponsored by the IRE Albuquerque, New Mexico, May 1959).

Viterbi, A. J., Acquisition and tracking behavior of phase locked loops. External Publ. No. 673. Jet Propul. Lab., Pasadena, California, July 1959.

Viterbi, A. J., The effect of sinusoidal interference on phase-locked loops. Sect. Rep. 8–583. Jet Propul. Lab., Pasadena, California, December 16, 1959.

1960

Davies, G. L., A narrow-band tracking filter and frequency multiplier. Tech. Note TD 50. Royal Aircraft Establ., May 1960.

Philco Res. Div., Space Commun. Tech., Acquisition and tracking system for space vehicles. Vol. I, 84–86, November 29, 1960.

Robinson, E. M., and Woods, C. R., Acquisition capabilities of phase-locked oscillators in the presence of noise. Tech. Inform. Ser., No. R60 DSD 11. Gen. Elec. Co., September 15, 1960.

Sapp, D. H., A synchronous detection system utilizing a new method of frequency and phase control. Master's Thesis, Univ. of Pennsylvania, Philadelphia, Pennsylvania, June 1960.

Weaver, C. S., Preliminary studies of adaptive processes applied to phase locked loops. Philco Corp., WDL-TR-1334. September 15, 1960.

1961

Brown, L. R., Experimental determination of signal-to-noise relationships in PCM/FM and PCM/PM transmission. Interim Rep. NASA N62–13483. Electro-Mech. Res. Inc., Sarasota, Florida, October 20, 1961.

Clarke, C. E., Tellier, J. C., and Urban, S. J., Phase lock studies. RPA 621–6–1, January 1 to March 31, 1961. Philco Res. Div., Vol. I, April 28, 1961.

Clarke, C. E., Golay, M. J. E., and Urban, S. J., Phase-lock Studies. RPA 621–6–2, April 1 to June 30, 1961. Philco Res. Div., Vol. II, August 18, 1961.

Clarke, C. E., Golay, M. J. E., and Urban, S. J., Phase lock studies. RPA No. 621–6–3, Philco Res. Div., Vol. III, October 27, 1961. 3rd Quart. Rep. covering July 1–September 30, 1961.

Develet, J. A., Fundamental sensitivity limitations for second-order phase-lock loops. Rep. 8616–0002–NU–000. Space Technol. Lab., Los Angeles, California, June 1961.

Easterling, M., A long-range precision ranging system. Tech. Rep. 32–80. Jet Propul. Lab., Pasadena, California. Presented at URSI Meeting, Washington, D. C., May 1961.

Fralick, S., Mumma, J., and Develet, J., The analysis of advanced synchronous telemetry techniques (U). Final Rep. TRW Space Technol. Labs., Los Angeles, California, October 1961 (AD–631 857).

Huylar, J., Lawhorn, R., and Weaver, C. S., Study of adaptive processes applied to phase-locked loops. Tech. Doc. Rep., AD–609 242. Philco Corp., Palo Alto, California, December 1961.

Martin, B. D., The Mariner planetary communication system design. Tech. Rep. 32–85. Jet Propul. Lab., Pasadena, California, May 15, 1961.

Rowbotham, J. R., Phase-locked loop study. Final Rep., N68–80467. Motorola, Inc., Scottsdale, Arizona, June 15, 1961.

Stephenson, J. M., Analysis of phase locked loops. Western Develop. Labs. WDLTR 1599, AD–448 450. Philco Corp., Palo Alto, California, November 1961.

1962

Boyer, R., Digital control of a second-order linear AFC system with a large time delay. N65–24984, Jet Propul. Lab., Pasadena, California, August 30, 1962.

Develet, J. A., The influence of time delay on second-order phase-lock loop acquisition range. Rep. No. 9332, 6–9. Space Technol. Lab., Inc., Los Angeles, California, September 1962.
Develet, J. A., Coherent FDM/FM Telephone Communication. Tech. Rep. No. 8614-6004–NU–000. Space Technol. Lab., Los Angeles, California, January 1, 1962.
Dollard, P. M., and Jacobs, I., Weak-signal communication techniques (U). AD–299 219. Bell Telephone Lab., Inc., Whippany, New Jersey, April 1962.
Jennings, R. R., and Miller, D. C., A low noise correlation frequency tracker. AD–412 630. Naval Avionics Facility, Indianapolis, Indiana, February 1962.
Klapper, J., and Rabinovici, B., A frequency synthesizer study for the proposed AN/GRC–103. RCA Rep. CR-62-419-8, AD 289562. RCA Corp., July 1962.
Martin, B. D., The pioneer IV lunar probe: A minimum-power FM/PM system design. N62-11501. Jet Propul. Lab., Pasadena, California, March 15, 1962.
Philco Corp., Doppler tracking loop optimization study. AD–408 920. Philco Corp., Palo Alto, California, December 1962.
Riedel, E. G., Jr., The effect of frequency tracking, the use of a phase lock loop, and predicted tracking on receiver sensitivity. AD-286920. Air Force Inst. of Tech., Wright-Patterson AFB, Ohio, August 1962.
Schilling, D. L., The response of an automatic phase control system to FM signals and noise. Rep. PIBMRI-1040-62, N63-12322. Polytech. Inst. of Brooklyn, Brooklyn, New York, June 1962.
Thomas, E. F., Investigation and analog simulation of the type two and type three phase-lock loop. AD-295096. Air Force Inst. of Tech., Wright-Patterson AFB, Ohio, December 1962.
Woodman, R. F., A phase-locked phase filter for the minitrack system. N62-15136. Nat. Aeronaut. and Space Admin., Goddard Space Flight Center, Greenbelt, Maryland, September 1962.

1963

Aetjen, R. M., Cancellation of Doppler frequency shift (U). AD-404 428. Air Force Cambridge Res. Labs. Bedford, Massachusetts, March 1963.
Baghdady, E. J., and Marshall, A. C., FM improvement Techniques. Final Rep., N63-22119. ADCOM, Inc., Cambridge, Massachusetts, February 1, 1963.
Develet, J. A., Statistical design and performance of high-sensitivity frequency-feedback receivers (U). AD-408 639. Aerospace Corp., Los Angeles, California, May 1963.
Heckert, J. P., Study of the phase-locked loop for doppler tracking (U). AD-287107. Philco Corp., Palo Alto, California, 1963.
Hill, E. R., Techniques for synchronizing pulse-code-modulated telemetry (U). AD-402 192. Naval Ordnance Lab., Corona, California, February 1963.
Hoffman, L. A., Receiver design and phase-lock loop (U). AD-459 435. Aerospace Corp., Los Angeles, California, May 1963.
Johnson, W. A., A general analysis of the false-lock problem associated with the phase-lock loop (U). Rep. No. TDR-269, AD-427 155. Aerospace Corp., El Segundo, California, October 1963.
McLaughlin, R. J., A lock-on probability analysis for the initial synchronization of phase-locked loops (U). AD-433 698. Tech. Rep. No. 372, Cruft Lab., Harvard Univ., Cambridge, Massachusetts, February 1963.
Roland, W. F., A threshold criterion for phase locked loop design (U). AD-481 321. Naval Postgraduate School, Monterey, California, 1963.

Svoboda, D. E., A phase-locked receiving array for high-frequency communications use. AD-464 374. Antenna Lab., Ohio State Univ. Res. Foundation, Columbus, Ohio, August, 1963.
Van Trees, H. L., Optimum power division in coherent communication systems. AD-406 882. Lincoln Lab., M.I.T., Lexington, Massachusetts, February 1963.
Viterbi, A. J., Phase-locked loop dynamics in the presence of noise by Fokker-Planck techniques. N63-14881. Jet Propul. Lab., Pasadena, California, March 29, 1963.
Viterbi, A. J., Phase-lock loop systems. *In* "Space Communications," (A. Balakrishnan, ed.) p. 123–142. McGraw-Hill, New York, 1963.
Yates, F. F., Communication link performance. N63-23576. Aerospace Corp., El Segundo, California, August 1963.
Yates, F. F., Phase-lock demodulation of sinusoidal FM (U). AD-425 734. Aerospace Corp., San Bernardino, California, November 1963.

1964

ADCOM, Inc., Advanced threshold reductions techniques study. 1st Quart. Rep., N65-30842. ADCOM, Inc., Cambridge, Massachusetts, 1964.
Aupperle, E. M., Locked instability and forced oscillations in automatic phase control systems. Tech. Rep., AD-463100. Cooley Electron. Lab. Univ. of Michigan, Ann Arbor, Michigan, December 1964.
Baker, T. S., Analysis of the synchronization of an automatic phase control system. AD-610 691. Cruft Lab., Harvard Univ., Cambridge, Massachusetts, November 1964.
Derusso, P. M., Michaels, L. H., and Tuel, W. G., Jr., Design and stability of phase locked loops. Final Rep., AD-612 036, N65-19493. Rensselaer Polytech. Inst., Troy, New York, December 1964.
Develet, J. A., Jr., Fundamental sensitivity limitations for second order phase-lock loops. AD-416 683. TRW Space Technol. Labs., Los Angeles, California, 1964.
Georgia Inst. of Tech., Active notch filter. AD-438 252. Atlanta Eng. Experiment Station, April 1964.
Hartl, P., The principle of the phase locked loop and its application in communication receivers for space travel. AD-458 074. Royal Aircraft Estab., Farnborough, England, October 1964.
Schilling, D. L., and Billig, J., A comparison of the threshold performance of the frequency demodulator using feedback and the phase locked loop. Rep. No. PIB MRI-1207-64. Polytech. Inst. of Brooklyn, Brooklyn, New York, February 28, 1964.
Walker, J. R., and Overlander, R., Oscillator, phase-locked. AD-462 369. Space and Inform. Syst. Div., North Ameri. Aviat. Inc., Downey, California, April 1964.
Woodman, R., A narrow-band tracking filter. N64-27254. Nat. Aeronaut. Space Administration, Goddard Space Flight Center, Greenbelt, Maryland, January 17, 1964.
Vaughan, G. R., Osborne, E. F., and Entwistle, G. S., Locked oscillator phase modulator. Appendix D, Final Rep., N65-29140. Defense and Space Center, Westinghouse Elec. Corp., Baltimore, Maryland, August 25, 1964.

1965

Abrams, B. S., Oberst, J. F., Berkoff, M., and Schilling, D. L., Phase locked loop threshold investigations. Rep. PIBMRI-1274-65. Polytech. Inst. of Brooklyn, Brooklyn, New York, June 1965.
Baird, C. A., Jr., A dual mode phase locked loop. N66-18640. Harry Diamond Lab., Washington, D.C., September 30, 1965.

5. PLL—Reports

Becker, H. D., Lawton, J. G., and Chang, T. T., Investigations of advanced analog communications techniques. AD-613 703. Final Rep. Cornell Aeronauti. Lab. Inc., Buffalo, New York, March 1965.

Beery, W. M., Frequency stabilization of frequency-shift-keyed transmissions (U). AD-459 332. Nat. Bur. of Stand., Boulder, Colorado, February 1965.

Bratt, P., Pull-in performance of first-order phase-locked loops. N65-19589, AD-611878. MITRE Corp., Bedford, Massachusetts, February 1965.

ITT, Methods and techniques study report, three-oscillator combiner. AD-617 144. ITT Federal Labs., Nutley, New Jersey, June 1965.

McIntyre, R., Transportable satellite communications terminal X-band receiving facility. Final Tech. Rep., AD-618 907. Sylvania Electron. Syst., Williamsville, New York, June 1965.

Schilling, D. L., Billig, J., and Kermisch, D., Error rates in FSK using phase locked loop demodulator. PIBMRI-1254-65. Polytech. Inst. of Brooklyn, Brooklyn, New York, February 11, 1965.

Smith, P. G., *et al.*, A Study of the effects of interference on narrow-band phase lock loops. Final Rep. N66-10691. Res. Triangle Inst., Durham, North Carolina, October 15, 1965.

Stratemeyer, H. P., A low-noise phase-locked-oscillator multiplier. N66-10399. NASA, Goddard Space Flight Center Short-Term Freq. Stability Symp., 1965.

Summer, H. M., An analysis of the effects of differential phase feedback on a Type 2 feedback control system as applied to phase-lock receivers. Tech. Rep. No. 8, N66-85762. Elect. Eng. Dept., Auburn Res. Foundation, Auburn, Alabama, Inc., June 15, 1965.

Svoboda, D. E., Phase and amplitude control for arrays with increased directivity. AD-461 633. Ohio State Univ. Res. Foundation, Columbus, Ohio, March 1965.

Tausworthe, R. C., Minimizing VCO noise effects in phase-locked loops. N65-32470, pp. 287–289. Jet Propul. Lab., Pasadena, California, June 30, 1965.

Tausworthe, R. C., A new method for calculating phase-locked loop performance. Space Programs Summ., No. 37-31, Vol. IV, pp. 292–300. Jet Propul. Lab., Pasadena, California, February 1965.

Wilson, C. S., and Warren, W. B., A frequency measuring spectrum analyzer. N66-15724. Georgia Inst. of Tech., Atlanta, Eng. Experiment Station, November 1965.

1966

Cambi, E., A survey of the phase-lock loop. N66-30867. European Space Vehicle Launcher Develop. Organ., Paris, France, May 1966.

Compton, R. T., Jr., The effect of a pure time delay on the stability of a phase-lock loop. N66-26382, AD-630473. Ohio State Univ. Res. Foundation, February 1966.

Gardner, F. M., Kent, S. S., and Dasenbrock, R. D., Theory of phaselock techniques. N66-10515, Resdel Eng. Corp. 1966.

Luby, D. C., Gill, W. J., Ballard, E. J., and Spilker, S. S., Demodulation of angle-modulated telemetry signals. Vol. 1, Advanced Demodulation Techniques, AD 639 787; Vol. 2, Review of Demodulation Methods, AD-639788, August 1966.

Motorola Inc., Vehicle tracking receiver design. N66-38787. Motorola, Inc., Scottsdale, Arizona, August 3, 1966.

Smith, L. J., Use of phase-lock loop control for driving ultrasonic transducers. N66-33455. NASA Lewis Res. Center, August 1966.

Tausworthe, R. C., Theory and practical design of phase-locked receivers, Volume 1. Rep. No. 32–819, N66-17323. Jet Propul. Lab., Pasadena, California, February 15, 1966.

Williams, T. R., A note on phase-locked loops in space communications. Goddard Summer Workshop, 1966, N67-22749, 111–119.

1967

Electrac Inc., Study report for the development of techniques to automatically acquire the carrier of AM or PM signals. N67-35322. Electrac, Inc., Anaheim, California, June 15, 1967.
Olsen, D. P., Equivalence of PLL systems and a discriminator followed by a nonlinear feedback filter. N67-30908. Electron. Syst. Res. Lab., Pudue Univ., Lafayette, Indiana, June 1967.
Techn. Hochschule, Hanover, A contribution for optimization demodulation of disturbed frequency modulation signals (in German). West Germany, 1967.

1968

Carden, F. F., Jones, T. J., Martin, C. R., and Merrill, M. D., The nonlinear transient behavior of second, third, and fourth order phase-locked loops. N69-13453. Elec. Eng. Dept., New Mexico State Univ., 1968.
Carden, F. F., Hintz, T. B., and Kelly, L. R., The FDM demodulating characteristics of nonlinear phase-locked loops. N69-13536. New Mexico State Univ., University Park, New Mexico, 1968.
Gunigal, T. E., and Santarpia, D. E., A digital voltage-controlled oscillator for phase-locked loops. N68-24389. NASA Goddard Space Flight Center, Greenbelt, Md., May 1968.
Lindenlaub, J. C., and Olsen, D. P., A study of the extended linear range phase lock loop. Rep. TR-EE68-27. Pudue Univ., Lafayette, Indiana, August 1968.
Seay, T. S., Short-term oscillator stability specifications for phase-locked loops. N68-26668, AD669090. Lincoln Lab., M.I.T., Lexington, Massachusetts, April 29, 1968.
Smith, B. M., Some aspects of phase-locked loop behavior in the presence of noise. Ph. D. Thesis, Univ. of Adelaide, Australia, May 1968.
Viskanta, V. Z., Compound phase-locked loop receiver. N68-34007. TRW Syst. Group, Redondo Beach, California, 1968.
Lindsey, W. C., Nonlinear analysis and synthesis of generalized tracking systems. Part I, USCEE317 December 1968; Part II, USCEE342, April 1969. Univ. of Southern California, Los Angeles, California.

1969

Osborne, P. W., Threshold analysis of phase locked loops. Ph.D. Thesis. Polytech. Inst. of Brooklyn, Brooklyn, New York, 1969; Univ. Microfilms No. 69-20355.
Osborne, P. W., and Schilling, D. L., Threshold analysis of phase locked loops. Rep. No. PIBEE69-002. Polytech. Inst. of Brooklyn, Brooklyn, New York, 1969.

1970

Segrue, D. F., The first order phase locked loop. MS Proj. Rep. Polytech. Inst. of Brooklyn, Brooklyn, New York, June 1970.
Chalkley, H. E., False lock in sampled-data phase lock loops. Ph.D. Thesis, Virginia Polytech. Inst., 1969; Univ. Microfilms No. 69-9925.

6. FMFB, Multiple Loops, Comparisons, etc. — Reports

1961

Hamilton, A. R., Threshold improvement in FM detection by use of feedback. Rep. CRR-246. Collins Radio Co., Cedar Rapids, Iowa, August 1961.

Spilker, J. J., Jr., and Magill, D. T., The delay-lock discriminator—An optimum tracking device. Tech. Rep. LSMD-894802. Missiles and Space Div., Lockheed Aircraft Corp., Sunnyvale, California, March 1961.

1963

Baghdady, E. J., and Marshall, A. C., FM improvement techniques. Final Rep., ADCOM, Inc., Cambridge, Massachusetts, February 1, 1963.

Van Trees, H. L., An introduction to feedback demodulation. N64-17371. Lincoln Lab., M.I.T., Lexington, Massachusetts, August 16, 1963.

Wojnar, A., An analysis and synthesis procedure for feedback FM systems. Tech. Rep. 415. Res. Lab. of Electron. M.I.T., Cambridge, Massachusetts, September 30, 1963.

1964

Battail, G., Reception threshold for frequency modulation. NASA Rep. N64-31562, September 1964; translated into English from *Ann. Telecommun.* **19,** No. 1–2, 1–28 (1964).

Duncan, J., FM demodulator threshold reduction. NASA-CR-57496. Montana State College, Bozeman, Montana, September 1964.

Schilling, D. L., and Billig, J., A comparison of the threshold performance of the frequency demodulator using feedback and the phase locked loop. N65-31810. Microwave Res. Inst., Polytech. Inst. of Brooklyn, Brooklyn, New York, February 28, 1964.

1965

Abbate, J. V., and Schilling, D. L., Optimum demodulation of phase and frequency modulated signals by Bayes criterion. PIBMRI-1253-65. Polytech. Inst. of Brooklyn, Brooklyn, New York, June 1965.

Ghais, A. F., and Wachsman, R. H., SNR behavior of coherent phase demodulators. NASA-CR-80890. ADCOM, Inc., Cambridge, Massachusetts, September 1965.

1966

Luby, D. D., Ballard, E. J., Gill, W. J., and Spilker, J. J., Jr., Demodulation of angle-modulated telemetry signals, Vols, I and II. Defense Doc. Center No. AD639787 and 639788. Philco Company, Palo Alto, California, August 1966.

1967

Filippi, C. A., Advanced threshold reduction techniques study. NASA CR-682. ADCOM Inc., January 1967.

Suter, C. F., Jr., Performance of a combination phase and frequency lock system. NOLTR-67-21, AD-655 812. Naval Ordnance Lab., White Oak, Maryland, February 1967.

1968

Lindenlaub, J. C., and Olsen, D. P., A study of the extended linear range phase lock loop. Rep. TR-EE68-27. Purdue Univ., Lafayette, Indiana, August 1968.

1969

Guida, A., and Schilling, D. L., Optimum frequency modulation receivers. Polytech. Inst. of Brooklyn, Brooklyn, New York, 1969.

Lindsey, W. C., Nonlinear analysis and synthesis of generalized tracking systems. Part I, USCEE317, December 1968, and Part II, USCEE 342, April 1969. Univ. of Southern California, Los Angeles, California.

Ziemer, R. E., Experimental comparison of Costas and PLL demodulators in RFI environments. Rep. X-520-69-355. Goddard Space Flight Center, September 1969.

Author Index

Numbers in parentheses are reference numbers and indicate that an author's work is referred to, although his name is not cited in the text. Numbers in italics show the page on which the complete reference is listed.

Aaronson, G., 267(8), 279(8), 284, 288(8), 290, 291, 292, 293, 294, *301*
Abrams, B. S., 154, *169*
Acampora, A., 2, *3*, 37(15), *47*, 229(6), 256(13), 257(15), 259, 262, *264*, 267(8), 279(8), 284, 288(8), 290, 291, 292, 293, 294, *301*
Allen, D. B., 44(47), *48*
Arguimbau, L. B., 25(3), *46*
Avins, J., 25(2), *46*

Baghdady, E. J., 110(8), *112*, 152(7), *169*, 201, *224*
Bedrosian, E., 40(23), *47*, 181, 182, *224*
Benjaminson, A., 300(44), *302*
Bennett, W. R. 266(1), 267(7), 268(7), 270, *300*, *301*, *303*
Berkoff, M., 154, *169*
Billig, J., 201, *224*
Bode, H. W., 13(6), *22*
Boor, S. B., 44(47), *48*
Bremmer, H., 4(3), *21*
Brown, L. R., 267(4), *301*

Bucher, T. T. N., 40(24), *47*
Byrne, C. J., 297(30), *302*

Cafissi, R., 44(44), *48*
Camenzind, H. R., 44(48), *48*
Carlson, A. B., 268(15), *301*
Carson, J. R., 40(19), *47*
Chaffee, J. G., 2, *3*, 194(5), *224*
Chestnut, H., 7(4), 11(4), 13(4), *22*, 85, 124(1), *168*
Clapp, J. K., 297(29), *302*
Cohn, J., 35(10), *46*
Corrington, M. S., 40(28), *46*
Costas, J. P., 296(23), *301*
Creutz, J., 271(14), 285(14), 287, 288, 289, *301*
Crosby, M. G., 35(9), 36(9), *46*

Davenport, W. B., 14(7), *22*, 128(3), *168*
Davey, J. R., *303*
Davis, B. R., 201, *224*

de Bellescize, H., 1(1), *3*
Develet, J. A., 1, *3*, 38(18), *47*, 154, *169*, 201, *224*

Ito, S., 2, *3*, 225(2), *264*
Izatt, J. B., 40, 41, 43, *47*, 181, *224*

Edson, W. A., 44(37), *47*
Enloe, L. H., 2, *3*, 40(33), 42, *47*, 53(1), 71, 72, 74, *75*, 170, 196, 201(1), 206, 207, 213, 214, 217, *224*, 308, *338*, *354*
Entwistole, G. S., 298(37), *302*
Evers, A. F., 297(27), *302*

Jaffe, R., 1, *3*
Jones, J. J., 266, 268, *301*

Kent, S. S., 43, 44(35), 46(35), *48*, 299(41) *302*
Klapper, J., 36(13), 40(25), *47*, 261(16), 262(16), *264*, 267(8, 10), 268(13), 271(13, 14), 272(13), 275, 277(13), 278, 279(8), 280, 282, 284, 285(14), 287, 288(8), 289, 290, 291, 292(22), 293, 294, 297(25), *301*, *302*
Kobayaski, S., *224*
Koll, V. G., 267(9), 268(9), 273(16), *301*, *303*
Kotelnikov, V. A., 266(2), 268(2), *300*
Krishnan, S., 45, 46(50), *48*

Fancourt, K. G., 25(1), *46*, 186, 187, 188, *224*
Frankle, J., 36(14), 38(14), 39, *47*, 225(4, 5), 249(4), 250(4), *264*, 267(8), 279(8), 284, 288(8), 290, 291, 292, 293, 294, *301*
Frazier, J. P., 103, *112*, 298, *302*
Fredendall, G. L., 1(2), *3*, 300(45), *303*
Frutiger, P., 2(13), *3*, 201(10), *224*
Fry, T. C., 40(19), *47*

Gagliardi, R. M., 225(1), *264*
Gardner, F. M., 43, 44(35), 46(35), *47*, *48*, 80(2), 92, 104, 105, 106, *112*, 299(41), 300(42), *302*
Gardner, W. A., 44(49), *48*
Giger, A. J., 194(5), *224*
Goblick, T. J., *346*
Graszkowski, J., 44(43), *48*
Grebene, A. B., 44(48), *48*
Guillemin, E. A., 254(11), *264*
Gupta, S. C., 152(6), *169*

Langseth, R. E., *109*
Lambert, R. F., *109*
Lawton, J. G., 268(11), *301*
Lefrak, F. H., 37(15), *47*, 197(7), 198, 203(7, 13), 204(7, 13), *224*, 299(38), *302*
Leon, B. J., 40(22), *47*
Lewis, F. D., 297(29), *302*
Lindenlaub, J. C., 256(14), *264*
Lindsey, W. C., 2(9), *3*, 263, *264*, 296(24), *302*
Lucky, R. W., *303*
Luna, A., 44(44), *48*

Harmon, I., 297(26), *302*
Heinemann, H., 225(5), *264*
Heitzman, R. E., 135(4), 154, *168*
Hiroshige, K., 298, *302*

McLaughlin, P., 267(8), 279(8), 284, 288(8), 290, 291, 292(22), 293, 294, *301*
Malling, L. R., 44(41), *48*

Mayer, R. W., 7(4), 11(4), 13(4), *22*, *85*, 124(1), *168*
Mazer, W. M., 267(5), 270, 296, *301*
Mazo, J. E., 273(16), *301*
Medhurst, R. G., 40(30), *47*
Menkes, E. D., 297(26), *302*
Meyerhoff, A. A., 267(5), 270, 296, *301*
Middleton, D., 40(31), *47*
Miller, R. L., 297(28), *302*
Moore, H., 203(13), 204(13), *224*
Morita, M., 2, *3*, 225(2), *264*
Moschytz, G. S., 297(35), *302*
Mullen, J. A., 40(31), *47*
Murphy, J. V., 300(46), *303*

Newton, A., 2, *3*, 37(15), *47*, 203(13), 204(13), *224*, 225(5), 257(15), 259, 262, *264*, 292(22), *301*
Noda, K., 44(46), *48*
Norwood, M. H., 44(45), *48*
Nyquist, H., 279, *301*

Oberst, J. F., 154, *169*
Oliver, W., 331(3), *338*
Osborne, E. F. 298(37), *302*
Ozolins, L., 203(13), 204(13), *224*, 299(39), *302*

Page, J., 103, *112*, 298, *302*
Panter, P. F., 40(26), *47*
Papoulis, A., 4(2), 10(2), 14(8), *21*, *22*, 276(19), *301*
Pelchat, G. M., 44(47), *48*
Peter, M., 300(43), *302*
Pierce, B. O., *107*, *348*

Rabinovici, B., 297(25), *302*
Real, R. R., 44(42), *48*

Rechtin, E., 1, *3*
Reich, H. J., 44(39), *48*
Rice, S. O., 26(4), 35(4, 11, 12), 36(4, 13), 40(23), *46*, *47*, 181, 182, 196, *224*, 266(1), 275(17), 276(17), *300*, *301*
Richman, D., 1(3), *3*, 297(32), 300(32), *302*
Ringdahl, I., 276(18), *301*
Robinson, L. M., 254(12), 256, 257(12), *264*
Root, W. L., 14(7), *22*, 128(3), *168*
Rowe, H. E., 40(32), *47*

Saito, S., *224*
Salz, J., 267(7, 9), 268(7, 9), 270, 273(16), *301*, *303*
Sann, K. H., 44(40), *48*
Schader, J. H., 299(39), *302*
Schilling, D. L., 1(7), 154, *169*, 201, *224*, 276(18), *301*
Schwartz, M., 28(5, 6), 36(5), *46*
Seeley, S. M., 25(2), *46*
Seifert, W. W., 4(1), *21*, *96*
Shaft, P. D., 267(6), 268(6), *300*
Shannon, C. E., 38(17), *47*
Skwirzynski, J. K., 25(1), *46*, 186, 187, 188, *224*
Smith, B. M., 110(7, 9), *112*
Smith, E. F., 268(12), *301*
Solem, R. J., 152(6), *169*
Springett, J. C., 299(40), *302*
Steeg, C. W., 4(1), *21*, *96*
Stein, S., 266, 268 *301*
Stewart J. L., 40(29), *47*
Stiffler, J. J., 250(8), 252(8), *264*
Stone, R. F., 225(3), *264*
Strandberg, M. W. P., 300(43), *302*
Stratemeyer, H. P., 297(31), *302*
Stumpers, F. L. H. M., 34, 35, 40(21), *46*, *47*

Tausworthe, R. C., 80(1), 99(1), 103, *112*
Thomas, C. M., 154, *169*
Truxal, J. G., *96*
Turin, G. L., 266(3), 268(3), *301*

Uhran, J. J., 256(14), *264*

van der Pol, B., 4(3), *21*, 40(20), *47*
Van Trees, H. L., 1(6), *3*
Van Valkenburg, M. E., 7(5), 9(5), *22*
Vaughn, G. R., 298(37), *302*
Veiga, G. M., 261(16), 262(16), *264*
Viterbi, A. J., 2(8) *3*, *80*(3, 4), 84(3), 91, 94, 98(3, 5), 99, 100, 102, 103, 105, 107(3, 4), 108, 109, *112*, 135(4), *168*, 249, *264*, 288(21), *301*

Walsh, C. P., 32, *46*
Warner, A. W., 44(38), *47*
Warren, W. B., Jr., 297(36), *302*
Weiner, D. D., 40(22), *47*
Weldon, E. J., Jr., *303*
Wendt, K. R., 1(2), *3*, 300(45), *303*

Zinn, M. K., 37, *47*, *344*

Subject Index

Adder, 328
AM-to-PM conversion, 311
Amplifier
 dynamic range, 328
 specification, 158, 161, 164, 167, 218, 221, 223, 237, 238, 246, 247
Amplitude response, 9
Analog-signal tests, 326–334
Angle-modulation, derivation of improvement equations, 339–343
Angular feedback demodulators, 38
Articulation testing, 324
Autocorrelation function, 15–17
Automatic frequency control (AFC), 49–50, 322
 hold-in range tests, 322
 pull-in range tests, 322

Bandwidth, Carson's rule, 31
Baseband filter, noise bandwidth, graphically, 330
Baseband intrinsic noise ratio, 332
Binary FM, 268–295, *see also* Digital FM
 baseband AC coupling, 281
 baseband DC coupling, 281
 bit error rate for constant CNR, 277
 decision, 270

demodulation with angular-feedback demodulators, 283–295
design problem, 291
deviation index, 270
effect of center frequency offset, 281
encirclement noise, 274
error rates for constant energy ratio, 279
example, 277–283
with impulsive noise, 290
intersymbol interference, 285
limiter-discriminator demodulation, 274–283
with offset in center frequency, 277
optimum deviation index, 280
optimum PLL bandwidth, 285
phase difference, 270
PLL bandwidth, 284
PLL demodulation, 268
predetection bandwidth for PLL, 290
predetection filter, 284
probability of error, 274–295
rate of aiding spikes, 285
receiver bandwidth, 279
rectangular transitions, 268
spike-rate minimization, 285
spikes per bit, 278
staircase, 278
type of PLL, 284

Binary systems, 269
Bit error rate (BER), 268, 325
Bit transitions, 267
Bode plot, 13

Carrier
  phasor and noise representation, 26, 33
  reference extractor, 297
Carson's rule bandwidth, 38
CCIR loading formula, 166, 222
Channel, 335
  capacity, 346
Coherent, AM reception, 1
Coherent harmonic signal, 297
Coherent PSK, 268, 296
Coherent ranging system, 299
Component tests, 304–313
Compound-loop demodulator, 225–264, *see also* FMFB-PLL, etc., and ERPLD
  baseband filter, 228–230, 238, 247
  general design considerations, 227–230
  internal detector, 225–227
  internal IF, 228, 234, 238, 241, 243, 245, 246
  open-loop transfer characteristic, 227
  threshold, 225
Convolution integral 9,
Correlation techniques, 268
Cross-talk, 181
  equivalent generator, 182
  spectral densities, 182
Crystal oscillator, 43–44, 300
Cycles slippings (or skipping), in PLL, 101, 104, 110, 132–133, 247–248

Delay, 73–75, 110–112, 147–149, 168, 209–212, 299, 305, 306–307
Demodulator
  correlation, 266
  ideal, 346
  performance boundary, 39, 346
  twin-filter, 266
Differentially-coherent PSK, 268
Digital FM, 265–295, 325, *see also* Binary FM
  block diagram of system, 266
  causes of errors, 274
  cycle slippings, 272–279
  decision circuits, 336
  demodulator, 267
  error mechanism, 269–274
  error rates, 266–269
  general principles, 265
  phase continuity, 266
  phase steps, 272
  postdetection processor, 267
  predetection filtering, 267
  receiver, 266
  reception with angular-feedback demodulators, 265
  required signal power, 268
  signal angle, 273
  spectra, 266
  system, 266
  test setup, 335
  tests, 325
  transmitter, 266
  word analyzers, 336
Digital-signal tests, 334–338
Discriminator, 23–37, 66, 73, 186–191
  balanced type, 26, 283
  characteristics, 312
  distortion factors, 186–191
  equivalent baseband response, 37, 344–345
  equivalent distortion generator, 187
  finite detection bandwidth, 66
  IF leakage, 312
  operational principles, 25–33
  optimized, 188
  output, 33
  peak-to-peak separation, 67, 73, 187, 218
  response to impulse, 344
    to noise, 26–36
  sensitivity, 26, 73, 221, 223, 245, 312
  tests, 312
Distortion, 123, 181–186
  contours, 126–127
  equivalent generator, 123, 128, 177–183, 186–187
  FDM, 128
  harmonic, 126
  intermodulation, 125
  due to limiter-discriminator, 186–191
  noise-loading tests, 331
  output point, 131, 195
  phase detector, 131, 162
  spectrum, 129
  techniques to reduce, 131–132, 193–195
  tests, 329, 331–333

Subject Index

types of, 125
VCO, 126–130, 191–195
Dynamic range, receiver, 337

Encirclement of origin, 33, 271
Energy ratio, 268, 325
  vs CNR, 279, 326
Equivalent baseband filter, 52, 351–352
Equivalent noise bandwidth, 19–21
  integrals, 21
Equivalent noise generator, 56, 83, 250–254
ERPLD, 225, 257–263, *see also* Extended-range PLL
  alternate implementations, 260
  block diagrams, 258
  closed-loop response, 261
  equivalent filter realization, 258
  equivalent noise generator, 260
  linear analysis, 261
  linearity, 257, 259
  monotonic range, 257
  noise bandwidth, 257, 261
  optimization, 262
  phase error, 261
  predetection filtering, 261
  response functions, 259
  sensitivity, 258
  threshold, 258, 261
Error mechanism, 269–274
Error signal, 12
Errorless transmission, 268
Excess phase shift, 74, 110
Extended-range phase detector, 250–260
  by carrier signal waveshaping, 250
  by postdetection synthesis, 252
Extended-range PLL, 247–263, *see also* ERPLD
  generalized Tanlock comparator, 257
  loss-of-lock analysis, 249
  loss-of-lock rates, 250
  model with generalized phase detector, 249
  noise, 250
  by phase feedback, *see* ERPLD
  Tanlock system, 254–257
  threshold reduction, 250
Evaluation procedures, 304–338

FDM-FM, 30, 115, 122, 128, 135, 141–147, 165, 172, 180, 182, 186, 195, 197, 203, 215, 219, 221–223, 331–333
  experimental threshold data, 197
FDM signal tests, 324
FDM signals, 341–343
Feedback theory, 11–14
Filter, 10, 328
  noise bandwidth, 328
  radius of gyration, 35
  rectangular, 10
FM carrier
  effect of off tuning, 351
  impulse train, 252
  sawtooth, 252
  shaping, 251
  square-wave, 252
FM carrier filtering, 40–42
  approximation to response, 40
  baseband equivalent circuit, 40
  in digital FM, 40
FM demodulator, distributed feedback type, 226
FM detection, minimum attainable threshold, 227
FM feedback loop (FMFB), 49–75, 170–247, *see also* FMFB with delay
  AFC tests, 318
  amplifier, 218
  baseband filter, 219
  closed-loop response, 60–64, 202
    tests, 322
  closed-loop stability, 59
  closed-loop transfer function, 59–64
  conditions of low distortion, 177
  DC tracking, 58
  demodulator, 64–65
  describing equations, 50–52
  design for low threshold, 201–212
  discriminator distortion, 186–191
  distortion-limited region, 177–195
  double-pole IF filter, 194
  dynamic instability, 201
  effect of excess delay, 73–75
    of off-tuning, 62–64, 66–69
  effective noise bandwidth, 200
  element limitations, 66–67
  equivalent distortion generator, 179
  examples, 220–223
  experimental optimization, 319
  factors affecting threshold, 195–201
  for FDM, 178, 221

fed-back noise, 71
feedback factor, 71, 219
   check, 318
feedback threshold, 197
   characteristics, 203, 210
   criterion, 199
filter parameters, 210
FM improvement region, 170–177
frequency deviation, 199
fundamental limitations, 68–69
harmonic distortion, 179
high CNR conditions, 56
IF center frequency, 217
IF design rules, 203
IF filter response, 52
IF input spectrum, 180
IF network distortion, 181–186
IF tests, 318
initial setup tests, 317
input noise, effects due to, 54–57, 69–73
input off-tuning, 71
limiter-discriminator, 218
linear equivalent circuit, 53–56
loop baseband filter, 202
magnitude and phase response, 60–64
minimum noise bandwidth, 203
minimum threshold CNR, 204, 208, 210
noise bandwidth, 197
   with off-tuning, 71, 73
noise penalty, 174
nonlinear characteristic of, 191
nonlinear operation, 65–73
NPR, 178
open-loop response, 59, 63
   tests, 322
open loop tests, 319
optimum feedback factor F, 204, 208, 210
optimum zero constant, 204
output distortion, 177
output noise, 172–177
output NPR, 185
output SNR (or NPR) at threshold, 209
performance curve, 170–172
for PM, 174
poles and zeros, 59
postdemodulator filter, 219
pre-FMFB filter, 199
regions of operation, 170–172
residual noise, 177
signal-to-distortion ratios, 190

single-channel system, 173
SNR correction factors, 173–177
   for speech modulation, 206–209, 220
   step-by-step design procedure 215–223
   for test-tone at top baseband frequency, 207
tests, 317–322
threshold, 69–73
above threshold, 65
threshold CNR, 219
threshold-distortion tradeoff, 320
threshold-limited region, 195–215
threshold mechanism, 195
threshold SNR, 219
top-channel NPR, 185
tracking objectives, 49–50, 64–65
for TT modulation, 179
VCO, 191, 218
VCO jitter above threshold, 198, 200
VCO noise and drift, 212–215
wide-band basebands, 203
worst-channel NPR, 219
zero constants, 219
FM modulator, 298
FM reception, 23–42
   below-threshold region, 23
   distortion-limited region, 23, 40–42
   improvement region, 23, 28–33
FMFB with delay, 209–212
   feedback threshold characteristics, 210
   minimum threshold, 210
   noise bandwidth minimization, 210
   open-loop transfer characteristic, 209
   optimum feedback factor, 210
FMFB-ERPLD, 225, 262
   design parameters, 263
   internal detector, 263
   SNR, 262
   threshold CNR, 262
FMFB-FMFB, 225, 239–247
   closed-loop noise bandwidth, 243
   design, 239
      parameters, 243
   example, 244–247
   functional block diagram, 239
   inner-loop design, 242
   internal FMFB loop, 241
   linear equivalent model, 239
   outer loop design parameters, 241

output SNR, 241
postdemodulator filter, 244
system design curves, 241
threshold CNR, 242
threshold curves, 239
FMFB-PLL, 225, 231–239
  design, 231
    parameters, 231
    example, 234–239
    feedback factor F, 231
    functional block diagram, 231
    inner-loop parameters, 236
    linear equivalent model, 231
    outer-loop parameters, 236
    output SNR, 231
    threshold CNR, 234
    threshold curves, 231
    zero constants, 231
Fourier transform, 5
Frequency divider, regenerative, 297
Frequency feedback loop, *see* FM feedback loop
Frequency multiplier, 297
Frequency response, 9–11
  lowpass equivalent, 11
Frequency-shift keying (FSK), 266, *see also* Digital FM, Binary FM
Frequency synthesizer, 296

Gain-bandwidth limitations, 74
Gaussian distribution, 16
Gaussian filter, threshold characteristics, 196
Gaussian noise, 38
Generalized tracker, 2

Ideal demodulator boundary, 38
IF amplifier tests, 311
IF filter, 69, 310
  inadequate compensation, 69
  nonlinear modulation response, 69
  tests, 310
IF operating frequency, 66, 217
Impulsive FM, 296
Incidental FM, 297
Information rate, 346
Input mixer, tests, 309
Input-output relationship, 9
Integrate-dump circuit, 267, 270
Intersymbol interference 274

Laplace transform, 4–6
  bilateral, 5
  table, 6
  unilateral, 5
LD, *see* Discriminator
Limiter
  insufficient limiting, 66
  tests, 311
Limiter-discriminator, *see* Discriminator
Linear system, 8
Loop filter, tests, 307
Loop tests, 304
Loss-of-lock impulses (LLI), *see* Cycle slippings

Matched filter, 267
Maximum likelihood receiver, 268
Mean-square value, 16
Mixer saturation, 66
Modified first-order loop, 284, 295
  open-loop response, 284
Multiple-loop demodulator, *see* Compound-loop demodulator

Network, 4–11
  minimum-phase, 8
  order of, 7
  passive, 8
Noise, 14–21, 29, 26–36, 37–39, 325
  angle, 27–28, 33, 36, 271
  averages 14–16
  bandwidth, 20
    with delay, 75, 110
  character, 325
  characteristics of FDM, 122
  correction factor, 118–122, 173–177
  doublet type, 273
  ergodic, 14, 16
  frequency division multiplex, 30, 115, 172, 341–343
  immunity of FM and PM, 29
  narrowband, 18–19
  origin encirclements, 33, 271
  performance equations, 29–31, 115, 172, 339–343
  performance of various demodulators, 38–39
  PLL FM output, 119
  PLL PM output, 120
  postdemodulator selectivity, effect of, 117, 174

390 Subject Index

properties, 14–21
representation, 14–21
shaped by single tuned circuit, 276
single-channel speech, 29, 115, 172, 339–341
spectrum, 35
spike, 271
stationary, 14
Noise bandwidth, minimum, 139, 353
Noise power ratio (NPR) 30, 115, 128–130, 172, 185, 324, 341–343
vs CNR tests, 329
Noncoherent FSK, 268
Non return-to-zero (NRZ) baseband, 283
Nyquist's signaling rate, 279

Operator p, 8
Optimum loop bandwidth, 248
Orthogonal FSK, 268
Oscillator stabilization, 300

PCM timing, 300
Peak factor, 136, 203, 222
Performance characteristic for FM, typical, 24, 114, 172
Phase detector, 44–46, 76, 95, 123, 158, 248, 308
 balanced type, 45
 characteristic, 248
 defining equation, 44–45
 distortion, 123–126
 doubly-balanced type, 45–46
 extended-range, 250–260
 mode of operation, 45
 monotonic range, 46, 248
 multiplier-type, 46, 250
 optimum characteristic, 250
 saturation, 95
 sawtooth characteristic, 250, 254
 sensitivity, 45–46, 158, 257–259, 308
 shape of characteristic, 248
 square-wave characteristic, 250
 tests, 308
 truncated sine characteristic, 250
Phase error
 critically-damped loop, 295
 first-order loop, 292
 frequency step, 292
 second-order loop, 292
 square-wave input, 294

Phase-locked Klystron modulator, 299
Phase-locked loop (PLL), 76–169, *see also* PLL with frequency offset
 above-threshold, 113–132
 acquisition time, 99, 103
 amplifier, 158
 applications, 295–303
 as bandpass filter, 298
 basic nonlinearity, 80
 channel-threshold spread, 135
 circuit design, 155–168
 closed-loop tests, 317
 closed-loop transfer function, 86–90
 compensating filter, 162
 conditions of low distortion, 122
 cycle-skipping, 104–105
  parameters, 133
  rate, 110, 132, 285–289
 DC tracking, 84–85
 demodulator for FM, 94–95
 derivable from FMFB, 76
 describing equations, 77–80
 design
  for FM demodulation, 113
  for low threshold, 138–168
 design curves, 159
 design equations, 144
 design examples, 160–168
 digital FM tests, 317
 distortion-limited region, 122–132, 156
 dynamic tracking equation, 79
 effect of delay, 110, 147, 165
 element limitations, 95
 entire baseband threshold, 144
 equivalent distortion generator, 149
 error responses, 153
 error signal, 77
 essential elements, 76
 factors affecting threshold, 132–138
 filter components, 158
 first-order, type-one, 86–87
 FM (speech) demodulator, 160
 FM improvement region, 113–122
 Fokker–Planck techniques, 107
 frequency division multiplex, 115, 165
 frequency offsets, effects of, 79, 102, 151, 285–289
 functional block diagram, 155
 Gaussian modulation, 136
 of higher-order, 152

Subject Index

hold-in range, 85
inherent nonlinear characteristics, 95
input filter, 137, 261
linear equivalent circuit, 80–84
linear SNR-CNR region, 113–122
loop order, 85–86
loop type, 85–86
loss-of-lock, 104
   impulses, 109
maximum input frequency step, 104
maximum sweep rate, 103
mean–square VCO phase noise, 108
minimum noise bandwidth with delay, 110
minimum phase error, 141
model with equivalent noise input, 83
noise bandwidth, 143, 159
   minimum, 139
noise performance, 82–84, 105–110
   at high CNR, 94, 115
nonlinear differential equation, 78, 96
nonlinear noise mechanism, 105
as notch filter, 298
nth order, 152–154
open-loop response, 85, 87
open-loop tests, 315
optimum bandwidth, 110, 139, 143, 148, 151, 285
oscillator frequency jitter, 122
output noise power density, 133
output SNR, 115, 145, 159
peak phase error, 105–106, 136–137, 151–152, 293–295
performance and design equations, 156–157
performance tests, 313–317
phase detector, 80
   distortion, 123–127
   outputs, 101
   sensitivity, 158
   tests, 308–309
phase-error minimization, 138–141
phase margin, 152, 315
phase-plane trajectories, 103
predetection bandwidth, 137–138, 142–143, 289
pull-in range, 85, 99, 102
pull-in time, 99, 102
pull-out frequency, 106
rate of LLI, 109, 248, 295

reference extraction, 151
regenerative mode, 109
regions of operation, 113
residual noise, 122
root loci, 91
second-order
   type-one, 87, 89–90
   type-two, 86, 88
single-channel signals, 115, 160–164
single-channel threshold, 143
spike-noise model, 134
stability, 86, 90–91
static describing equation, 79
step-by-step design, 154–168
summary of design, 155–159
synchronism, 104
synchronized condition, 78
synchronous model, 81
system design, 154–168
table of basic characteristics, 93
   of design and performance, 156–157
test-tone (TT) modulation, 136–137, 146
of third order, 90, 152
threshold, 105–110, 156
above threshold, 115, 156
threshold CNR, 134, 159
   minimization, 146
threshold condition, 133
threshold criterion, 136
threshold impulse, 109
threshold-limited region, 132–154
threshold noise model, 132
threshold NPR with loop delay, 149
threshold-producing noise, 132
threshold SNR (or NPR), 159
top-channel threshold, 144
total phase error, 110, 136
total spike rate, 109, 132
tracking objectives, 77
transient response, 91–94, 292–295
   tests, 317
variations in loop gain, 80
VCO center frequency, 158
VCO noise and drift, effect of, 149–152
worst-channel NPR, 159
worst-channel threshold, 165
Phase modulator, 257, 298
Phase plane, 96–104
   analysis, 96–104
   critical trajectory, 99

first-order type-one loop, 98–99
graphical construction, 97–98
limit of frequency lock, 100
portrait, 99
second-order type-one loop, 102–103
second-order type-two loop, 99–102
Phase-reference extraction, 296
Phase response, 9
Phased array antenna control, 299
PLL, see Phase-locked loop
PLL with frequency offset, 151, 285–288
cycle slipping rate, 287
minimum spike rate, 285
optimum bandwidth, 285
rate of aiding cycle slippings, 288
Poisson distribution, 17, 276
Pole-zero cancellation, 230
Poles, 8
Power response, 18–19
Power spectral density, 17–19
colored, 18
white or flat, 18
Pre-FMFB filter, 199
Probability density function, 15–17
Probability theory, 15
PSK demodulation, 296
Psophometrically weighted SNR, 166, 222

Quasi-random digital baseband, 271

Random signals, 14–21, see also Noise
Receiver, dual-purpose, 265
Response
closed-loop, 12
open-loop, 12
Return signal, 12

Second-order loop, with complex zeros, 260
Shannon's limit, 38
Single-channel speech, 115, 141–146, 160–164, 172, 207–212, 220–221
Single-channel speech model, 339
Single-channel tests, 125, 324
Single pole 68, 181, 196, see also Single-tuned circuit
IF filter, 68–69
off-tuning, 69
radius of gyration, 196
threshold impulse rate, 196

Single-tuned circuit 40, 62–63, 68, 351, see also Single pole
baseband-equivalent response, 41, 63, 351
distortion data, 40–42, 181–182
off-tuning, 62–63, 351
response of lowpass equivalent, 41
Small noise, 35
SNR
vs CNR tests, 329
psophometrically weighted, 166, 222
Speech modulation, see Single-channel speech
Spike noise, 33–36
source of, 35
spectral distribution, 35
Spike rate, 35
carrier offset, effect of, 36
minimum, 248
modulation, effect of, 36
Spikes, 17, 275
aiding, 275
opposing, 275
Stability tests, 13
Standard deviation, 16
Synchronization, television receivers, 1
System tests, 304, 323–338

Tanlock system, 254–257
Test equipment, 333
Test methods, 327
Testing procedures, 304–338
Threshold
approximation to, 35
compound loops, 225–262
discriminator, 33–39
experimental optimization, 319
FM feedback loop, 69–73, 195–223
model, 35
onset of, 35
phase-locked loop, 105–110, 132–168
point, 33
tests, 323
Threshold impulses
(ThI), 109, 247
tradeoff between LLI and ThI, 248
Tone tests, 324, 329
Transfer function, 8
Trigonometric identities, 21
Two-tone modulation, 124, 331

Subject Index

Varactor diode, 348
Variance, 16
Video signal tests, 325
Voice signal tests, 324
Voltage-controlled oscillator (VCO), 42–44, 191, 304–307, 348
  crystal, 44
  delay, 306
  distortion, 126–130, 191–195, 348
  hysteresis, 304
  incidental amplitude modulation, 305
  jitter, 215, 297
  linearity, 194
  multivibrator, 43
  nonlinear characteristic, 44
  sensitivity, 306
    calibration, 326
  specifications, 43–44
  spurious components, 304
  tests, 304–307
  types of, 43
  varactor-controlled, 43, 127, 348

Wiener–Khintchine theorem, 17
Word generator, 334
Worst-channel performance, 165, 329

Zeros, 8

ELECTRICAL SCIENCE

A Series of Monographs and Texts

Editors

Henry G. Booker
UNIVERSITY OF CALIFORNIA AT SAN DIEGO
LA JOLLA, CALIFORNIA

Nicholas DeClaris
UNIVERSITY OF MARYLAND
COLLEGE PARK, MARYLAND

Joseph E. Rowe. Nonlinear Electron-Wave Interaction Phenomena. 1965

Max J. O. Strutt. Semiconductor Devices: Volume I.
  Semiconductors and Semiconductor Diodes. 1966

Austin Blaquiere. Nonlinear System Analysis. 1966

Victor Rumsey. Frequency Independent Antennas. 1966

Charles K. Birdsall and William B. Bridges. Electron Dynamics of Diode Regions. 1966

A. D. Kuz'min and A. E. Salomonovich. Radioastronomical Methods of Antenna
  Measurements. 1966

Charles Cook and Marvin Bernfeld. Radar Signals: An Introduction to Theory and Application.
  1967

J. W. Crispin, Jr., and K. M. Siegel (eds.). Methods of Radar Cross Section Analysis. 1968

Giuseppe Biorci (ed.). Network and Switching Theory. 1968

Ernest C. Okress (ed.). Microwave Power Engineering:
  Volume 1. Generation, Transmission, Rectification. 1968
  Volume 2. Applications. 1968

T. R. Bashkow (ed.). Engineering Applications of Digital Computers. 1968

Julius T. Tou (ed.). Applied Automata Theory. 1968

Robert Lyon-Caen. Diodes, Transistors, and Integrated Circuits for Switching Systems. 1969

M. Ronald Wohlers. Lumped and Distributed Passive Networks. 1969

Michel Cuenod and Allen E. Durling. A Discrete-Time Approach for System Analysis. 1969

K. Kurokawa. An Introduction to the Theory of Microwave Circuits. 1969

H. K. Messerle. Energy Conversion Statics. 1969

George Tyras. Radiation and Propagation of Electromagnetic Waves. 1969

Georges Metzger and Jean-Paul Vabre. Transmission Lines with Pulse Excitation. 1969

C. L. Sheng. Threshold Logic. 1969

Dale M. Grimes. Electromagnetism and Quantum Theory. 1969

Robert O. Harger. Synthetic Aperture Radar Systems: Theory and Design. 1970

M. A. Lampert and P. Mark. Current Injection in Solids. 1970

W. V. T. Rusch and P. D. Potter. Analysis of Reflector Antennas. 1970

Amar Mukhopadhyay. Recent Developments in Switching Theory. 1971

A. D. Whalen. Detection of Signals in Noise. 1971

J. E. Rubio. The Theory of Linear Systems. 1971

Keinosuke Fukunaga. Introduction To Statistical Pattern Recognition. 1972

Jacob Klapper and John T. Frankle. Phase-Locked and Frequency-Feedback Systems: Principles and Techniques. 1972